# Mine Water Treatment – Active and Passive Methods

Christian Wolkersdorfer

# Mine Water Treatment – Active and Passive Methods

Christian Wolkersdorfer
South African Research Chair for Acid Mine Drainage Treatment
Tshwane University of Technology (TUT)
Pretoria, South Africa

ISBN 978-3-662-65772-0 ISBN 978-3-662-65770-6 (eBook)
https://doi.org/10.1007/978-3-662-65770-6



This Springer imprint is published by the registered company Springer-Verlag GmbH, DE, part of Springer Nature.
The registered company address is: Heidelberger Platz 3, 14197 Berlin, Germany

*I dedicate this book to my friend Prof Dr Paul Younger who sadly passed away too soon in April 2018.*

*Only those who will risk going too far can possibly find out just how far one can go.*
Thomas Stearns Eliot

*I love him who desires the impossible.*
Johann Wolfgang von Goethe
*Faust II*

*"But I don't want to go among mad people," Alice remarked.*
*"Oh, you can't help that," said the Cat: "we're all mad here. I'm mad. You're mad."*
*"How do you know I'm mad?" said Alice.*
*"You must be," said the Cat, "or you wouldn't have come here."*
Lewis Carroll
*Alice's Adventures in Wonderland*

*Here I am conscious that I have fallen far short of the possible. Simply because my powers are insufficient to cope with the task.—May others come and do it better.*
Ludwig Wittgenstein
*Tractatus Logico-Philosophicus*

# Preface

There is a large body of literature on mine water treatment, and more is added annually as the worldwide problems associated with contaminated mine water are becoming known. While we usually prefer to see an increase in the size of an artist's oeuvre, this is of limited benefit with the ever-growing number of specialist publications on mine water and water treatment methods. Therefore, rightly, the question arises: why another book? This can be answered simply: because, to date, there are not many books providing a comprehensive summary of the knowledge on this topic.

As the German court rulings on the Rammelsberg and Meggen metal mine cases show, treating contaminated mine water is a very long-term task, but not an "eternity liability" (7. Senat des Bundesverwaltungsgerichts 2014; Anonymous 1995). However, the question of "how long is very long term" cannot yet be conclusively assessed (Beckmann 2006; Spieth 2015; Wolfers and Ademmer 2010). I deliberately refrain from discussing the issue here, as given its complexity, even ten pages could not illuminate the question in all its aspects in every single country where mine water treatment exists (and I do not even dare to tackle the question of how long "eternity" is, but I would say around 25–50 years). Irrespective of this, it is essential to know the various possibilities for treating mine water to be able to adapt such a plant to the respective pollutant loads and legal framework conditions. This is only possible if we know the combined oeuvre of colleagues who deal with mine water treatment, which is the topic of this book. I would have liked to have written a cookbook or a *Circa Instans* to enable the practitioner to plan a plant – you know, like those books with DINs and ISOs and process sketches and rules of thumb to be able to plan and size a plant. However, during the literature review, I quickly realised that this would have required a completely different approach. It was more important for me to summarise the literature for you so that it would be possible to carry out planning based on the available body of knowledge. As a result, the work has unfortunately become a little theory biased, which is why I ask for your forbearance.

As the technology of mine water treatment has changed fundamentally in some areas over the past decades, the present work could have been even more comprehensive and detailed (if your publication is not included here, this is not intentional, but simply due to the large number of publications on the subject). The portfolio of pollutants to be removed is also steadily increasing as they are recognised as being problematic for the environment. One example is the ubiquitous PCBs (polychlorinated biphenyls), which have been recognised as a particular problem in German mine waters in recent years (Schabronath 2018) but have been perceived as such in the USA since the late 1980s (Bench 2000). However, I have chosen to set out only the broad outlines of each process, and then refer to the literature. Some of these are in languages other than English. I would recommend that you use the same method that was used for the rough translation of this book, namely the DeepL Translator (www.DeepL.com), and it will be easy for you to follow a German, Chinese, or French text. The chapter lengths in this book differ due to individual procedures described; however, this does not signify that one method is less or more important than the other. It merely means that there was more to write about one method than another. Some of the information on case studies that I give here has been entrusted to me by colleagues in the spirit of the Chatham House rule. I thank them all for sharing their data with me on this premise.

The target groups for this book are all those who wish to familiarise themselves with mine water treatment technologies using simple descriptions. Specifically, I have in mind mining engineers, engineers, geologists, geoecologists, biologists, government officials, environmental activists, students, remediation contractors, and journalists. You do not necessarily need a comprehensive education in physics and chemistry. However, I assume a basic understanding of the chemical, physical as well as biological processes that occur around us or are important in the context of water or mine water remediation. The comprehensive bibliography with over 1000 entries should enable you to acquire supplementary information at any time.

However, this book also addresses you, the reader who is doing monitoring as part of your daily routine. Finally, the book is aimed at engineering consultants who plan mine water treatment plants and want to obtain an overview of the technologies before digging into the literature. I have also included examples from my 32 years of experience in the mining environment, basically focusing on "pitfalls": What can be done wrong? Some of these examples could be understood as a critique of the existing scientific system, in which often only the quantity of publications is relevant – not their quality.

For several years now, the GARD (Global Acid Rock Drainage) Guide has provided a relatively comprehensive guide on how to prevent or treat contaminated mine water (Verburg et al. 2009). Many details of the procedures presented in this book can also be looked up in that guide. Sometimes, the GARD Guide is criticised for being "industry-oriented" – but the guide was created precisely with this in mind: to show tried and tested methods for preventing or cleaning up contaminated mine water. Membrane processes therefore take up just less than 500 words in the GARD Guide; these gaps have now been filled in this book.

A gap in *this* book is biometallurgy, which is now also entering the literature as geobiotechnology. It seems to be experiencing a revival at present, although many of the key features were already developed in the 1970s–1990s (Lundgren and Silver 1980; Paños 1999). There is no doubt that biometallurgical processes play a major role (Rohwerder et al. 2003) and their importance will increase considerably in the coming decades if mine water and mining wastes can be successfully converted from waste to raw material. However, under the keyword "biometallurgy" there is still no large-scale plant that treats mine water and makes the metals available as a raw material. The same applies to the keyword "Circular Economy"– unfortunately, we have not yet reached the point where any valuable materials from mine water or the residues from mine water treatment could be fed into the raw material cycle. Therefore, no further information on this topic is given in this book, although I do not fully agree with the opinion of Fritz Haber, who wrote about gold extraction from seawater: "I have given up looking for this dubious needle in a haystack" (Haber 1927, p. 314). At the beginning of 2022, Rio Tinto started a worldwide initiative, asking researchers to submit projects for raw material extraction from mining-influenced water (MIW) – let's see what ultimately transpires.

It is possible that mining for non-energy raw materials is experiencing a renaissance in some traditional mining countries such as Germany (Mischo and Cramer 2020), even though many of the projects that showed promise just a few years ago have now been discontinued. This would require a reorientation in mine water treatment technology, because the requirements for a clean environment are much higher today than in the days when there were hundreds of operating mines in central Europe. Let me first mention the deposits of rare earth metals in Delitzsch (Storkwitz: Seltenerden Storkwitz AG 2013, p. 23), copper in Lusatia (Seidler 2012), fluorspar (Niederschlag fluorite-barite vein deposit: Rauner 2011) or tin in the Ore Mountains (Tellerhäuser project of Saxore Bergbau GmbH on tin, zinc, silver, indium, and iron: Sebastian 2013, p. 151, Roscher 2021), which are currently mined or were explored some time ago. Modern and efficient technologies for the treatment of mine water at these and all other sites around the world will help to strengthen the acceptance of mining among the population to obtain the social licence to operate. However, an open information policy, sensible wording and transparency also go a long way towards changing the negative perception of mining by strengthening an understanding of what is happening in mining and in relation to mine water. Perhaps blogs, press releases or websites that criticise the mining sector or even suspect conspiracies would lose their explosive power as a result.

With the end of coal mining in Germany in December 2018 (Fischer 2016; Tönjes 2016), the topic of mine water treatment moved to the forefront of discussions about eternal tasks. It will take quite some time before the natural flooding of abandoned mines will be completed (Baglikow 2012; Terwelp 2013), and only then the quality of the mine water that will reach the ecosphere will be precisely known. Therefore, for the time being, no further information on the topic of German hard coal mine flooding is given in this book.

In conclusion, however, let us not be under the illusion that we will one day succeed in dispensing with the treatment of mine water. Walter Ficklin used to say "Drainage happens..." (*1937–†1993), and George Vranesh even called mine water "the common enemy" (Vranesh 1979). There will be neither sufficiently good in situ processes, nor will mining cease one day, and nor will using ISL (in situ leaching) or ISR (in situ recovery) processes entirely solve the problem of contaminated mine water. None of us, no matter how we feel about the mining industry, is ready to give up using one metal or another in our daily lives. Even Stone Age people extracted raw materials and traded them "worldwide" (Holgate 1991; Shepherd 1993), although the resulting environmental problems – if any – were comparatively minor, but this was mainly due to the size of the "operations." There would only be an end to mining if we lowered ourselves to the level of Anthropoidea. Since few of us would probably want to do that, we need to try to be proactive in addressing the problems and put procedures in place as early as possible from which responsible stewardship of nature can be derived. At least the large corporations usually have the money and attentive shareholders to largely eliminate the environmental problems they cause – on the other hand, the global small-scale mining industry, which is hardly noticed (Fig. 1), often lacks the information and the financial resources to avoid or treat mine water (Sierra Florez and Gaona 2021). As long as gold is mined by Brazilian small-scale gold miners (called garimpeiros) or rare earth metals come via China from the Congo (keyword coltan), where they are mined by children and women, and everyone lacks the financial means for water treatment, we will have to deal with the problem of mine water.

**Fig. 1** Artisanal sapphire miners at the Ilakaka river near Isalo National Park, Madagascar. Suspended solids and stream bed modifications are impairing the river's ecosystem (© Andreas Hafenscher)

Conrad Matschoss, the editor of the German Agricola translation from 1928, wrote about Georgius Agricola in his preface (Agricola 1928 [1557], p. IX):

> Agricola made it the purpose of his writings to inspire the youth to the study of nature. He devoted himself passionately and with all his soul to the study of nature, and he placed science higher than wealth, fortune, and positions of honour. We need this spirit of passionate devotion to science more than ever for the advancement of mankind.

I hope that this book, in the spirit of Agricola, will help to improve mine water treatment (if you are not familiar with a term, consult the glossary in the Appendix). Furthermore, I hope that it will inspire one or the other reader to "export" or optimise a known method. If this happens, I consider my goal achieved – Glückauf!

In the South African summer (Pretoria), the Central European winter (Tyrol), and the sultry South Atlantic (St. Helena)
2020 (German version)
In hot Namibia, the Kalahari, the Karoo, and refreshing Cape Town
2022 (English translation)

Pretoria, South Africa Christian Wolkersdorfer

# Acknowledgements

First, thanks to Christin Jahns, who convinced me in several telephone calls to Austria to write the initial German version of this book on mine water treatment. A special word of thanks to the Saxon State Office for Environment, Agriculture and Geology (LfULG) in Dresden/Saxony and to Frank Sander, who provided the idea for this book by funding a study as part of the German–Czech EU project Vodamin. Without their support, this book would not have been possible.

Thanks to my colleague Wu Qiang of the Chinese Mining University in Xuzhou, who gave me a recommendation of Chinese literature on the subject – 谢谢你们的帮助。Bob Hedin generously provided photos of his pigments, Jeff Skousen of his passive treatment equipment, Kathy Karakatsanis provided the principal picture for the mine water treatment system in eMalahleni as well as for the Optimum Coal Mine, and Eberhard Janneck contributed pictures of his schwertmannite. Charles Cravotta III selflessly provided tables and figures, and Naďa Rapantova and Jiří Wlosok translated the English abstract into Czech. Irina Levchuk helped me understand a foreign language text, and Olga Oleksiienko pondered how we could get to the East Ukrainian–Russian war zone without coming to harm while visiting a mine water treatment plant (ultimately, we did not visit that plant). Evan T. Williams II and Gary Antol Sr. spent a Saturday morning locating the Thompson well at the Vesta No. 5 colliery in Pennsylvania (thank you, Mr Zuckerberg). Thanks to everyone who responded to my e-mail inquiries or who allowed me to use a photo that was taken by them, such as Martin Jochman, or that featured them. In particular, I would like to thank the colleagues who are involved with the F-LLX process. With no technology I have written as many e-mails as I have with this one to obtain information. So, thanks to Todd Beers, Dave Cercone, and John McArdle for sending me their material – I am most grateful for your generosity. All photos without copyright notices are my own.

Thanks to Stephanie Preuß, Sebastian Müller, and Martina Mechler at Springer-Verlag, who were responsible for the German manuscript and Mounika Gujju, Simon Shah-Rohlfs, and Naresh Veerabathini for handling the English translation. Thanks to Ingrid G. Buchan, who carefully copy-edited the English translation of this book.

Above all, I would like to thank all my friends, colleagues, students, bosses, employees, and co-workers in Clausthal, Freiberg, and Munich in Germany, on Cape Breton Island in

Canada, in Finland, and in South Africa, who have accompanied me over the past three decades. Without their tireless support and without the mistakes they made together with me, this work would not have existed.

I thank all my friends at the International Mine Water Association who provided answers to questions that I could not answer myself. First and foremost, I wish to thank Bob Kleinmann, who encouraged me when I could not go on! But I am also very grateful for the assistance of our former presidents Adrian Brown and John Waterhouse and the treasurers Lee Atkinson, Jennifer Geroni, and Alison Turner.

Thanks to Ines, who gave me a completely new perspective on mine water for 4 days in the summer of 2012. Thanks to Kathleen and Mike as well as to all employees of the Momo Coffee Roastery in Freiberg/Saxony, who played a substantial role in making it possible to finish the German version of this text. Without your coffee, tea, and cake, some of the chapters would not have turned out so well! Thank you to the people of Jamestown on St. Helena who hosted me for a week, especially Toni, Ivy and Nim, and Jane from Anne's Place. I thank Kay Florence, whose confidence for some time encouraged me not to lose sight of this text, and who allowed me to use illustrations I had made for her dissertation. *Ngiyabonga* to my chicken farmer and former postdoc Themba Oranso Mahlangu who took the trouble to fill in all the missing DOIs (before we found the excellent online tool www.crossref.org/simpleTextQuery). Thanks to Zicki who always goes the extra mile to support me and takes care of all the administrative tasks. *Kiitos* also to Riko Bergroth, from Muurla in Finland, who patiently listened to me as we set up our first containerized VFR (vertical flow reactor) in Finland. I thank Anne Weber, who provided me with a list of names of the remaining pit lakes in Lusatia. Thanks to all those who took the trouble to improve the initial German text, first Bernd Schreiber, Kathrin Kranz, and Elke v. Hünefeld-Mugova, but especially to my colleagues Georg Wieber, Frank Wisotzky, Michael Paul, Thomas Walter, Broder Merkel, Esther Takaluoma, Petra Schneider, Oliver Totsche, Felix Bilek, Uwe Grünewald, and the Franconian Roland Haseneder. I was able to incorporate most of your comments into the text – the responsibility for any remaining errors remains with me, of course.

I hope that one day we will succeed in jointly bringing a process to market maturity that will take mine water treatment up a notch – and if not, then at least we have had fun experimenting.

Finally, I would like to thank Ulrike, Karoline, and Franziska. Even if you did not always feel it – you have always been the motivation for all my projects.

# Abstract | Zusammenfassung | Shrnutí

## English Abstract

The initial reason for this compilation of mine water treatment processes was the German-Czech joint project VODAMIN. The aim of this project was to investigate the effects of mining activities on the quality of groundwater and surface waters. In terms of work package four, treatment methods for mine water were to be presented and, if possible, evaluated. The aim of this publication is therefore to address a range of methods for assessing mine water. Currently known methods and systems for the treatment of mine water are presented. Firstly, an overview of mine water in general will be given, as well as how it can be classified and how correct sampling should be carried out. This is followed by a compilation of the known methods for the treatment of mine water, based on the internationally accepted classification into active and passive treatment methods, (monitored) natural attenuation and in situ measures. At the end, there is a presentation of alternative uses for abandoned mines, aimed at mine water. Interspersed, glossary-like contributions highlight possible pitfalls of sampling or management.

To this end, some 3000 publications on the subject of mine water treatment were carefully studied and, based on this, a compilation of all the important processes currently known will be given. A comprehensive bibliography with over 1000 entries allows you to obtain further information. Since access to literature varies from person to person, there are always several citations for each process, so that in any case at least one publication should be found for reference.

Up until now, it has only been possible to a limited extent to plan the exact design of a mine water treatment system based on water analysis data. Therefore, laboratory tests and pilot plants are often necessary to determine the optimal plant configuration for the full-scale plant. Future research and development should therefore aim at an optimised understanding of the process to get closer to the goal of integrated mine water treatment. In this regard, the "Internet of Mine Water" could also make a contribution (see Sect. 1.1), as it interconnects all the relevant components of mine water management. Reliable and correct sampling procedures with flow rates and on-site parameters are essential for planning a mine water treatment plant. Although there is no generally accepted procedure, company standards that are largely based on national or international standards have become established. Some parameters must always be determined for each sampling procedure, whereas others can be determined as needed.

A total of twelve groups of active treatment processes that are currently being used or are at advanced stages of development are presented. These include neutralisation, electrochemical processes, and applications of membrane-based technologies, as well as some methods that are less well known to date. Special variants, which are often not used because of patenting reasons, but differ only minimally from other processes, are not discussed. Eleven of the passive treatment processes are presented. These are, for example, limestone drains or channels, constructed wetlands, permeable reactive walls, and reducing and alkalinity producing systems. Methods that have only been used at one or two sites or are at a very early stage of development are largely disregarded. Three alternative methods, which show considerable potential for development, are associated with in situ measures in open pits and underground mines. These alternative methods include (monitored) natural attenuation, modification of mining conditions, or the biometallurgical processes that are at an early stage of development. Of the in situ methods, the various in-lake processes and the reinjection of residues are described in more detail.

Finally, a compilation of possible alternative uses follows. These are divided into the use of the abandoned mines themselves and the use of the residues resulting from mine water treatment.

## German Zusammenfassung

Anlass dieser Zusammenstellung von Reinigungsverfahren für Grubenwasser war das deutsch-tschechische Gemeinschaftsprojekt VODAMIN. Dieses hatte sich zum Ziel gesetzt, die Auswirkungen der Bergbauaktivitäten auf die Wasserqualität von Grund- und Oberflächenwässern zu untersuchen. Innerhalb des Arbeitspaketes 4 sollten Reinigungsverfahren für Grubenwasser vorgestellt und wenn möglich bewertet werden. Ziel dieser Publikation ist daher, auf eine Reihe von Verfahrensweisen zur Beurteilung von Grubenwasser einzugehen. Es werden die derzeit bekannten Methoden und Systeme zur Reinigung von Grubenwasser vorgestellt. Zunächst wird in einem Überblick dargelegt, was Grubenwasser ist, wie es klassifiziert werden kann und wie eine korrekte Probenahme zu erfolgen hat. Danach folgt eine Zusammenstellung der bekannten Methoden zur Reinigung von Grubenwasser, basierend auf der international üblichen Einteilung in aktive und passive Verfahren, (kontrollierte) natürliche Selbstreinigung sowie In-situ-Maßnahmen. Am Ende findet sich eine Darstellung von alternativen Nutzungsmöglichkeiten für aufgelassene Bergwerke, die auf das Grubenwasser abzielt. Eingestreute, glossenartige Beiträge beleuchten mögliche Fallstricke der Probenahme oder des Managements.

Dazu wurden etwa 3000 Publikationen zum Thema Grubenwasserreinigung sorgfältig studiert und darauf basierend eine Zusammenstellung aller wichtigen, derzeit bekannten Verfahren gegeben. Ein umfassendes Literaturverzeichnis mit über 1000 Eintragen erlaubt es dem Leser, weitergehende Informationen einzuholen. Da der Zugang zu Literatur individuell verschieden ist, gibt es zu jedem Verfahren stets mehrere Zitate, sodass in jedem Fall wenigstens eine Publikation zum Nachlesen gefunden werden sollte.

Bislang ist es nur eingeschränkt möglich, anhand der Wasseranalyse eine exakte Bemessung des Reinigungssystems zu planen. Daher sind vor einer Vollinstallation oftmals Laborversuche und Pilotanlagen notwendig, um die optimale Anlagenkonfiguration zu ermitteln. Künftige Forschungen und Entwicklungen sollten daher auf ein optimiertes Prozessverständnis abzielen, um dem Ziel einer integrierten Grubenwasserreinigung näherzukommen. Dazu könnte auch das „Internet of Mine Water" beitragen, in dem alle relevanten Komponenten des Grubenwassermanagements zusammengefasst sind.

Zur Planung einer Grubenwasserreinigungsanlage ist eine zuverlässige und korrekte Probenahme mit Volumenströmen und Vor-Ort-Parametern unerlässlich. Obwohl es keine allgemein anerkannte Verfahrensweise gibt, haben sich Firmenstandards eingespielt, die weitgehend auf nationalen oder internationalen Standards beruhen. Einige Parameter sind grundsätzlich bei jeder Probenahme zu ermitteln, wohingegen andere nach Bedarf ermittelt werden können.

Insgesamt werden zwölf Gruppen von aktiven Reinigungsverfahren vorgestellt, die derzeit angewendet werden oder in fortgeschrittenen Entwicklungsstadien sind. Darunter befinden sich die Neutralisation, elektrochemische Verfahren, Membrananwendungen und einige bislang weniger bekannte Methoden. Auf spezielle Varianten, die meist aus patentrechtlichen Gründen nicht zum Einsatz kommen und sich nur minimal von anderen Verfahren unterscheiden, wird dabei nicht eingegangen. Von den passiven Verfahren werden elf vorgestellt. Dabei handelt es sich beispielsweise um Carbonatkanäle, konstruierte Feuchtgebiete, permeable reaktive Wände und reduzierende Alkalinitätssysteme. Verfahren, die bislang nur an einem oder zwei Standorten zum Einsatz kamen oder sich in einem sehr frühen Entwicklungsstadium befinden, bleiben dabei ebenfalls weitgehend unberücksichtigt. Drei alternative Methoden, die ein erhebliches Entwicklungspotenzial aufweisen, leiten über zu In-situ-Maßnahmen in Tagebauen und Untertagebergwerken. Zu den alternativen Methoden zählen (kontrollierte) natürliche Selbstreinigung, Änderung der Abbaubedingungen oder die im Anfangsstadium stehenden biometallurgischen Verfahren. Von den In-situ-Methoden werden die unterschiedlichen Inlake-Verfahren sowie die Ruckspülung von Reststoffen näher beschrieben.

Abschließend folgt eine Zusammenstellung von alternativen Nutzungsmöglichkeiten. Diese sind aufgeteilt in Nutzung der aufgelassenen Bergwerke selbst und Nutzung der Reststoffe, die bei der Grubenwasserreinigung anfallen.

## Czech Shrnutí

Předložená publikace uvádí řadu postupů pro hodnocení důlních vod a představuje dosud známé metody a systémy čištění důlních vod. V první části je uveden obecný přehled znalostí o důlních vodách, jejich klasifikaci a vhodných způsobech vzorkování. Následuje souhrn známých metod čištění důlních vod, rozdělený na základě mezinárodní klasifikace na aktivní a pasivní metody, přirozenou atenuaci a *in-situ* metody. Na závěr jsou představeny alternativní způsoby využití uzavřených důlních děl se zaměřením na využití důlních vod. Text je navíc doplněn o příklady zdůrazňující potenciální nástrahy způsobů vzorkování a správy.

Svým objemem jsou důlní vody považovány za jeden z největších zdrojů vod přímo souvisejících s antropogenní činností, ne vždy jsou však kontaminovány. I přesto jsou nečištěnými kyselými důlními vodami či důlními vodami s obsahem kovů kontaminovány či potenciálně ohroženy tisíce kilometrů vodních toků, rozsáhlá území v přírodních rezervacích a nespočet aquiferů. Pro minimalizaci negativních dopadů na ekosféru a antroposféru je proto nezbytné čistit důlní vody tak, aby byl dopad na životní prostředí minimální nebo žádný. Z tohoto důvodu jsou důlní vody čištěny na přijatelnou kvalitu všude tam, kde jsou dostupné finanční prostředky.

Možnosti plánování systému čištění vod, který je navržen pouze na základě výsledků chemické analýzy vod, jsou zatím omezeny. Před plným provozem systému je potřeba nejdříve provádět laboratorní experimenty a následně využít testování pilotních technologií a určit tak optimální konfiguraci celého systému. Budoucí výzkum a vývoj by proto měl být zaměřen na komplexní pochopení dílčích procesů za účelem zřízení integrovaného systému čištění důlních vod. K dosažení tohoto cíle by mohl přispět i "Internet of Mine Water", prostřednictvím kterého mohou být vzaty v úvahu všechny složky procesu hospodaření s důlními vodami.

Pro navržení systému čištění důlních vod je nezbytná správná metodika vzorkování. Protože neexistuje žádná všeobecně uznávaná metodika, vyvinula řada firem normy, jež jsou z velké části založeny na národních nebo mezinárodních směrnicích. Během vzorkování musí být některé parametry stanovovány při každém odběru, zatímco jiné mohou být stanoveny podle potřeby. Publikace shrnuje nejběžněji stanovované parametry a správnou metodiku vzorkování a měření parametrů *on-site*.

Je zde popsáno celkem dvanáct druhů aktivních metod čištění, jež jsou v současné době používány, nebo jsou v pokročilé fázi vývoje. Zahrnuty jsou neutralizace, elektrochemické procesy, membránové aplikace i některé méně známé metody. Specifické druhy metod, které se jen nepatrně liší od jiných a jejichž využití je většinou omezeno kvůli patentové ochraně, nejsou zahrnuty. Z pasivních metod je zde popsáno jedenáct druhů. Patří k nim vápencové drenáže, umělé mokřady, propustné reaktivní bariéry a redukční a alkalické systémy. Metody, jež byly testovány pouze na jednom nebo dvou místech, nebo jsou zatím v rané fázi vývoje, nejsou popsány. Dále jsou uvedeny tři alternativní metody se značným rozvojovým potenciálem. Mezi alternativní metody patří přirozená atenuace, modifikace těžebních metod nebo biometalurgické procesy, které jsou v současné době v rané fázi vývoje. Navazuje výčet *in-situ* metod využitelných pro povrchové a hlubinné doly. *Z in-situ* metod jsou charakterizovány různé *"in-lake"* procesy a zpětná injektáž reziduí z procesu čištění vod.

V závěru jsou uvedeny způsoby alternativního využití důlních děl. Ty jsou rozděleny jednak na využití samotných uzavřených důlních děl, jednak na využití zbytkových materiálů z procesu čištění důlních vod.

# Contents


**1 Introduction** ... 1

1.1 Sidenotes – or Experiences After More Than a Decade of Literature Review ... 1

1.2 Definition of Terms ... 7

1.2.1 Problems with the Definition of Terms ... 7

1.2.2 Active Mine Water Treatment ... 11

1.2.3 Base Capacity ($k_B$; Acidity; m-Value) ... 12

1.2.4 Mine ... 13

1.2.5 Bioreactor ... 14

1.2.6 Circular Economy ... 14

1.2.7 Mine Water Discharge – “Decant” ... 15

1.2.8 First Flush ... 15

1.2.9 Mine Flooding ... 18

1.2.10 Mine Water (Mine Drainage, Mining Influenced Water) ... 19

1.2.11 Constructed Wetlands for Mine Water Treatment ... 20

1.2.12 Coagulation and Flocculation ... 21

1.2.13 Net Acidic or Net Alkaline Mine Water ... 21

1.2.14 Passive Mine Water Treatment ... 22

1.2.15 Treatment Wetlands for Municipal Wastewater ... 23

1.2.16 Phytoremediation ... 23

1.2.17 pH Value ... 24

1.2.18 Acid Capacity ($k_A$; Alkalinity; p-Value) ... 25

1.2.19 Acid Mine Drainage ... 26

1.2.20 Sorption, Adsorption, Coprecipitation, Surface Complexation and Other Such Reactions ... 26

1.2.21 Heavy Metal ... 29

1.2.22 Base Metal ... 29

1.3 Formation of Mine Water and Buffer Mechanisms ... 29

1.4 Classification of Mine Water ... 36

**2 Preliminary Investigations** . . . . . 41
2.1 Introductory Remarks . . . . . 41
2.2 Mine Water Sampling . . . . . 47
2.2.1 Checklists and Notes. . . . . 47
2.2.2 Note on Occupational Health and Safety . . . . . 49
2.2.3 Sampling Methods . . . . . 50
2.2.4 Quality Control . . . . . 55
2.2.5 Measuring Instruments and Sampling . . . . . 57
2.2.6 Sample Names . . . . . 59
2.2.7 Dissolved and Total Concentrations . . . . . 60
2.2.8 Documentation . . . . . 63
2.3 Essential On-Site Parameters. . . . . 64
2.3.1 Introductory Note . . . . . 64
2.3.2 Electrical Conductivity (Specific Conductance). . . . . 66
2.3.3 Base Capacity ($k_B$; Acidity). . . . . 68
2.3.4 Acid Capacity ($k_A$; Alkalinity) . . . . . 69
2.3.5 Flow and Loads. . . . . 70
2.3.6 pH Value . . . . . 77
2.3.7 Iron Concentration . . . . . 80
2.3.8 Manganese Concentration. . . . . 81
2.3.9 Aluminium Concentration . . . . . 82
2.3.10 Redox Potential ($E_h$, ORP) . . . . . 83
2.3.11 Oxygen Saturation . . . . . 85
2.4 Water Analysis . . . . . 86
2.5 Lime Addition or Column Tests. . . . . 86
2.6 Active or Passive Mine Water Treatment? . . . . . 88
2.7 The Endless Mine Water Treatment Plant . . . . . 90
2.8 Pourbaix Diagrams (Stability Diagrams, Predominance Diagrams, $E_h$-pH-Diagrams, "Confusogram" *sensu* P. Wade). . . . . 92

**3 Active Treatment Methods for Mine Water** . . . . . 95
3.1 Introduction . . . . . 95
3.2 Neutralisation Process . . . . . 97
3.2.1 Principles and Historical Development. . . . . 97
3.2.2 Low Density Sludge (LDS) Process . . . . . 105
3.2.3 High Density Sludge (HDS) Process . . . . . 107
3.2.4 It Is Easier to Live in a Box . . . . . 110
3.3 Electrochemical Processes. . . . . 112
3.3.1 Electrocoagulation . . . . . 112
3.3.2 Electrosorption (Condensation Deionisation). . . . . 116
3.3.3 Electrodialysis/Membrane-Based Electrolysis . . . . . 117

3.4 Membrane-Based Processes. 118
3.4.1 Introduction. 118
3.4.2 Microfiltration. 123
3.4.3 Ultrafiltration. 124
3.4.4 Nanofiltration 124
3.4.5 Reverse Osmosis (RO) 125
3.4.6 SPARRO Process (Slurry Precipitation and Recycle Reverse Osmosis) 128
3.4.7 Forward Osmosis (FO) 129
3.5 Precipitation Methods for Uncommon Contaminants 130
3.6 Ettringite Precipitation 131
3.6.1 SAVMIN™ Process 131
3.6.2 Other Procedures. 134
3.7 Schwertmannite Process 134
3.8 Bioreactors (Fermenters) 135
3.9 Ion Exchange 138
3.10 Sorption 141
3.11 Advanced Oxidation 144
3.12 Flotation Liquid-Liquid Extraction (F-LLX; VEP; HydroFlex™ Technology) 145
3.13 Eutectic Freeze Crystallisation 147

**4 Passive Treatment Methods for Mine Water** 151
4.1 Note 151
4.2 What Is Passive Mine Water Treatment? 155
4.3 Limestone Drains and Channels 158
4.3.1 Classification of Limestone Drains and Channels 158
4.3.2 Anoxic Limestone Drain (ALD). 158
4.3.3 Oxic Limestone Drain (OLD). 162
4.3.4 Open Limestone Channel (OLC) 163
4.4 Constructed Wetlands. 164
4.4.1 Prologue (I Have Always Wanted to Write This) 164
4.4.2 Aerobic Constructed Wetland (Reed Bed) 166
4.4.3 Anaerobic Constructed Wetland (Anaerobic Wetland, Compost Wetland). 169
4.5 Reducing and Alkalinity Producing Systems (RAPS); Successive Alkalinity Producing Systems (SAPS); Sulfate Reducing Bioreactor, Vertical Flow Wetlands 172
4.6 Settling Pond (Settling Basin, Settlement Lagoon) 173
4.7 Permeable Reactive Walls 177
4.8 Vertical Flow Reactor (VFR). 179
4.9 Passive Oxidation Systems (Cascades, "Trompe") 182
4.10 ARUM (Acid Reduction Using Microbiology) Process 186

**5 Alternative Methods for the Management of Mine Water** ... 189
5.1 Thoughts on Alternative Treatment Methods and Their Application ... 189
5.2 Natural and Monitored Natural Attenuation ... 193
5.2.1 Natural Attenuation ... 193
5.2.2 Monitored Natural Attenuation ... 195
5.3 Change in Mining Methods ... 197
5.4 Biometallurgy, Geobiotechnology, Biomimetics or Agro-metallurgy ... 198

**6 In Situ and On-site Remediation Measures** ... 203
6.1 Introductory Remark ... 203
6.2 In-lake Processes ... 204
6.2.1 Introduction ... 204
6.2.2 In-lake Liming ... 204
6.2.3 Stimulated Iron and Sulfate Reduction in Lakes ... 204
6.2.4 Electrochemical and Electro-biochemical Treatment ... 206
6.3 Chemical Treatment Measures to Reduce Pollutants ... 207
6.3.1 Treatment of Acidic Lakes ... 207
6.3.2 On-site Chemical Treatment Measures ... 212
6.4 Reinjection of Sludge, Treatment Residues or Lime ... 213
6.5 Remediation of Contaminated Watercourses ... 216
6.6 In situ Remediation of Uranium-Containing Mine and Seepage Water ... 217
6.7 Mixing of Pyrite-containing Substrates with Alkaline Material ... 219
6.8 Closure of Drainage and Mine Adits ... 222

**7 Post-mining Usage of Mine Sites or Residues of the Treatment Process** ... 227
7.1 Post-mining Usage of Remediated Sites ... 227
7.2 Treatment Residues as Recyclable Materials (Circular Economy) ... 234

**8 Finish** ... 243

**Correction to: Introduction** ... C1

Lagniappe – Bonus Chapter ... 245

References ... 253

Index ... 311

# Notes from the Author

## Legal Notice

A large part of the mentioned processes for mine water treatment is protected by patent. The absence of a reference to the patent does not signify that the process could be freely used by anyone. This also applies to registered trade names, which are not always indicated. In case of doubt, the international database of the European Patent Office, the Internet search pages of Google Patents or, if necessary, a patent attorney (usually the most expensive option) should be consulted. Furthermore, the author does not take responsibility if a system does not work, although it was built based on the information in this book. It is the responsibility of each designer of a mine water treatment system to additionally consult the literature cited or to consult an expert. If, on the other hand, a system mentioned in this book works perfectly, I will gladly take credit for it.

## Company Names

Any use of trade, company, or product names is for descriptive purposes only and does not imply my or the publisher's endorsement. I do not have any commercial relationships with any of the companies listed nor have received any products or services free of charge for the specific purpose of presenting them here. For some of the companies mentioned, I did consultancy work.

## Note on Gender Mainstreaming

I have refrained from using the feminine and masculine forms throughout to make the text easier to read for my readers. Of course, all examples, explanations, or potential errors generally refer to all three genders.

## Texts and Illustrations from Previous Publications

In a few cases, I have used sentences or illustrations from my earlier publications without referencing them in detail. Wherever it was linguistically possible, however, I have clearly reworked the texts. All these publications are listed in the bibliography. The text of this publication is based on an expert opinion for the LfULG (Sächsisches Landesamt für Umwelt, Landwirtschaft und Geologie) in Dresden, Saxony, Germany. Its publication by Springer Nature was approved with Ref: 46(13)-4331/142/27, B 559 in May 2014.

# List of Figures

Fig. 1.1 Flow diagram for the "Internet of Mine Water". (From Wolkersdorfer 2013) . . . 6
Fig. 1.2 Terms and pH values commonly used in connection with acid capacity and base capacity ($k_A$, $k_B$) in mine water. The three wavy symbols each represent a mine water whose $k_B$ and $k_A$ values are determined using a base (NaOH) or an acid (HCl, $H_2SO_4$). The inflection point in the acidic range can vary between 3.7 and 5.1, depending on the mineralisation, whereas that in the alkaline range is either 8.2 or 8.3, depending on the specification. Concentrations should be expressed in mmol $L^{-1}$ and not in mg $L^{-1}$ $CaCO_3$ as is often the case. The p- and m-values, whose designations are derived from the equivalence point of phenolphthalein and methyl red, are only of historical relevance . . . . . . . . 13
Fig. 1.3 Characteristic first flush scenario using the example of the Rothschönberg adit in Saxony. (Modified from Wolkersdorfer 2008); Cessation of mining in 1969 (Jobst et al. 1994) . . . . . . . . 16
Fig. 1.4 First flush at the Buttonwood adit, one of the main mine water discharges in the Wyoming Basin of eastern Pennsylvania, USA. (Modified after Ladwig et al. 1984, and supplemented with data from the USGS water database. For the period after 1980, only four individual measurements were available, so that no minimum and maximum areas could be represented; from Wolkersdorfer 2008) . . . . . . . . 16
Fig. 1.5 Stalactite of melanterite above the puddle where Alpers and Nordstrom measured the mine water to have a pH of −3.6. The mine water in the 2-L-beaker has a pH of −0.7 and a temperature of 35 °C. (Pers. comm. Kirk Nordstrom 2019; photo: Kirk Nordstrom & Charly Alpers). Scale: Plastic beaker . . . . . . . . 25
Fig. 1.6 Schematic representation of abiotic pyrite and marcasite oxidation. (Based on information in Kester et al. 1975; Singer and Stumm 1970; Stumm and Morgan 1996) . . . . . . . . 30

Fig. 1.7 Drops of acid mine drainage with pH 2 at a gallery roof in the former Glasebach mine in Straßberg/Harz, Germany. Second level: area of the pyrite passage. (Image width approx. 20 cm) . . . . . 31
Fig. 1.8 Red coloured water in a section of the Rio Tinto River in Spain. (Photo: Marta Sostre). . . . . 33
Fig. 1.9 Ficklin diagram with mine waters from Colorado, USA. (Modified and supplemented from Ficklin et al. 1992). All waters investigated by Ficklin in the vicinity of mines are shown as a shaded background. "Metals" is a synonym for "metal concentrations" of the ordinate; FND: Fanie Nel Discharge, Witbank, South Africa . . . . . 36
Fig. 1.10 Expanded Younger-Rees diagram (Rees et al. 2002; Younger 1995, p. 106; 2007, p. 98). I. Acidic spoil leachates, tailings drainage, and shallow oxygenated workings in pyrite-rich strata; II. Majority of fresh, shallow, ferruginous mine waters; III. Previously acidic mine waters, since neutralised; IV. Deep-sourced pumped, saline, mine waters; and V. Field in which few mine waters plot . . . . . 37
Fig. 2.1 Possible development of mine water quality in a mine where the pyrite concentration exceeds that of the carbonates. The second pH increase is due to different buffer processes and the fact that the pyrite is largely weathered. (Modified after Younger et al. 2002) . . . . . 44
Fig. 2.2 Sampling waste from a less environmentally conscious colleague. (Cape Breton Island, Nova Scotia, Canada – but not only found there) . . . . . 48
Fig. 2.3 Particles in water and filtering techniques. The limits of 0.02, 0.20 and 0.45 μm refer to membrane filters. (Adapted from Ranville and Schmiermund 1999, p. 185; supplemented and modified from Stumm and Morgan 1996, p. 821) . . . . . 52
Fig. 2.4 Ion balance calculated by different methods. Top: without speciation; middle: without speciation, protons taken into account; bottom: with speciation and protons. (After Nordstrom et al. 2009) . . . . . 56
Fig. 2.5 Decrease in electrical conductivity due to scaling and biofouling on an online probe. Over the course of 9 weeks, the electrical conductivity decreases by 0.7 mS $cm^{-1}$. For comparison, the initial individual measurements taken with a hand-held meter. An evaluation of the measurement in the context of a salt tracer test was not possible . . . . . 58
Fig. 2.6 Forms in which metal species occur, using copper species as an example and potential filtration methods. (Modified after Stumm and Morgan 1996) . . . . . 60
Fig. 2.7 Paul Younger† during on-site determination of iron concentrations (August 2000) . . . . . 65

Fig. 2.8 Result of a chemical-thermodynamic calculation with PHREEQC (PHREEQC 3.5.0–1400). Result of the electrical conductivity calculation highlighted in colour . . . . . . . . . . 67
Fig. 2.9 Flow measurement with an acoustic digital flow meter in a mine water influenced stream (Cape Breton Island, Nova Scotia, Canada). The operator correctly stands downstream of the meter and holds the meter far away from the body to avoid interfering with the flow as far as possible . . . . . . . . . . 71
Fig. 2.10 Incorrectly installed measuring weir at the former IMPI pilot plant in Middleburg, Gauteng Province, South Africa. The sharp edge on the water side of the weir plate is missing . . . . . . . . . . 72
Fig. 2.11 Incorrectly constructed triangular weir for a mine water outlet of the 1 B Mine Pool on Cape Breton Island, Nova Scotia, Canada. The weir plate is too thick (2 cm), is made of non-stainless material, and lacks the sharp edge on the water side. The weir was in use for a period of about 1 year to measure flows with "high accuracy" . . . . . . . . . . 73
Fig. 2.12 Calibrated flow at a measuring weir of the Metsämonttu mine, Finland, and comparison with the flow from the Kindsvater-Shen equation . . . . . . . . . . 74
Fig. 2.13 Flow measurements in an open channel using the 4-lamella 3-point method and preferential flow in the channel's lower right corner (the figure is a section against the direction of flow; velocity in m $s^{-1}$). The single-point lamella method in the middle of the channel gave readings that were 20% too low compared to the salt dilution method. The water level in the channel on the measurement day was 15 cm (Gernrode/Harz, Germany; outlet Hagenbachtal adit, 15th October 2003) . . . . . . . . . . 76
Fig. 2.14 Photometric on-site analysis of total iron and $Fe^{2+}$ . . . . . . . . . . 81
Fig. 2.15 Aluminium concentrations in sequentially filtered mine water from the 1 B Mine Pool in Canada. Of the 970 ± 184 µg $L^{-1}$ Al filtered with a 0.45 µm pore size filter, 750 ± 148 µg $L^{-1}$ is still present in colloidal form (B183 well, Cape Breton Island, Canada) . . . . . . . . . . 82
Fig. 2.16 Limestone and compost experiment (horse manure) with acid mine drainage (100 L tank, approx. 5 kg limestone or manure). A stable pH of 7 is reached after only 2–8 h, as much of the acidity comes from the $CO_2$ in the mine water. Kinetic: water was stirred regularly; static: water was not stirred. Error in pH measurement 0.01 . . . . . . . . . . 87
Fig. 2.17 Relationship between the investment costs of an active mine water treatment plant and the volume of mine water to be treated. (After Morin and Hutt 2006) . . . . . . . . . . 89

Fig. 2.18 Graphical decision aid for and against active or passive water treatment. The boundaries between the processes must be considered blurred. On the ordinate there is an upper area with less contaminated mine water (lower concentration of iron, lower base capacity and flow) and a lower area with more heavily contaminated mine water. $k_B$ in mmol $L^{-1}$, Fe and Zn in mg $L^{-1}$ (ERMITE Consortium et al. 2004; Younger 2002b) . . . . . . . . . . 89

Fig. 2.19 Classification of mine water from different uranium mines based on quotients of U, As, and Ca. Blue squares are seepage and infiltration waters, orange circles mine waters from uranium mine A. For comparison, the diamonds show mine waters from two other uranium mines (B and C). . . . . . . . . . 91

Fig. 2.20 Scatter diagram of uranium and arsenic mass concentrations at uranium mine A. The discharge limits and monitoring concentrations of the mine water treatment plant are shown in green. The blue box corresponds to the uranium and arsenic mass concentrations of the seepage and infiltration waters based on the statistical evaluation given in Fig. 2.19. Initially, the discharge limit for arsenic was set at 0.2 mg $L^{-1}$. By increasing the discharge limit for arsenic to 0.3 mg $L^{-1}$, lower volumes of infiltration and seepage waters are treated . . . . . . . . . . 92

Fig. 2.21 Pourbaix diagram showing the development of mining influenced water from the solid, immune phase to the dissolved, corroded species and the insoluble, passivated phase after mine water treatment. (Supplemented and modified after Pourbaix 1966, p. 314 ff.; Pourbaix 1973, p. 19). . . . . . . . . . 93

Fig. 3.1 Flow diagram for the most relevant, mainly active mine water treatment methods. (Modified and supplemented after Jacobs and Pulles 2007; Younger et al. 2002) [1,2]different decision paths. . . . . . . . . . 96

Fig. 3.2 Pumps for chemical feed in an active mine water treatment plant (Straßberg/Harz, Germany) . . . . . . . . . . 97

Fig. 3.3 pH-dependent solubility of metal hydroxides. (Modified and supplemented after Cravotta 2008, original data obtained from Charles A. Cravotta III, pers. comm. comm. 2013). . . . . . . . . . 103

Fig. 3.4 pH-dependent oxidation rate of Fe(II). (Modified from Singer and Stumm 1969, with data from Stumm and Lee 1961). . . . . . . . . . 103

Fig. 3.5 Examples of pH-dependent sorption of metal cations to iron hydroxides ($[Fe_{tot}] = 10^{-3}$ mol, $[Me] = 5 \times 10^{-7}$ mol; $I = 0.1$ mol $NaNO_3$). (Adapted from Dzombak and Morel 1990; Stumm and Morgan 1996, p. 543) . . . . . . . . . . 104

Fig. 3.6 Principle of conventional mine water treatment (low density sludge process). "Hydrated lime" is used here to represent any usable alkaline material. . . . . . . . . . 106

Fig. 3.7 Aeration of mine water in a low density sludge plant (mine water treatment plant Schwarze Pumpe, Lusatia, Germany). . . . . . . . . . . . . . . . 107
Fig. 3.8 Sludge dewatered by filter presses. (Acid rock drainage treatment plant Halifax Airport, Nova Scotia, Canada; image width approx. 1 m) . . . 108
Fig. 3.9 Principle of the classic high density sludge process with partial recirculation of the sludge. "Hydrated lime" is used here to represent any usable alkaline material. . . . . . . . . . . . . . . . . . . . . . . . . . . 108
Fig. 3.10 Sludge collection tank of the former active Horden mine water treatment plant (County Durham, England, United Kingdom) . . . . . . . . . 109
Fig. 3.11 Schematic drawing of the OxTube of SansOx Oy. (Modified from the company brochure; see also Fig. 4.25) . . . . . . . . . . . . . . . . . . . . . . . . 111
Fig. 3.12 Principle of electrocoagulation. (Modified after Vepsäläinen 2012) . . . . . 113
Fig. 3.13 Possible arrangements of electrodes in electrocoagulation (After Liu et al. 2010). Blue arrows indicate the flow of water . . . . . . . . . . . . . . 114
Fig. 3.14 Electrocoagulation of mine water on a laboratory scale. *Left*: Raw water; *right*: mine water after electrocoagulation. In reality, these plants are at least the size of garden sheds . . . . . . . . . . . . . . . . . . . . . . . 114
Fig. 3.15 Block diagram of the mine water treatment plant with electrocoagulation at the *Glubokaya* (Глубокая) coal mine near Donetsk (Донецьк) in the Donets Basin in Free Ukraine. (Modified from Kalayev et al. 2006). *SS:* suspended solids, *DR:* dry residue without suspended solids . . . . . . . . . . . . . . . . . . . . . . . . . . . . . . . . . . . . 115
Fig. 3.16 Separation limits of different membrane filters and selection of retained constituents. Spacing of the membranes in the graph corresponds to the logarithmic pore width. Retention of constituents based on Schäfer et al. (2006, p. 2) . . . . . . . . . . . . . . . . . . . 119
Fig. 3.17 Number of publications on water treatment using membrane technology between 1990 and 2020. On the right-hand axis, the number of those that are relevant to mine water. (Source: Clarivate Analytics Science Citation Expanded [SCIE]) . . . . . . . . . . . . . . . . . . . . 121
Fig. 3.18 Ultrafiltration unit of the eMalahleni mine water treatment plant in Gauteng Province, South Africa . . . . . . . . . . . . . . . . . . . . . . . . . . . . 124
Fig. 3.19 Simplified principle of osmosis (forward osmosis) and reverse osmosis. In osmosis, ions "migrate" through a semi-permeable membrane along a concentration gradient, thereby increasing the pressure on the concentrate side. In reverse osmosis, on the other hand, pressure is exerted on the side of the concentrate and the water ions diffuse ("migrate") through the membrane to the side of lower concentration . . . . . . . . . . . . . . . . . . . . . . . . . . . . . . . . . . . . . 126
Fig. 3.20 Reverse osmosis unit of the eMalahleni mine water treatment plant in Gauteng Province, South Africa. (See also Fig. 7.7). . . . . . . . . . . . . . . 127

Fig. 3.21 Flow chart for the SPARRO process. (Modified after Juby and Schutte 2000; Juby et al. 1996) . . . 128
Fig. 3.22 Schematic diagram of forward osmosis. (Modified from Cath et al. 2006; adapted from the idea of Weichgrebe et al. 2014) . . . 130
Fig. 3.23 Current process diagram of the Schlema-Alberoda, Germany, mine water treatment plant including residue disposal. (After Fig. 3.4.2.1–23 in Badstübner et al. 2010) . . . 131
Fig. 3.24 Stability diagram of ettringite in the system $CaO/Al_2O_3/SO_3/H_2O$; projected into the $Al(OH)_4$ plane not visible in the graph, $\{Al(OH)_4^-\} = 0$. (Modified after Hampson and Bailey 1982, Fig. 1; optimal pH range of 11.6–12.0 added after Smit 1999) . . . 132
Fig. 3.25 SAVMIN™ pilot plant at Western Utilities Corporation's Mogale gold mine in South Africa. (From Roger Paul: "Presentation to the Parliamentary Committee on Water and Environmental Affairs" – 7/8 September 2011) . . . 133
Fig. 3.26 Schwertmannite on growth elements (Left: image width approx. 2 m) and microbially produced Schwertmannite (Right: image width approx. 15 cm). (Photos: Eberhard Janneck) . . . 135
Fig. 3.27 Used and recommended pH values for selective precipitation of metals and arsenic as sulfides. (Adapted from Kaksonen and Şahinkaya (2012) with data from Govind et al. (1997); Hammack et al. (1994a); Kaksonen and Puhakka (2007); Tabak et al. (2003)) . . . 136
Fig. 3.28 Various synthetic resins for ion exchangers (LANXESS Deutschland GmbH) . . . 138
Fig. 3.29 Industrial ion exchange system for wastewater treatment from copper electrolysis at Montanwerke Brixlegg/Tyrol, Austria . . . 139
Fig. 3.30 Number of publications on water treatment with sorbents between 1990 and 2020.On the right axis, the number of those relevant to mine water. (Source: Clarivate Analytics Science Citation Index Expanded [SCIE]) . . . 142
Fig. 3.31 Comparison of sulfate removal between F-LLX and other treatment methods. (After Monzyk et al. 2010) . . . 145
Fig. 3.32 Former water treatment plant for eutectic freeze crystallisation at the Soshanguve campus of the Tshwane University of Technology, Pretoria, South Africa. (HybridICE® process) . . . 148
Fig. 3.33 PROXA Water Eutectic Freeze Crystallization plant at the New Vaal Colliery, an open pit mine of Anglo American Coal in South Africa. (Photo: Jochen Wolkersdorfer) . . . 149

Fig. 4.1 Part of the passive mine water treatment system for the tailings pile at Enos Colliery (Pike County, Indiana, USA). The operation of the wetland is described in Behum et al. (2008). (The photo composite shows the Canal Aerobic Wetland in the 6.5 ha constructed Enos East wetland) . . . . . . . . . . 152
Fig. 4.2 Periodic table of elements for the passive treatment of mine water. Explanations are given in Table 4.2 . . . . . . . . . . 155
Fig. 4.3 Sedimentation and macrophyte basin and – no longer necessary – activated carbon filter bed and aerobic wetland at the Urgeiriça uranium mine, Portugal. Mine water ($Q \approx 90\ L\ min^{-1}$) flows from right to left, and pH is raised from about 6 to 8 on average. Iron, uranium and radium are retained in the system . . . . . . . . . . 157
Fig. 4.4 Flow diagram for passive mine water treatment processes. (modified after Hedin et al. 1994a; Skousen et al. 1998, 2000; PIRAMID Consortium 2003) . . . . . . . . . . 159
Fig. 4.5 Anoxic limestone drain during the construction phase. The liner sits in the open drain which is already filled with limestone at the rear end. (Photo: Jeff Skouson) . . . . . . . . . . 160
Fig. 4.6 Open limestone channel with iron hydroxide precipitates at the mine water discharge of the former Dominion No. 11 coal mine on Cape Breton Island, Nova Scotia, Canada . . . . . . . . . . 163
Fig. 4.7 Former open limestone channel at the Mina de Campanema/Minas Gerais, Brazil. The limestone clogging is clearly visible . . . . . . . . . . 165
Fig. 4.8 First European anaerobic constructed wetland Quaking Houses (County Durham, England, United Kingdom). Plant cover is absent, as the sludge had been replaced after 10 years of operation . . . . . . . . . . 165
Fig. 4.9 Inlet (left) and outlet (right) of the aerobic constructed wetland Neville Street (Cape Breton Island, Nova Scotia, Canada). Water flows into the constructed wetland in an ochre and metal-rich state (left) and exits in a crystal-clear state with a substantially reduced metal load (right) . . . . . . . . . . 167
Fig. 4.10 Sampling locations of Table 4.4 in the anaerobic constructed wetland Westmoreland County, Pennsylvania, USA. (After Hedin et al. 1988) . . . . 170
Fig. 4.11 Illustration of the overall process for sulfate reduction and formation of monosulfides and pyrite in sediments. (Modified after Berner 1972) . . . . . . 170
Fig. 4.12 Channels caused by gas upwelling on the surface of the organic substrate covered with iron oxyhydrates in the RAPS Bowden Close, County Durham, England. Diameter of the channels in the centimetre range; image width in the centre about 3 m . . . . . . . . . . 172

Fig. 4.13 Site plan of the Bowden Close RAPS, County Durham, England. (Modified from Wolkersdorfer et al. 2016) . . . . . . . . . . . . . . . . . . . . . . . . 174
Fig. 4.14 RAPS II at the Bowden Close site, County Durham, England, looking west. Typical wetland plants are growing by succession on the substrate of limestone and horse manure. Outside the image on the right is RAPS I and on the left is the constructed wetland . . . . . . . 174
Fig. 4.15 Settling pond at the former Straßberg/Harz, Germany, mine water treatment plant. The green colour is mainly caused by suspended limestone. Since the relinquishment of this treatment system, the pond is overgrown with plants . . . . . . . . . . . . . . . . . . . . . . . . . . . . . . . . 175
Fig. 4.16 Settling pond at the Neville Street passive mine water treatment system, Nova Scotia, Canada, before and after installation of baffle sheets. (Details in Wolkersdorfer 2011) . . . . . . . . . . . . . . . . . . . . . . . . . . 176
Fig. 4.17 Schematic diagram of the settling process in a simple settling pond. (Modified after Crittenden et al. 2012) . . . . . . . . . . . . . . . . . . . . . . . . . . . 177
Fig. 4.18 Remediation options for groundwater contamination: (**a**) untreated contamination; (**b**) reactive wall; (**c**) pump and treat; and (**d**) funnel-and-gate system. (After Starr and Cherry 1994) . . . . . . . . . . . 178
Fig. 4.19 Schematic diagram of a vertical flow reactor (modified after Florence et al. 2016). Prolotroll (*Stollentroll*, Adit Troll) by Walter Moers (2000), in *The 13½ Lives of Captain Bluebear* © Verlagsgruppe Random House GmbH . . . . . . . . . . . . . . . . . . . . . . . . . . . . 180
Fig. 4.20 Containerised vertical flow reactor (VFR) at the former Finnish polymetallic Metsämonttu mine with ventilation of the mine water (Sansox Oy OxTube). The outlet of the reactor with measuring weir for the flow is located at the rear end of the 20″-container (water volume: 24 $m^3$) . . . . . . . . . . . . . . . . . . . . . . . . . . . . . . . . . . . . . . 181
Fig. 4.21 Oxidation cascade of the ventilator adit on the Leitzach river, Upper Bavaria, Germany. Flow rate about 2 $m^3$ $min^{-1}$; width of the channel about 1 m. In the meantime it was completely rebuilt . . . . . . 183
Fig. 4.22 Oxidation cascade of the Neville Street passive mine water treatment plant, Nova Scotia, Canada, during a tracer test with uranine (sodium fluorescein). Flow rate about 1 $m^3$ $s^{-1}$; height of the individual cascade stages about 0.6 m . . . . . . . . . . . . . . . . . . . . . . . . . . . . . . . . . 183
Fig. 4.23 Results of two aeration experiments of mine water from Finland and South Africa. The mine water in Kotalahti, Finland, has sufficient buffer capacity to buffer the protons formed during iron hydrolysis, which is lacking in the mine water from the West Rand, South Africa, so that the pH value decreases . . . . . . . . . . . . . . . . . . . . . . . . . . . . . . . . . . 184
Fig. 4.24 Five different stair- and weir-type aeration cascades. (Modified after Geroni 2011, Fig. 4.5) . . . . . . . . . . . . . . . . . . . . . . . . . . . . . . . . . . 184

Fig. 4.25 OxTube in 90° angle installation on the Metsämonttu vertical flow reactor. The oxygen saturation in the mine water could thus be raised from about 40 to 60% . . . 186
Fig. 4.26 Macrophyte basin of an ARUM system at the former Cunha Baixa uranium mine, Portugal. Iron, uranium and radium are retained in the system . . . 187
Fig. 5.1 Natural attenuation using the example of a mine water adit in Bavaria, Germany (Phillipstollen on the Leitzach). Left: Outflow from the collapsed adit; middle: after a flow path of 5 m; right: after a flow path of 20 m. Shortly after the inflow into the Leitzach receiving watercourse, no more iron can be visually detected, as it has almost completely precipitated as iron oxyhydrate . . . 190
Fig. 5.2 William Simon (left, with walking stick), coordinator of the Animas River Stakeholders Group (Colorado, USA), explains the work of a group of volunteers in the remediation of water bodies contaminated by acid mine drainage from former gold mining operations to interested colleagues from the International Mine Water Association (IMWA) . . . 191
Fig. 5.3 Decision tree for monitored natural attenuation. (After ERMITE Consortium et al. 2004) . . . 196
Fig. 5.4 Left: Collection trench of copper-containing solution from heap leaching; right: Preparation of a heap for microbiologically assisted acid leaching of copper-rich ores (Chilean El Salvador copper mine of the Corporación Nacional del Cobre de Chile – Codelco) . . . 199
Fig. 5.5 Heap leaching at the Finnish Terrafame nickel mine (formerly Talvivaara) . . . 199
Fig. 5.6 Copper rich mine water with copper mineral precipitates from the Kilian adit, Marsberg, Germany (width of image 20 cm) . . . 200
Fig. 6.1 Enclosures (“macrocosms”) of varying sizes in residual lake 111, Lusatia, Germany. (Photo: Peter Radke, LMBV) . . . 205
Fig. 6.2 Principle of electro-biochemical water treatment (modified according to company brochure Inotec Inc., Salt Lake City, USA) . . . 207
Fig. 6.3 Neutralisation and $CO_2$ addition station at Lake Scheibe. (Modified from Strzodka et al. 2016, Figure: GMB GmbH) . . . 210
Fig. 6.4 Injection of slaked lime into Lake Scheibe, Germany, in November 2010 via hose lines and a land-based injection station. (Image width about 500 m; Photo: Peter Radke, LMBV) . . . 211
Fig. 6.5 Diffuse discharge of mining-influenced groundwater. Left: Cadegan Brook (Cape Breton Island, Nova Scotia, Canada; image width approx. 1 m); right: Gessenbach (Gessental near Ronneburg, Thuringia, Germany; image width approx. 2 m) . . . 217
Fig. 6.6 Mixing overburden and lime to prevent acid formation in the dumps. (Modified from Huisamen 2017; after Wisotzky 2003) . . . 220

Fig. 6.7 Partial damming of a drainage gallery. (After Foreman 1971) . . . . . . . . . 222
Fig. 6.8 Example of an underground dam. (After Lang 1999) . . . . . . . . . . . . . . . . 223
Fig. 6.9 Examples of dammed drainage galleries. (Adapted from Halliburton Company 1970; Scott and Hays 1975). . . . . . . . . . . . . . . . . 223
Fig. 6.10 "Königl. Verträglicher Gesellschaft Stolln" at the Rote Graben near Freiberg/Saxony, Germany, from the nineteenth century, closed with a grid. Part of the World Heritage Site *Montanregion Erzgebirge/Krušnohoří* . . . . . . . . . . . . . . . . . . . . . . . . . . . . . . 225
Fig. 7.1 Biosphere project "Eden" in a former china clay mine in Cornwall, United Kingdom. (Photo: Jürgen Matern; Wikimedia Commons, CC-BY-3.0) . . . . . . . . . . . . . . . . . . . . . . . . . . . . . . . . . . . . . . 228
Fig. 7.2 The InterContinental Shanghai Wonderland Hotel in Songjiang, which opened in November 2018. Architects: JADE + QA (JADE + Quarry Associates), ECADI – East China Architectural Design & Research Institute and Atkins. (Photo: Martin Jochman, JADE + QA). . . 228
Fig. 7.3 Premier Coal crayfish breeding in Western Australia. Left: Breeding tank; right: crayfish *(Cherax tenuimanus)*. . . . . . . . . . . . . . . . . . . . . . . . 230
Fig. 7.4 Signage near the main geothermal boreholes of the flooded Springhill, Nova Scotia, Canada, coal mines. . . . . . . . . . . . . . . . . . . . . . . 230
Fig. 7.5 Mine water treatment process at the plant in Dębieńsko, Poland. (After Ericsson and Hallmans 1996) . . . . . . . . . . . . . . . . . . . . . . . . . . . 236
Fig. 7.6 Waste rock piles with ochre from a passive mine water treatment after dewatering and drying (**a**) and dried as well as burnt ochre (**b**) (Photos: Bob Hedin) . . . . . . . . . . . . . . . . . . . . . . . . . . . . . . . . 237
Fig. 7.7 Mine water treatment plant eMalahleni. From top left to bottom right: Pre-treatment (leaching basin in the background on the left and coal mine on the right); reverse osmosis; building made of gypsum; mineral water produced in the plant. (Gauteng Province, South Africa) . . . 239
Fig. 7.8 Flow chart of the eMalahleni and Kromdraai mine water treatment plants in Gauteng Province, South Africa. (Modified from Gunther and Mey 2008). . . . . . . . . . . . . . . . . . . . . . . . . . . . . . . . . . . . . . . . . 240

# List of Tables


Table 1.1 Journals and online sources that publish negative or unsuccessful results of experiments 4
Table 1.2 SI unit prefixes. Please note that the prefix is capitalised only from $10^6$ (mega); all others are lower case – i.e.: kg, not Kg (Bureau international des poids et mesures (BIPM) 2019, p. 121), because K stands for the unit Kelvin 11
Table 1.3 Definition of terms related to surface and precipitation reactions and selected literature 27
Table 1.4 Prediction of maximum iron concentrations in mine water from newly flooded British deep mines as a function of the total sulfur concentration of the mined seams (Younger 2002a) and comparison with measured concentrations from Mpumalanga, South Africa. (Pers. comm. Altus Huisamen) 32
Table 1.5 Classification of mine water based on the proposal of the Federal Water Pollution Control Administration (Hill 1968; Scott and Hays 1975); units in mg $L^{-1}$; pH without unit; and acidity in mg $L^{-1}$ $CaCO_3$ equivalents 38
Table 1.6 Visual pH meter scale. Colour indications are based on the Munsell scale. (After Younger 2010) 40
Table 2.1 Mine water treatment technologies depending on the parameter to be treated, which can be considered BAT/BATEA (Hatch 2014; Pouw et al. 2015) 42
Table 2.2 Selection of numerical, chemical-thermodynamic codes, arranged according to the actuality of the version 44
Table 2.3 On-site parameters in the Rehbach stream at the former passive mine water treatment plant Lehesten, Germany in May 2008 46
Table 2.4 Recommended water quantities and pre-treatments for standardised mine water sampling 51

Table 2.5 Recommended scope of analysis for different types of mine water investigations; parameters in bold should be determined on-site; for parameters in italics, total concentrations as well as those of individual species should be determined; additional parameters should be analysed depending on geological conditions or presumed history of the mine 54
Table 2.6 Expressions that should be used for analysis in mine water, using iron as an example 61
Table 2.7 Essential parameters to be recorded when sampling mine water (details in Sect. 2.3 and Table 2.5) 64
Table 2.8 Calculation of pH averages. The correct average pH value is 3.41 and not 3.63 (pH values of a mine water near Carolina in Mpumalanga Province, South Africa; online calculator: www.wolkersdorfer.info/ph) 79
Table 2.9 Coefficients for the calculation of the redox potential according to Eq. 2.8. Further coefficients and an online calculator can be found at www.wolkersdorfer.info/orp 83
Table 3.1 Compilation of treatment processes for the neutralisation of mine water 98
Table 3.2 Characteristics of low- and high density sludge processes from seven water treatment plants using the neutralisation process 99
Table 3.3 Selected alkaline materials suitable for mine water neutralisation 100
Table 3.4 Commonly used flocculants and coagulants 101
Table 3.5 Amount of alkali required to raise the pH of mine water at Horden Colliery (County Durham, England) to between 8 and 8.5 (from Croxford et al. 2004, Table 4). Meanwhile, the plant has been replaced by a passive water treatment system (Davies et al. 2012) 105
Table 3.6 Compilation of minimum pH values required to precipitate metals from mine water and to achieve concentrations below 1 mg $L^{-1}$ (without considering sorption or coprecipitation) 106
Table 3.7 Selected characteristic properties of membrane processes as well as electrodialysis (compiled from the sources mentioned in the text) 119
Table 3.8 List of ions that can substitute lattice sites (diadochic substitution) in ettringite ($Ca_6Al_2[(OH)_{12}|(SO_4)_3]\cdot 26\ H_2O$) 132
Table 4.1 Classification of passive and natural processes for mine water treatment 152
Table 4.2 Oxidation and reduction conditions in "conventional" passive treatment systems 156
Table 4.3 Empirical treatment rates $R_A$ of aerobic ("alkaline") and anaerobic ("acidic") constructed wetlands to be used for the area calculation of constructed wetlands (Hedin et al. 1994a) 169

Table 4.4 Change of mine water chemistry in the course of a anaerobic constructed wetland in Westmoreland County, Pennsylvania, USA (Hedin et al. 1988) when sampled on 7th October 1987 (sampling locations in Fig. 4.10); units in mg $L^{-1}$, pH without unit, $k_B$ in mmol $L^{-1}$ 169
Table 5.1 Selected key parameters of the Saxon, Germany, mine adit database (updated to 2009): *n:* Number, $\bar{x}$: mean value. Number *n* refers to all individual analyses of the parameter, not to the number of sampling points; mean of the pH based on the proton activity (www.wolkersdorfer.info/pH_en) 192
Table 6.1 Selection of acidic lakes whose water quality has been altered by the addition of chemicals or mine water treatment sludges 208
Table 7.1 Cost comparison of a passive mine water treatment plant with marketing the recovered iron oxide and a conventional low density sludge plant (after pers. comm. Bob Hedin 2008). *Final storage costs if the sludge is not marketable 238

# Introduction

# 1

## 1.1 Sidenotes – or Experiences After More Than a Decade of Literature Review

During the research for this book, I noticed that the terms "innovative", "unique", "first time" or "no or little waste" often appear in connection with many processes. The associated publications or final reports then state "successfully treated". However, as soon as it is supposed to move from laboratory scale to pilot plant or even industrial scale, many processes remain in their infancy and never make it to application. Often, I found that an "innovative" process has already been published by other authors or that patent reasons prevent a process from being optimised by others. My aim with this book is to show what has already been done and where you can read details of various treatment methods (best practice, however, I will largely avoid, because copy-and-paste will get you nowhere, especially in the mine water sector). Don't try to reinvent the wheel, but – if you are involved in research – identify the weak point of a method and try to improve it – James Watt didn't invent the steam engine either, as many believe, but merely optimised a partial aspect, which he had even copied from another method (look up "centrifugal governor" or "process intensification"). And Johannes Gutenberg did not invent printing. Try to become a James Watt or a Johannes Gutenberg of mine water treatment as a researcher, or, alternatively, if you don't want to or can't (if you are pressed for time, you can safely skip the following paragraph): When you have developed a supposedly new method, write the following:

> With the innovative, environmentally friendly MyTREatmeNT® process, it has been possible for the first time to treat mine water cost-effectively, with the waste volume reduced to a minimum and the costs comparable to those of conventional methods. In the pilot plant test, the mine water from the A mine was successfully treated up to the discharge limits specified by

The original version of this chapter has been revised. The correction to this chapter can be found at https://doi.org/10.1007/978-3-662-65770-6_9



C. Wolkersdorfer, *Mine Water Treatment – Active and Passive Methods*, https://doi.org/10.1007/978-3-662-65770-6_1

the authorities. According to a cost-benefit analysis, the new process has the potential for use on an industrial scale and can substantially reduce treatment costs. Further research is needed to obtain the optimal conditions for a pilot plant and commercialisation of the process.

(As I write these lines, I get an e-mail asking me to review an article about arsenic removal from drinking water – I have the feeling the authors copied my sentences above). Put the name of your process instead of MyTREatmeNT, get the name trademarked and the process patented by slightly modifying an existing process. Arrange for a sufficient number of press releases, invite radio and television, put a website online and open a Facebook account to which you invite all your business friends. Apply for a research project (not less than €250,000) jointly with a well-known research institution or a lesser-known industrial company and publish at least three papers each year that need to be insignificantly different in content as long as the titles are different from each other. Distribute them as widely as possible in journals with high impact factors and come up with a new project after 3 years. Adjust your data as you see fit, leave out unpopular values, or make a typo – the reviewers won't notice because they are overloaded with reviews and can't find time to check your data in depth. Above all, make incomprehensible, perhaps even oulipotent, nested sentences peppered with foreign words so that everyone thinks you are highly educated – or try sorptive extraction of unobtainium (Misra 1990). Use lots of nouns and acronyms (which you'd best reintroduce), invent a word that doesn't yet exist in dictionaries (e.g. OuEaMiPo), and your method will sound innovative; fill your texts mainly with equations, whereby you should give preference to differential equations (I would like to give you a good negative example, but the search engine of your choice will probably lead you, as a companion within the Turing galaxy, quite quickly on the right track, and I don't want to offend anyone personally). Then, the wider you spread your publications, and the more often you regurgitate your results, the more successful you will become in your métier. On no account should you try to deal with too many different topics; this would contradict the utilitarian nature of your publishing style and would only benefit your career and your reputation to a limited extent. In my literature searches for the GARD Guide, only less than 500 out of an initial 5000 articles were found to be relevant or unduplicated (www.wolkersdorfer.info/gard/refbase/index.php). It is up to you to decide whether you want to interpret this to mean that only a tenth of all publications are actually required.

Only the high density sludge (HDS) method seems to me to be a truly innovative process that has had no predecessor in mine water treatment or the industrial treatment of water. Three years of intensive research went into the process, and the only article published on it does not even meet the standards required of technical articles – it does not have a single citation. And the inventor even relinquished his patent rights after a few years – perhaps one of the many reasons why the process caught on? Take Paul Kostenbader, Wilhelm Röntgen or even Jonas Salk as an example, who replied, when asked who held the patents for the polio vaccine he had invented: "Well, mankind, I would say. There is no patent. Can you patent the sun?" (Oshinsky 2005, p. 211). Wilhelm Röntgen, who even refused to accept the title of nobility that had been offered to him, explained to the engineer Max Levy of the German AEG company "that he was absolutely of the opinion, in accordance with the good tradition of German professors, that his inventions and discoveries belonged to the general public and should not be reserved for individual enterprises by means of patents, licensing agreements and the like. He was clear about the fact that with this statement he renounced to draw monetary advantages from his invention" (Glasser 1995, pp. 88, 277).

Finally, in my research I have come across publications that quite simply repeat what researchers in other countries have already done in exactly the same way. This can almost be described as impertinent. Of course, everyone is free to repeat already published experiments in a modified form or in a case-specific way. However, if the experiment is basically just a copy of what is in another publication, and if that is then published across multiple journals with only minor changes, then I consider that behaviour to be decidedly questionable ethically, if not uncollegiate. Especially when it becomes apparent that these researchers seem to systematically repeat and publish almost all the experiments of another research group without producing much that is new or their own – then I am at a loss for words. Like now ...

Furthermore, it was striking in my research that authors are very selective in reporting their data. This fact often prevents the comparison of techniques or costs, which was also outlined by Makhathini et al. (2020). Schäfer and Schwarz (2019) have examined this "publication bias" in more detail in an article that is worth reading. Papers are given or articles are published, in my view, at different times, sometimes by different authors with almost identical text, and when comparing the data in the publications, supposedly disagreeable measurements are omitted or values are suddenly printed incorrectly, or no errors are given for the values (an impressive example of five publications compared with each other can be found here: www.wolkersdorfer.info/selektiv). Call it what you will, for example alternative facts. From a scientific point of view, this is hardly understandable and does little to promote acceptance of mining in circles that are already critical of mining. I am not talking about "rude affairs" here, but it must be the goal of all of us to present reliable, resilient, verifiable, and, above all, correct results. We need to get to the point where our values – as in physics, medicine or chemistry – come with error bars or point out the margins of error. Hardly any of the publications I have worked on met these requirements (and I am not excluding my publications that were written in this tradition without error bars). We should get into the habit of admitting errors and publishing unsuccessful experiments or data.

There are journals that specialise specifically in "failed" experiments or research. Harvard University (Institute for Quantitative Social Science – IQSS) has set up a website to document not only successful experimental data but also negative experimental results for posterity (www.dataverse.org). It would be nice if papers relating to mine water treatment could also be found there, so that we can learn from them in the future (Table 1.1). Van Emmerik et al. (2018) present another aspect: Scientists are increasingly refraining from field experiments, claiming that they involve greater "risks" of obtaining negative results and that the publication process takes longer. Instead, modelling is preferred to

**Table 1.1** Journals and online sources that publish negative or unsuccessful results of experiments

| Title | ISSN | Subject area |
|---|---|---|
| The Dataverse Project | www.dataverse.org | Open source research data repository software |
| *Journal of Contradicting Results in Science* | 2278–7194 \| Online | Open to the public |
| *Journal of Interesting Negative Results in Natural Language, Processing and Machine Learning* | 1916–7423 \| Online | Speech processing, machine learning |
| *Journal of Negative Results in BioMedicine* | 1477–5751 | Biomedicine |
| *Journal of Negative Results: Ecology & Evolutionary Biology* | 1459–4625 | Ecology, evolution |
| *Journal of Pharmaceutical Negative Results* | 2229–7723 | Pharmacy |
| *Journal of Unsolved Questions* | 2192–0745 | Open to the public |
| *Nature Precedings* | 1756–0357 | Publication discontinued |
| *The All Results Journals: Biol* | 2172–4784 | Biology |
| *The All Results Journals: Chem* | 2172–4563 | Chemistry |
| *The All Results Journals: Nano* | 2444–0035 | Nanotechnology |
| *The All Results Journals: Phys* | 2174–1417 | Physics |
| The *Journal of Irreproducible Results* – Official Organ of the Society for Basic Irreproducible Research | 0022–2038 | Open to all, anecdotal |

achieve a faster throughput of publishable articles (see also the comment in Sect. 1.2.20). In my research, this is exactly what I came across quite a few times, namely that researchers do not want to publish the results because, from their point of view, they were not successful. However, perhaps others could learn from these data, "pinnacles" of mine water research, so to speak.

How can you become a James Watt or Johannes Gutenberg of mine water treatment? By first reading, reading, and reading again – yes, it's sometimes a drudge to do so. Find weak points in the existing methods and concentrate on eliminating these (keyword "process intensification") or improve the co-treatment of mine water with other waste streams (Makhathini et al. 2020). Two aspects still pose an unsolved challenge after more than 50 years of intensive research: What should we do with the sludge produced by neutralisation (Ødegaard 2004), and how can we further utilise the highly concentrated brines generated by membrane processes as well as ion exchangers. For example, the Coal Research Bureau wrote in 1971 – without the slightest change to date: "There is currently no practical use for coal mine sludge nor is there any practical method for the recovery of by-products" (Coal Research Bureau 1971, p. 1). Or are you one of those bold, elucubrating fantasists? If so, ask yourself this question right now: can we develop a method that leaves neither sludge nor high-salinity solutions behind? Please do not repeat what others have done before you. Therefore, I challenge you:

> To everyone who thinks differently: the rebels, the idealists, the visionaries, the lateral thinkers, those who do not allow themselves to be pressed into any scheme, those who see things

> differently. They don't obey any rules and they don't respect the status quo. We can quote them, contradict them, admire or reject them. The only thing we can't do is ignore them because they change things, because they advance humanity. And while some think they're crazy, we see them as geniuses. Because those who are crazy enough to think they can change the world are the ones who do it.

And who said that? Steve Jobs and the computer company Apple in its advertising campaigns between 1997 and 2002 (Isaacson 2011, p. 329). Are you reading this text on an iPad or iPhone right now? Then you know which way to go forthwith. A colleague from the USA has summed up in his blog what is often lacking in the mining industry:

> As important as mining is to society, techniques and equipment that were first developed in the early 1900s are still standard in many modern mining facilities today. Mining is one of the last holdouts of dirty, inefficient industry that's just waiting to be revolutionized by new breakthrough clean technology. Latest innovations and cost reductions in clean technologies (cleantech) hold promise for making mining more profitable, safer and better for the planet.
>
> While there are clear benefits to mining companies implementing new technologies, there is risk involved with new technology. New technology – like bioremediation of mine tailings (the often toxic output from mining processes) or electrochemical water treatment – has historically struggled to find footholds in mining because companies generally don't like taking the risk of adopting new, unproven technology until others have. That attitude is now changing, as companies are increasingly motivated by dramatic new economic benefits promised by new green mining breakthroughs (Dallas Kachan on November 9, 2013: www.kachan.com/making-green-mining-less-oxymoron).

Should we establish why only a few new methods of mine water treatment find their way from pilot scale to industrial scale? If we are repeatedly confronted with the demand to consider in our recommendations for mine water treatment plants only methods that have shown their practical effectiveness over the past *X* years, how are we to bring new methods to market maturity? Some countries have always shown the courage to dare and try out new things. Or to put it another way: who invented the telephone? So if you do not like Watt or Gutenberg, then become a Reis of mine water treatment. And if you happen to head a mining company: Have the courage of an Artur Fischer and take a chance.

One example of a future development is the Internet of Mine Water (IoMW – mine water internet; Fig. 1.1). Already, the Internet of Things is seen as one of the upcoming disruptive technologies for technological progress (Carayannis et al. 2018; Losavio et al. 2019; S. R. I. Consulting Business Intelligence 2008). In the IoMW, all components relevant to water management in a mine, such as sensing devices (sensors), gate valves, sampling vessels, or measuring points, receive RFID radio tags. According to the idea of the Internet of Things, these components will be readable, recognisable, localisable, addressable and controllable via the Internet (S. R. I. Consulting Business Intelligence 2008, Appendix F). It does not matter whether the connection of the components to the IoMW is wireless, via a radio network, RFID sensors, or by means of a local network (More et al. 2020). All these data can then be used to forecast the water chemistry of a mine water treatment plant for a certain period of time (More and Wolkersdorfer 2021).

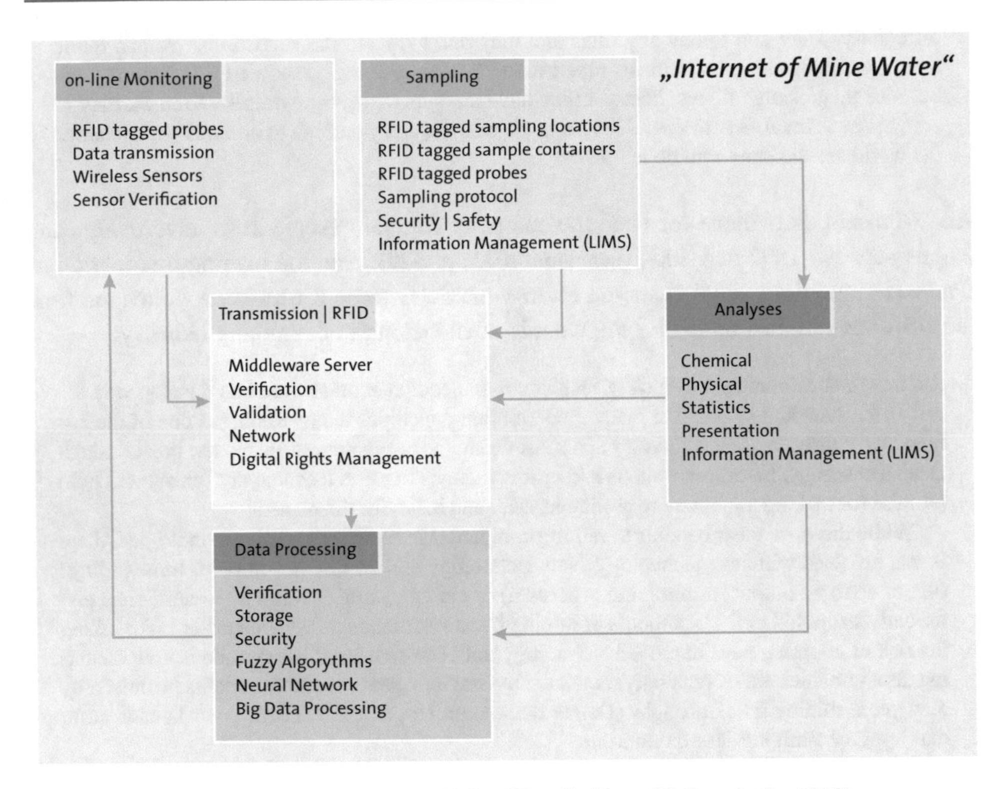


**Fig. 1.1** Flow diagram for the "Internet of Mine Water". (From Wolkersdorfer 2013)

Anyway – armed with all the research courage, you can finally start developing your new process and eventually pass on your memeto to us. Before doing so, however, you should have read at least *some* of the reference works on mine water treatment systems to avoid unnecessary replications (to date, there are hardly any reference works in languages other than English):

Herewith a selection of reference works, sorted alphabetically:

in English

- Hatch (2014): Study to Identify BATEA for the Management and Control of Effluent Quality from Mines
- Jacobs and Pulles (2007): Best Practice Guideline H4: Water Treatment
- Lorax Environmental (2003): Treatment of Sulfate in Mine Effluents
- Senes Consultants Limited (1994): Acid Mine Drainage – Status of Chemical Treatment and Sludge Management Practices
- Skelly and Loy and Penn Environmental Consultants (1973): Process, Procedures, and Methods to Control Pollution from Mining Activities
- Skousen et al. (1998): Handbook of Technologies for Avoidance and Remediation of Acid Mine Drainage.

- Society for Mining Metallurgy and Exploration (1998): Remediation of historical mine sites – technical summaries and bibliography
- The Pennsylvanian Department of Environmental Protection (1998): Coal mine drainage prediction and pollution prevention in Pennsylvania
- U. S. Environmental Protection Agency (1983): Design Manual – Neutralization of Acid Mine Drainage
- U. S. Environmental Protection Agency (2000): Abandoned Mine Site Characterization and Cleanup Handbook
- U. S. Environmental Protection Agency (2014): Reference guide to treatment technologies for mining-influenced water

in Turkish:

- Karadeniz (2005): Asit Maden (Kaya) Drenajında Aktif ve Pasif Çözüm Yöntemleri

in German:

- Wolkersdorfer (2021): Reinigungsverfahren für Grubenwasser

or in Chinese (where, incidentally, there are four words for mine water: 矿井水, 矿坑水, 的矿井水 and 矿井水的):

- 梁天成 (2004): 矿井水处理技术及标准规范实用手册。
- 何绪文and 贾建丽著 (2009): 矿井水处理及资源化的理论与实践。

With this in mind, "Glückauf!"; enjoy reading or studying – and let's start our joint journey through mine water treatment right away!

## 1.2 Definition of Terms

### 1.2.1 Problems with the Definition of Terms

Unlike, for example, chemistry, physics or mathematics, there are hardly any uniform definitions for the numerous terms that occur in connection with mining and water. This may primarily result from the fact that colleagues from a wide range of disciplines come together to work on mining influenced water. Above all, hydrogeologists, limnologists, biologists, chemists, botanists, mining engineers or engineers in general – and of course mine surveyors – with their technical terms should be mentioned. This is a good thing! But all these colleagues bring and use their own terms, some of which have evolved historically, when dealing with water in mining. When working on the glossary of the German AK Grubenwasser in the FS Hydrogeologie (Burghardt et al. 2017), which can currently be considered as the German standard work for the definition of relevant mine water terminology, or the definition for mine water in the Working Group 4.6 "Altbergbau" of the Engineering Geology Section in the German Geotechnical Society (DGGT) (2013), it became apparent how difficult it can sometimes be to define terms. In addition, many professionals in consulting firms have only limited access to literature databases or hardly any access to the latest publications at all. Let me give you an arbitrarily chosen example. The first reactive walls for mine water were constructed by the research group of David Blowes at Waterloo University (Blowes et al. 1995; Waybrant et al. 1995). In a publication on reactive walls in Lusatia, Germany, none of these relevant first publications are cited, and there is no reference to the investigations on reactive walls carried out by this working group for decades, apart from Benner et al. (2002). As already mentioned, this leads to unnecessary replications and to the fact that – as the example of Skado Dam/Lusatia, Germany shows – one falls into the same pitfalls as those of previous researchers.

I have not considered small nuances in the use of terms resulting from the various English dialects when defining them. Historically, too, there have been changes in the use of terms – if only because one or the other device or machine no longer exists today. The order of the next sections follows the order in the German book, to allow cross-references for those interested.

A classic example of a meaningless term with a fuzzy definition is the term "heavy metal". There are a total of 40 different definitions, and hardly any publication uses one definition correctly. When we just consider the specific density, the definitions of the density, above which a metal is a heavy metal, range between 3.5 and 7 g cm$^{-3}$. This results in numerous publications or reports with tables of analyses having the heading "heavy metals", in which the elements arsenic, aluminium, antimony, or sulfur appear. Arsenic and antimony, however, are semimetals or metalloids, aluminium with a specific density of 2.698 g cm$^{-3}$ can hardly be called a heavy metal, and sulfur is not a metal at all. The fact that Evangelou (1995, p. 97) makes the *faux pas* and discusses the metal aluminium in the subchapter "Heavy Metal Precipitation" should be noted here only in passing. Finally, in astronomy, all elements except hydrogen and helium are called "heavy metals" (Ridpath 2012, p. 302), because at the pressures in a sun, virtually all elements are present in metallic bonding. In addition, researchers often use the term as a "scare tactic" to express that mining influenced water is particularly "toxic" (see also Anonymous 2006). For this reason, the word "heavy metal" will not appear anywhere in this book except in this passage, in the index, and in the bibliography. In the Anglo-Saxon literature, terms such as "Contaminants of Potential Ecological Concern" (COPEC), "Potentially Toxic Element" (PTE), "Potentially Toxic Metal" (PTM) or "Metals of Concern" (MOC) have become common since the 1990s (the latter in Langmuir et al. 2005, p. 4), but these solve the problem only to a limited extent ("potentially toxic metal" seems to have been first used by Lewis et al. (1972), Sim and Lewin (1975), and Cherian and Goyer (1978)). In this book, we refer exclusively to metals, semimetals, or non-metals, or in the sense of Chapman (2012), "'Heavy metal' – cacophony, not symphony".

The inconsistent use of terms starts with the words “mining”, “pit” or “colliery”. In quite obvious cases, it is immediately clear whether a mining operation constitutes a “mine” or not. For example, for an underground coal mine. But what about a quarry that is partly underground, such as the German Wohlverwahrt-Nammen iron mine, which at first sight seems to be a quarry. In France, all quarries count as “mines”, and the “Directive 2006/21/EC of the European Parliament and of the Council of 15th March 2006 on the management of waste from extractive industries” deliberately chooses a very broad term to reflect the inconsistent definitions across the EU: “extractive industries” (European Commission 2006, Article 3(6)). Colloquially, the directive is referred to as the “Mining Waste Directive”, but one of the triggers for the directive was a quarry operation in Scandinavia (Getliffe 2002). The term therefore had to be changed.

If we now go one step further, the question arises: What is “mine water”? In fact, there is no uniform definition of the term in the international literature. Some colleagues consider any water produced in a mining operation as “mine water” and also include wastewater from ore processing. The journal *Mine Water and the Environment* regularly publishes articles dealing with wastewater from ore processing, as in some cases the chemical composition is quite similar to that of mine water *sensu strictu*. This brings us to the term “wastewater”. When is mine water wastewater? Or is mine water generally wastewater (legally, in some countries, it is not considered wastewater, but that is irrelevant here: Wolfers and Ademmer 2010). Sump water from open pit mines can undoubtedly be termed “mine water”, but the water quality is often so good that it could even be used as drinking water. In addition, there are numerous abandoned metal mines in Germany (pers. comm. Thomas Krasmann 2003, Stengel-Rutkowski 1993), or an abandoned gold mine near Sabie, South Africa, where the discharged mine water is fed into the drinking water supply. In some mines, e.g. Bad Gastein, Austria, the mine water is even used as healing water. Consequently, not all mine water is contaminated, and therefore the often observed negative press on mine water is not justified either. As the numerous publications on the classification of mine water show, it is still not possible to say unambiguously when a water is mine water and when it is not, especially since many classifications use only the inorganic chemistry without considering the genesis or the organic water constituents. What about acidic water from a quarry, such as the Großthiemig greywacke quarry in Brandenburg, Germany? Mine water or not mine water?

Next would be the question, what is acid mine drainage? In fact, many colleagues define acid mine drainage as water with a pH below 7. However, the pH is not relevant to the question of how to treat mine water or how contaminated it is. Therefore, as with the term heavy metal, caution should be exercised when using this term. Internationally, there is now a move away from using the term “acid mine drainage” towards using the term “mine leachate” or “mining influenced water” (MIW), because not only acid mine drainage, but also neutral or even alkaline mine drainage can cause considerable ecological and socio-economic problems. An example is the more or less pH neutral mine water of the German Niederschlema/Alberoda mine, which contained or still contains substantial concentrations of arsenic, sulfate and uranium (Paul et al. 2013). Many people in South Africa even refer to any mine water as acid mine drainage, against their better judgement, similar to Australia, where the acronym AMD stands for acid and metalliferous drainage, which encompasses the full spectrum of acid conditions through to pH neutral and saline water (pers. comm. Andy Davis, February 24, 2022).

Unfortunately, the EU Directive 2006/21/EC has put a spanner in the works, because it calls mine water "leachate". However, the definition in the directive is so broad that it would be nonsensical not to use the word mine water. Interestingly, Agricola – or rather his early New High German translator – does not use the word mine water anywhere in his standard work *De re metallica libris XII* (Agricola 1974 [1557]). In general, water plays only a minor role in his books, except when it comes to "water arts" – probably because at that time there were no reliable methods to treat contaminated mine water – it was mostly taken for granted. Around 1590, however, the "ochre marshes" were created at the Rammelsberg near Goslar, Germany, to remove most of the iron from the highly acidic and ferruginous mine water – although the original reason for this was the production of ochre for the paint industry (Brauer 2001). As late as 1955, the ochre sumps were extended to be able to improve clarifying the acid mine drainage. Let us hope, however, that not another 500 years will have to pass before we have a grip on mine water as a valuable material (that would be a circular economy after all!).

In the English literature, there are also clear regional differences in the use of English terms. For example, sometimes an open pit mine is called *open pit mine* (US English), and in other instances it is called *open cast mine* (mainly British English and non-English speakers) or *surface mine* (Australia). In addition, the neologisms used by colleagues from non-English speaking parts of the world sometimes make it difficult to define terms clearly. A dewatering system or pump can then become a "machine capable of moving water out of a mine to keep it dry" (from a summary of an article in *Mine Water and the Environment*).

If we now finally come to the terms "active" and "passive" mine water treatment, the chaos becomes almost perfect. In fact, there were and are extensive discussions about whether one or the other treatment plant should still be called "passive" or "semi-passive" or even "active". You too would laugh heartily at the contortions in some publications that use linguistic quibbles to make an active plant passive – or vice versa. In writing this publication, there was indeed some consideration of dispensing with this classification altogether, but since it has become generally accepted, I have finally retained it. The following definitions will certainly not meet with everyone's approval, nor can they be regarded as definitive; however, they certainly facilitate the reading of the present publication, since they are used in a largely uniform manner throughout. It is not intended to discuss whether a division into "active" and "passive" methods is useful or not. The definition and classification is used internationally and is therefore the basis of the book. It is true that every "active" method is "active", but not every "passive" method is really "passive", because maintenance and monitoring are still necessary. Since there are currently no better terms, I will use this classification for the time being.

Terms that are not defined in this chapter appear in the description of the various processes that can be used to improve the quality of mine water, or in the glossary. For the sake of understanding, only those terms that sometimes cause confusion at conferences or in the literature are defined. An issue that I encountered quite often is using multiple terms for a single concept. Two of many examples are the terms RAPS and SAPS describing the same passive treatment system or the many words used for aerobic constructed wetlands (e.g. reed bed). I tried to list all the terms I found in the relevant section, but you will surely find an example where I missed a term.

Before starting with the definition of terms, a table on SI unit prefixes, which sometimes seem to cause problems, is provided (Table 1.2). You won't believe how many Kelvin grams are around – it is hard to imagine what physical process that could describe. Details have been published by the International Bureau of Weights and Measures (Bureau international des poids et mesures 2019, pp. 31–33 or in English: pp. 143–145). Please note that SI units and the prefix are always placed recte, i.e. not in *italics*, before the unit. Also note that the unit symbol for the litre is a capital L to distinguish it from the numeral 1, which can occasionally look like a lowercase l in some character sets (such as this).

### 1.2.2 Active Mine Water Treatment

In active mine water treatment, chemicals, electrical energy or kinetic energy are used to remove undesirable substances from the mine water or to correct the pH value upwards or downwards by means of a wide range of processes. This usually requires extensive technical equipment, with the help of which the chemical and physical properties of the water are changed. By monitoring the water quality in conjunction with the addition of chemicals or aeration, it is possible to achieve almost any water quality specified by the operator or the authorities. Briefly: In active mine water treatment, the desired water quality is achieved by a constant supply of energy and chemical or biological reagents.

**Table 1.2** SI unit prefixes. Please note that the prefix is capitalised only from $10^6$ (mega); all others are lower case – i.e.: kg, not Kg (Bureau international des poids et mesures (BIPM) 2019, p. 121), because K stands for the unit Kelvin

| Factor | Name | Prefix | Factor | Name | Prefix |
|---|---|---|---|---|---|
| $10^1$ | deca | da | $10^{-1}$ | deci | d |
| $10^2$ | hecto | h | $10^{-2}$ | centi | c |
| $10^3$ | kilo | k (not K) | $10^{-3}$ | milli | m |
| $10^6$ | mega | M | $10^{-6}$ | micro | µ (not u or mu) |
| $10^9$ | giga | G | $10^{-9}$ | nano | n |
| $10^{12}$ | tera | T | $10^{-12}$ | pico | p |
| $10^{15}$ | peta | P | $10^{-15}$ | femto | f |
| $10^{18}$ | exa | E | $10^{-18}$ | atto | a |
| $10^{21}$ | zetta | Z | $10^{-21}$ | zepto | z |
| $10^{24}$ | yotta | Y | $10^{-24}$ | yocto | y |

### 1.2.3 Base Capacity ($k_B$; Acidity; m-Value)

There are an abundant number of publications on the definition of base and acid capacities (Sect. 1.2.18) in mine water and their determination or unit. There is major confusion, to say the least, and strictly speaking it would be time to agree on a uniform terminology. In the case of water, in which the base and acid capacity is essentially determined by the carbon species, the relationships are comparatively simple, and the definition of Stumm and Morgan (1996, p. 163 ff.) can be used. Unfortunately, not all authors adhere to the definition of the unit given by Stumm and Morgan (1996, p. 164) either: "The unit of [alk] or [ANC] is given in mol (moles of protons per liter) or in equivalents per liter." The unit "M" proposed by Stumm and Morgan (1996) is not correct according to the SI definition, because it must be correctly written "mol $L^{-1}$", since the symbol "M" is reserved for "mega" and "mile" respectively (Bureau international des poids et mesures 2019, pp. 31–104 or in English: pp. 143–209). Often authors use the unit mg $L^{-1}$ $CaCO_3$, where 50.04 mg $L^{-1}$ $CaCO_3$ corresponds to one millimole per litre. We will not go into the reasons for this here, but it is easier to calculate with moles than with mg $L^{-1}$, since an easier comparison of different species is possible (I know, it's confusing since the molar mass of $CaCO_3$ is 100.087 g $mol^{-1}$ – but the molar mass and equivalent weight of a compound are not always the same; the Ca in $CaCO_3$ has a charge of 2, and the acidity has a charge of only 1 – therefore the equivalent weight of $CaCO_3$ is its molar mass divided by 2, resulting in 50.04 g $mol^{-1}$). Details of the definitions are given by Kirby and Cravotta (2005a) – who largely use $CaCO_3$ eqv. as the unit. The term m-value, which comes from the inflection point of the indicator methyl red, has only historical significance and should be avoided – as should the error-prone titration with indicators.

The base capacity ($k_B$) of mine water is the property of the water to react with bases up to a predetermined pH (Fig. 1.2). Typically, the fixed pH values are 4.3 ($k_{B4.3}$) and 8.2 ($k_{B8.2}$); however, an exact determination is only possible based on the inflection point of the base titration and can vary within a few decimal points, especially at $k_{B4.3.}$ Kirby and Cravotta (2005a) therefore write of a pH of ≈ 4.5. A distinction is made between *juvenile acidity* and *vestigial acidity,* in accordance with Younger (1997, p. 460). Juvenile acidity arises primarily from (di)sulfide weathering (also referred to broadly as pyrite weathering) in the range of the varying water levels in a flooded mine. It also occurs above the mine water table in mines drained by adits. Hedin et al. (1994a) also indicate that in some systems metal acidity from weathering of secondary minerals contributes to juvenile acidity. Vestigial acidity is the acidity that enters the mine water from the dissolution of secondary minerals that have been exposed to weathering for a long time. These secondary minerals are often referred to as acid-generating salts (Bayless and Olyphant 1993), and because they are readily soluble in water, they contribute to the characteristic first flush effect (see Sect. 1.2.8 for details) of mine water after a mine is flooded.

Acidity is determined from the sum of all acidic components in the mine water, which can be metallic acids (e.g. hydrolysis of iron), inorganic acids (e.g. sulfuric acid) or organic

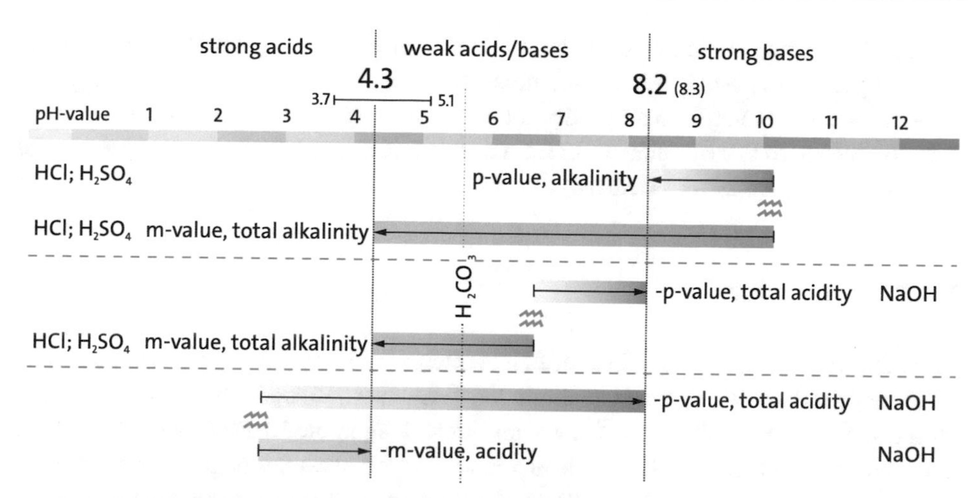


p-value: $K_A$ acid capacity 8.2 (phenolphthalein): alkalinity (strong bases)
m-value: $K_A$ acid capacity 4.3 (bromocresol green-methyl red): total alkalinity (weak/strong bases)
-p-value: $K_B$ base capacity 8.2 (phenolphthalein): total acidity (weak/strong acids)
-m-value: $K_B$ base capacity 4.3 (bromophenol blue): acidity (strong acids)

1 [mmol $L^{-1}$] = 1 [meq $L^{-1}$] = 50.04 [mg $L^{-1}$ $CaCO_3$]

**Fig. 1.2** Terms and pH values commonly used in connection with acid capacity and base capacity ($k_A$, $k_B$) in mine water. The three wavy symbols each represent a mine water whose $k_B$ and $k_A$ values are determined using a base (NaOH) or an acid (HCl, $H_2SO_4$). The inflection point in the acidic range can vary between 3.7 and 5.1, depending on the mineralisation, whereas that in the alkaline range is either 8.2 or 8.3, depending on the specification. Concentrations should be expressed in mmol $L^{-1}$ and not in mg $L^{-1}$ $CaCO_3$ as is often the case. The p- and m-values, whose designations are derived from the equivalence point of phenolphthalein and methyl red, are only of historical relevance

acids (e.g. tannin). It can be determined by titration with a base, usually NaOH (Wisotzky et al. 2018), or by approximate calculation from chemical analysis.

Caution should be exercised when the technical term "acidity" is used in connection with mine water, but only refers to mine water with pH values below 7. This should be avoided and called "acid" instead. In an unpublished report for the Straßberg/Harz mine from 1991 entitled "Report on laboratory investigations to reduce the heavy metal content and acidity in seepage water from the Straßberg mine" (Grüschow 1991), pH values are measured, but acidity, i.e. base capacity, is not measured at any point.

### 1.2.4 Mine

A mine is an operation in which raw materials are or were extracted. The raw materials can be metallic raw materials (e.g. copper, iron), non-metallic raw materials (e.g. fluorspar, graphite, barite, salt) or energy raw materials (e.g. coal, uranium). Mines can be underground (e.g. Metsämonttu, Finland), surface (e.g. Kromdraai, South Africa) or a combination of both (e.g. Wohlverwahrt-Nammen, Germany).

Installations for petroleum extraction are often subject to mining law but, strictly speaking, like geothermal wells, they are not mines. In some cases, however, the chemical composition of saline water from oil production or fracking (produced water or source water) may correspond to that of saline water from mines, and treatment is then analogous to that of mine water (Lane 2016; Veil 2013).

### 1.2.5 Bioreactor

Since the word bioreactor is so beautiful, it is often used in an inflationary manner, which is understandable. Be it for beer from the fermenter, Bionade from the fermenter, mushroom compost in the anaerobic wetland, cow dung in the RAPS (reducing and alkalinity producing system), sulfate-reducing reactors with sewage sludge or erythropoietin from biosynthesis – we are always talking about bioreactors in which the microbially catalysed processes take place.

Wildeman et al. (1993a, pp. 13–1) suggest calling a constructed wetland "bioreactor with a green toupee", Johnson and Hallberg (2002, p. 337) recommend calling anaerobic wetlands compost bioreactors rather than compost wetlands (though in most cases it is not compost at all, but manure), and Gusek (2002) refers to what others refer to as RAPS as bioreactors as well; a definition also found in Rees et al. (2004, p. 9). Dill et al. (1998, p. 338) state that "perhaps an extreme end of the bioreactors is the use of constructed wetlands to reduce sulfate levels". In Eloff et al. (2003, p. 102 ff.), the bioreactor is even called a "hydrogen contact reactor." Also in the book by Geller et al. (2013, p. 252 ff. for example) bioreactors are usually in situ processes in the lake, thus "passive" in the previous sense. The problem with nomenclature is also pointed out by Costello (2003, p. 15).

The term is most commonly used in the context of mine water to refer to vessels in which organic substrate is under reducing conditions and in which the sulfate of the mine water is converted to sulfide through microbial catalysis (Drury 1999; Hammack et al. 1994b). In this publication, therefore, let the term bioreactor be defined as follows: A bioreactor is a vessel in which organic substrate reacts with the sulfate and metals in the mine water with the participation of anaerobic bacteria. The organic substrate serves as a proton donor and the bacteria catalyse the process.

### 1.2.6 Circular Economy

The definition of the English term circular economy is not an exact translation of the long known German term "Kreislaufwirtschaft". Rather, this is understood as circular economy:

> A circular economy is an alternative to a (make, use, dispose) linear economy. In a circular economy, we keep the value in products and materials for as long as possible and minimise waste. When a product has reached the end of its life, resources are kept within the economy to be used again and again to create further value. In this context, it is essential to understand

an economy's 'societal metabolism', i.e. to quantify the amount of materials flowing in and out of the economy and how they are used, and particularly to see how many materials are recycled and used again as an input (European Innovation Partnership on Raw Materials 2016).

Consequently, in this sense, the reuse of mining waste for other purposes also falls under the term circular economy (Vidal-Legaz 2017).

### 1.2.7 Mine Water Discharge – "Decant"

In many publications we read about mine water "decant". What is the meaning of the word decant? According to the Merriam-Webster dictionary, decant means (i) to draw off (a liquid) without disturbing the sediment or the lower liquid layers, (ii) to pour (a liquid, such as wine) from one vessel into another, or (iii) to pour out, transfer, or unload as if by pouring. As we can see, in any case, decant refers to an *active process* in which a liquid is transferred gently from one container into another.

Therefore, when we refer to a mine water discharge, which is seldom a gentle or an active process, the word decant should be avoided. I also refrain from the term outfall, as this is usually associated with sewage water systems. Mine water should discharge or emanate from an adit or a seepage location, so that those real environmental effects caused by mine water are not hidden under a decant – which I usually associate with a good bottle of red wine.

### 1.2.8 First Flush

First flush (Younger 1997, p. 460) refers to the rapid increase in (contaminant) concentrations following the flooding of a mine and its subsequent decline (Figs. 1.3 and 1.4). Essentially, it results from the dissolution of easily soluble salts ("secondary mineral", efflorescent salts) by the rising mine water in the mine to be flooded and the groundwater penetrating from above. Its duration $t_f$ depends on the decrease of acidity $aci_{rem}$, which is controlled by buffering or solution processes, on the weathering rate $r_w$ of the acidic (secondary) minerals, the volume $V$ and the conductivities $K$ of the mine cavities as well as their hydraulic connection and the groundwater recharge $R_{GW}$. Loosely speaking, the mine workings are "kärchert" (loosely translated as cleaned by pressure washing) during the first flush, as undesirable substances are "washed out". As Younger (2000b, p. 61) was able to show, the duration of the first flush can be estimated as follows:

$$t_f = f\left(aci_{rem}, r_w, V, K, R_{GW}\right) \approx \left(3.95 \pm 1.2\right) \times t_r \quad (1.1)$$

where:

$t_f$ duration of the first flush

$t_r$ time elapsed for the flooding of the mine workings ("rebound time")

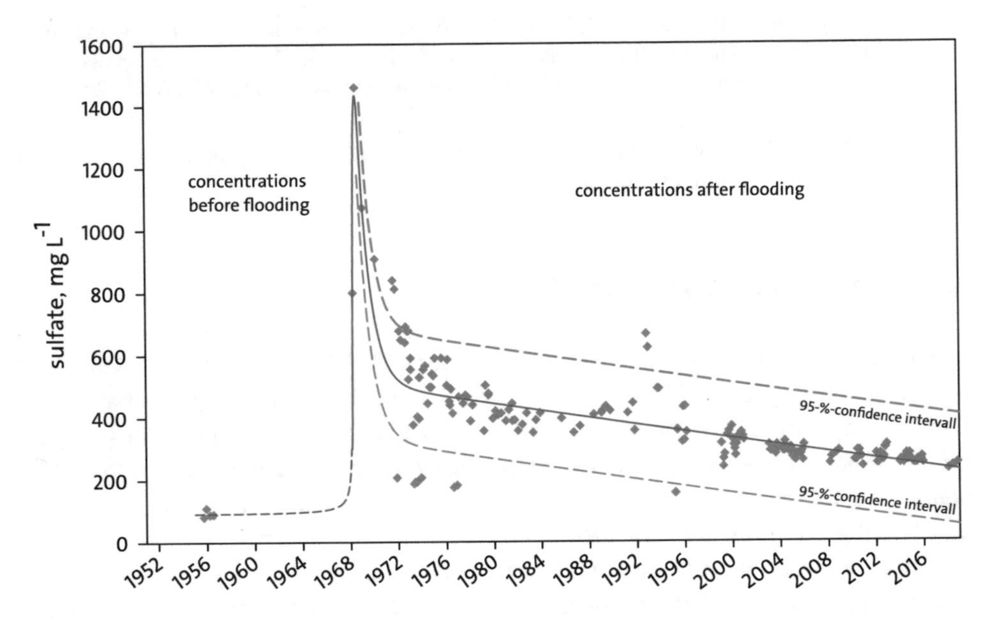


**Fig. 1.3** Characteristic first flush scenario using the example of the Rothschönberg adit in Saxony. (Modified from Wolkersdorfer 2008); Cessation of mining in 1969 (Jobst et al. 1994)

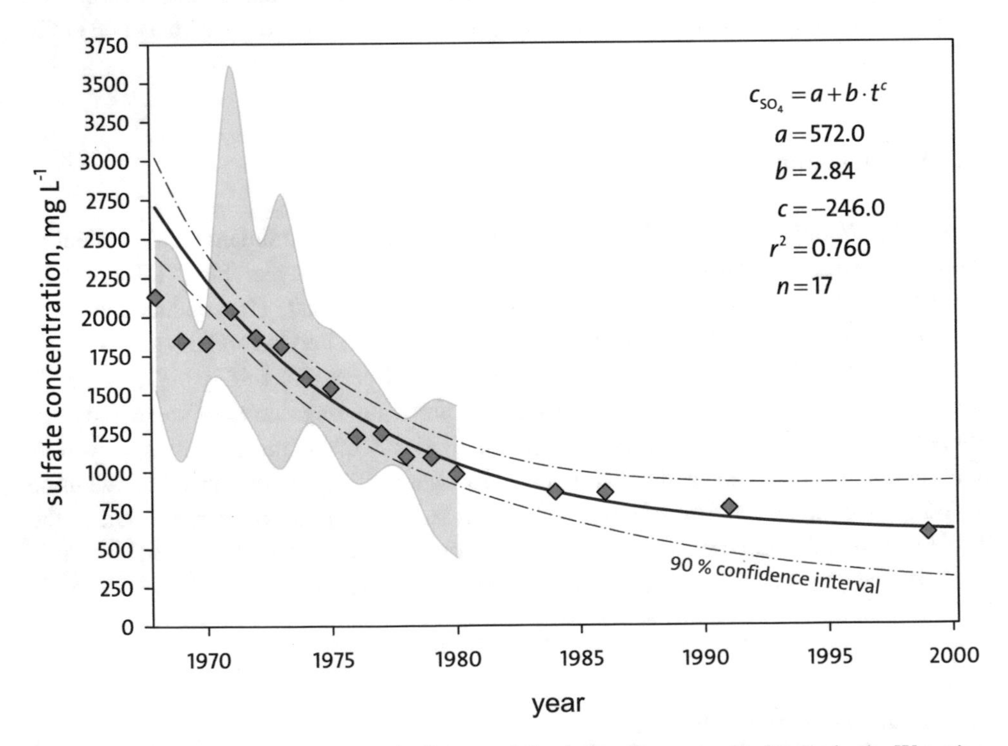


**Fig. 1.4** First flush at the Buttonwood adit, one of the main mine water discharges in the Wyoming Basin of eastern Pennsylvania, USA. (Modified after Ladwig et al. 1984, and supplemented with data from the USGS water database. For the period after 1980, only four individual measurements were available, so that no minimum and maximum areas could be represented; from Wolkersdorfer 2008)

This means that the first flush lasts approximately four times as long as the time it took for the underground mine to flood. However, an end to the first flush does not mean that afterwards the potential pollutant concentrations are so low that it would be possible to discharge the mine water into the receiving water courses without treatment. On the contrary, numerous individual cases show that mine water treatment may still be necessary after this period. Especially when a large part of the mine workings are located above the groundwater table, as for example at the German Reiche Zeche in Freiberg/Saxony or the Hohe Warte mine in Gernrode/Harz (Wolkersdorfer 2008), the continuous reproduction of juvenile acidity and the associated formation of acidic secondary minerals can extend the first flush over decades. Not all water constituents follow this trend, as they react differently in terms of chemical thermodynamics than, for example, iron or sulfate. These include arsenic, radium or hydrogencarbonate (bicarbonate).

Glover (1975, p. 65) described the first flush as follows: “Rule of thumb is that in any period of time equal in length to complete mine flooding, the iron concentration falls by 50%.” Commonly found in the literature in this regard is the explanation that the decrease in concentration would follow an exponential progression. This is of course correct, but there are some other functions that have exactly the same course, but due to the parameterisation allowed a better description of the chemical and physical processes. Renton et al. (1988), for example, prefer a Bateman function (in Systat Software 2002, ‘First Order Intermediate’), which, however, unlike the curve progression of the first flush, starts at concentration 0, then increases and finally decreases again.

Brusseau (1996), Hamm et al. (2008) and Paul et al. (2012) discuss the perfectly mixed flow reactor (PMFR) principle to describe the curve progress. In a perfectly mixed flow reactor, the influent water in the aquifer mixes completely, and the mixing time is relatively short compared to the mean residence time. Although the concept is appealing for flooded mines, the parameters of this principle need to be better defined to gain generality. Most notably, the PMFR does not include a term for new contaminant formation (juvenile acidity *sensu* Younger 1997), which is relevant to some mines. Paul et al. (2012) show that the decrease in uranium concentration at the German Schlema mine can be almost perfectly reproduced with the theoretical dilution curve. For Pöhla or Königstein, however, the curves can only be described to a limited extent (Paul et al. 2013). Hamm et al. (2008) defined a network of chemical reactors to describe the sulfate concentration of the Saizerais mine in Lorraine (France). They also succeeded in reproducing the curves relatively well with the model.

To date, only a few publications have accurately considered the curve shape of the first flush and discussed possible fitting of non-linear functions. The following potential equation descriptions, all of which have similar curves, are taken from Systat Software (2002, TableCurve2D). It is possible that an exact discussion of these responses underlying the curve shapes will lead to a better understanding of the first flush.

- First Order Decay A → B: Maximum concentration: *a;* Rate Constant: *b*
- First and Second Order Independent Decay A → B: Maximum concentration: *a*; First Order Rate Constant $k_1$: *b*; Second Order Rate Constant $k_2$: *c*

- Simultaneous First and Second Order Decay A → B: Maximum concentration: $a$; Rate Constants: First order $k_1$ is $b$, second order $k_2$ is $c$; this model assumes simultaneous first and second order reaction pathways.
- Simple Equilibrium A ↔ B: Maximum concentration: $a$; Rate Constants $k_f$ and $k_r$: $b$ and $c$
- Complex Equilibrium A ↔ B + C: Maximum concentration: $a$; Equilibrium Concentration: $b$; Net Rate Constant $k$: $c$
- Two First Order Independent Decay and Formation A → B, A → C: Maximum concentration: $a$; Rate Constants $k_1$ and $k_2$: $b$ and $c$
- Pulse Cumulative with Power Term: Transition Height: $a$; Transition Centre: $b$; Transition Width: Analytic Solution Unknown

### 1.2.9 Mine Flooding

In general, mine flooding is understood as the process of mine water rebound. This includes both passive flooding through the natural rebound in mine water after the cessation or reduction of dewatering, and active flooding. In the latter, the mine water rebound is accelerated, or the water quality improves by the supply of water from outside the mine. Flooding can also be controlled or uncontrolled, with the main elements of control being monitoring and the possibility of intervening in the processes of mine water rebound. As a rule, it is possible to actively intervene in the flooding process, either by adding external water or by keeping pumps in the mine workings in order to remove the water from the mine system, if necessary. In addition to the water level, external flooding can also affect the water quality in a mine. In terms of terminology, "flooding" is to be preferred to the word "mine water rebound", as it has long been established in the literature.

At the end of the flooding process, a hydraulically quasi-stationary state is established, which usually differs from that prior to mining, since the hydraulic conductivities in the rock have changed because of mining activities. This applies to both underground and open pit mines, whereby the hydraulic conditions in an open pit mine, in particular, are different from the natural situation prior to mining. The mine water can either be directed towards the receiving water courses via drainage tunnels, it can be pumped, or it flows to the deeper hydraulic potential via pathways in the rock. To minimise the discharge of contaminants, the aim will generally be to achieve a flooding level as high as possible, so that the oxygen supply to the primary minerals and thus their continued weathering is kept to a minimum. Gerth et al. (2006, p. 314) expressively summarise the principle behind mine flooding in one sentence: "Inherent in flooding itself are key aspects of principle solutions to control contaminant release."

Statements about potential flooding times are often required in advance of mine flooding. This allows the mine operator to design mine water treatment plants, to plan the construction of retention basins or to modify the design of inlets to receiving water courses. However, as Younger and Adams (1999, p. 39 ff.) have shown, these predictions are highly error-prone and are only reliable in very few cases. An initially more reliable method, which considers the water pressure dependent on the mine water level in the forecasts, was developed by Banks (2001) with the Mine Water Filling Model (MIFIM) for a Polish mine. He predicted four scenarios, two of which matched the actual flooding pattern quite well after the end of flooding. Self-critically, he writes: "One can speculate whether the apparent success of the re-run MIFIM model was due to insightful modeling and tightly constrained parameter selection. In reality, the choice of parameters was probably merely fortunate." (Banks et al. 2010). Westermann et al. (2017) show in detail which processes influence the flooding process and attempt to classify flooding processes to improve prediction in the future. Numerous examples of forecasting, based on German and international case studies, are also listed in Wolkersdorfer (2008, pp. 95–100).

### 1.2.10 Mine Water (Mine Drainage, Mining Influenced Water)

Mine water (also mine drainage) is all natural water that arises in connection with mining or quarrying operations and has either been in contact with the mine workings or open pit mine or arises in the course of pumping measures. The water either flows freely to the surface from adits or is pumped by means of suitable pumps via shafts or boreholes.

The term mine water is to be seen independently of its pH value or whether it is contaminated or uncontaminated, whereby "contamination" can have different meanings within the framework of valid legislative requirements (e.g. EU Water Framework Directive, Water Resources Act, Groundwater Ordinance, Surface Water Ordinance, State Water Acts). Many mine waters that have to be treated at one site due to their chemical composition could well be mineral water, medicinal water or even drinking water at another site if they have a similar composition. Also, the limit values for certain parameters differ considerably from region to region (Strosnider et al. 2020). According to Roschlau and Heintze (1975), mine water is composed of precipitation water and formation water, the latter being water inflows into the mine workings from an aquifer or water-bearing fissure or fault zones. A comprehensive definition is also given by Grothe (1962, p. 273 f.), in which he deals with various aspects of mine drainage and treatment.

According to this definition, water that occurs during raw material processing and beneficiation is not mine water but wastewater. However, provided that the chemical and physical properties of this wastewater are similar to those of mine water, the same treatment methods may be used.

Gusek and Figueroa (2009, p. 1) write on the definition of the term: "The term *mining influenced water* (MIW) was introduced in the first handbook in the series, *Basics of Metal Mining Influenced Water*. MIW is inclusive of a wide range of potential water-related issues that arise from the water–rock interactions that are common to mining operations—in contrast to traditional terms such as *acid rock drainage* and *acid mine drainage*, which refer to specific interactions and may imply a falsely narrow range of possible chemical characteristics. Simply defined, MIW is water that has been affected, adversely or not, by mining and metallurgical processes."

There are more and more publications that refer to *mining water* or *mine wastewater*. Often, however, the authors mean process water from the mining industry, but not mine water in the sense of the above definition. This increasingly causes irritation, since these process waters often have a different chemical composition from mine water and the methods presented there can only be transferred to mine water *s.s.* to a limited extent. Nevertheless, the impression is created that the methods are an approach to solve all problems with potentially contaminated mine water. There are even publications that refer to groundwater as wastewater (Iakovleva and Sillanpää 2013, Fig. 1) – in my opinion, this is not helpful when there is a need to develop consistent terminology in order to prevent miscommunication among stakeholders. In some areas there is a tendency to classify discharges of contaminated mine water as "natural sources" or to designate them as "ownerless" (e.g. the German North Rhine-Westphalia, Bavaria, Saxony). This may be a useful simplification from an economic or legislative point of view, but it is not entirely in line with the principles of the Water Framework Directive (WFD) for environmental protection. According to these, "all discharges [...] into surface waters are controlled according to the combined approach set out in this Article" (European Commission 2000, Article 10(1)), for which management plans and programmes of measures have to be drawn up. However, Spieth (2015) points out that "according to the European Commission's view to date, deterioration ... is only present [when] there is a change in the status classes determined under the Water Framework Directive. The contrary opinion, particularly in the German literature, according to which deterioration can already be assumed in the case of minor deviations from the status quo, is not convincing". If the aforementioned "natural sources" are referred to as mine water in this publication, this does not necessarily constitute a criticism of this approach, but is merely correct from a scientific or legislative point of view.

Directive 2006/21/EC of the European Parliament and of the Council of 15th March 2006 on the management of waste from extractive industries (European Commission 2006, Article 3(14)) refers to "leachate" in general terms. For the sake of completeness, this definition is given here: "'leachate' means any liquid percolating through the deposited waste and emitted from or contained within a waste facility, including polluted drainage, which may adversely affect the environment if not appropriately treated". For many reasons, I will not use this definition because it does not correspond to any of the definitions commonly used in hydrogeology, mining hydrogeology, or soil science. Also for practical reasons, I will not distinguish mine water, seepage water or groundwater, although I am aware that this may well be relevant for legal reasons. In the context of this book, water that occurs anywhere in the mine is called "mine water" or "mining influenced water".

### 1.2.11 Constructed Wetlands for Mine Water Treatment

There are two fundamentally different types of constructed wetlands for the treatment of mine water: aerobic and anaerobic constructed wetlands. These must be distinguished from constructed wetlands for municipal wastewater treatment (see definition of term in

Sect. 1.2.15) as in the latter, plant uptake of the nutrients in the wastewater takes place. Net acidic mine water must be treated with anaerobic constructed wetlands and net alkaline mine water should be treated with aerobic constructed wetlands – if a wetland proves to be a suitable option. Aerobic and anaerobic constructed wetlands are based on entirely different modes of operation, so they are not interchangeable. Basically, it would be like running a gasoline engine on diesel fuel – it doesn't work!

### 1.2.12 Coagulation and Flocculation

Coagulation refers to the agglomeration of colloids in mine water, usually associated with mine water treatment. This causes the colloids to become unstable and larger and precipitate out of the water. Since "natural" coagulation is a relatively slow process, polymer flocculants are added during mine water treatment to cause the colloids to coagulate faster and eventually flocculate. The two terms are often used interchangeably (Stumm and Morgan 1996, p. 822), specifically in water treatment (Valanko et al. 2020, p. 178). Energetically, colloids can still be separated from each other, whereas coagulated particles are subject to such strong bonds that they cannot be easily separated (Atkins 2006, p. 684).

### 1.2.13 Net Acidic or Net Alkaline Mine Water

Of all the terms relating to mine water treatment systems, those of *net acidity* and *net alkalinity* are probably the two most important. At the same time, they seem to be the most misunderstood, possibly due to the sometimes complicated presentation in Hedin et al. (1994a, pp. 27, 7), where these terms were introduced in the context of passive mine water treatment systems. However, the first use of these terms goes back further in time, with Agnew and Corbett (1969) using the term net acidity and McPhilliamy and Green (1973) using the term net alkalinity. In my view, therefore, it makes little sense for this successful concept to be rejected for the purpose of assessing the question of the correct method for the treatment of mine water.

It may be a little confusing at first that the authors Hedin et al. (1994a) use the term net acidity on page 7 and on further pages, but do not define it until page 27. This may have led many researchers to have an inadequate understanding of the underlying concept. In any case, there is no other way to explain why so many passive mine water treatment systems have been designed incorrectly or inadequately – both in the Anglo-Saxon-speaking world, but especially in the non-English-speaking world.

Hedin et al. (1994a, p. 7) state in this regard:

> When water contains both mineral acidity and alkalinity, a comparison of the two measurements results in a determination as to whether the water is net alkaline (alkalinity greater than acidity) or net acidic (acidity greater than alkalinity). Net alkaline water contains enough alkalinity to neutralize the mineral acidity represented by dissolved ferrous iron and Mn.

To calculate the net acidity or net alkalinity, the base and acid capacity (acidity/alkalinity) of the mine water must first be determined. This can be done by titration or by calculation. Then the following relationships apply (compare Sect. 1.2.3 and the summary at the beginning of the book):

Net Acidic: Acidity > Alkalinity or Base Capacity > Acid Capacity
Net Alkaline: Alkalinity > Acidity > or Acid Capacity > Base Capacity

For example, the mine water of the former fluorspar mine Straßberg/Harz, Germany, had a base capacity of 11.9 mmol $L^{-1}$ and an acid capacity of 2.1 mmol $L^{-1}$ in 2008 (Wolkersdorfer and Baierer 2013). Thus, we can calculate:

$$\text{Net acidity} = k_B - k_S = (11.9 - 2.1)\ \text{mmol L}^{-1} = 9.8\ \text{mmol L}^{-1} \tag{1.2}$$

Consequently, the mine water was net acidic at the time. In 2001, however, the mine water had a base capacity of 1.1 mmol $L^{-1}$ and an acid capacity of 1.9 mmol $L^{-1}$ (Neef 2004, Appendix 6.3). Thus, the result is:

$$\text{Net acidity} = k_B - k_S = (1.1 - 1.9)\ \text{mmol L}^{-1} = -0.8\ \text{mmol L}^{-1} \tag{1.3}$$

Consequently, the mine water was net alkaline at that time.

Kirby and Cravotta (2005a, p. 1927 f.), but also other authors, sometimes suggest other procedures or even reject the concept (e.g. Evangelou 1995; Lausitzer and Mitteldeutsche Bergbau-Verwaltungsgesellschaft mbH 2019, p. 6 f., Annex VII; Schöpke et al. 2001). Yet, for now, the above seems to me sufficient for the assessment of a mine water with regard to mine water treatment. The concept has proven itself in practice and is reliable for most practical applications. Moreover, it is a relatively simple equation in which the parameters can easily be determined.

### 1.2.14 Passive Mine Water Treatment

Passive mine water treatment *sensu stricto* uses only "natural" energy such as potential energy (height difference), solar energy (heat, photosynthesis) or biological energy (bacteria) to improve water quality. However, in many passive treatment systems, nutrients are added to activate the bacteria, or electrical energy is used to cause aeration of the mine water. In these cases, the term semi-passive mine water treatment is often used. There are also combined plants in which, as in Ynysarwed (South Wales, United Kingdom), a passive treatment plant is installed after active mine water treatment – or vice versa. Passive treatment does not mean that the operator of the system has to be "passive" – on the contrary: a passive mine water treatment system needs regular control, care and, if necessary, maintenance. A system where the operator is "passive" is called "*natural attenuation*". In one sentence: passive mine water treatment requires regular control of the structural designs in order to be able to use the naturally occurring processes that contribute to the improvement of the water quality.

### 1.2.15 Treatment Wetlands for Municipal Wastewater

The term "wetland" is a term also used in municipal wastewater treatment. Yet, as explained above, wetlands for municipal wastewater treatment and mine water treatment serve different functions as different water constituents must be removed. There are similarities in the construction of the treatment plants, which is why Vymazal (2011), who mainly gives a technical-historical account of constructed wetlands, also briefly mentions constructed wetlands for mine water (Vymazal 2011, pp. 62–64 f.). Care should therefore be taken when designing a constructed wetland for mine water treatment, as the guidelines for passive mine water treatment wetlands should be followed.

### 1.2.16 Phytoremediation

Phytoremediation involves the use of plants that take up potential pollutants from the soil or water and fix them in their biomass (Adams et al. 2000). These are generally passive measures in which the plants remove unwanted substances from the respective compartment. Planting on mine dumps, with the aim of removing potential soil contaminants, has also found its way into the literature as phytoremediation (Ernst 1996). Studies on this took place, for example, at the former Gessenhalde, Germany, between Ronneburg and Gera, where bioaccumulation of manganese occurred in particular (Phieler et al. 2015). Islam et al. (2016) were able to accumulate lead in *Arum* (perhaps the name similarity between the Latin name of *Arum arum* and the ARUM process – Acid Reduction Using Microbiology – developed by M. Kalin was intentional?). Phytoremediation thus differs from bioremediation, in which microorganisms play the decisive role (Davison and Jones 1990). These are even said to be able to degrade PCB in the root zone (discussion in Adams et al. 2000), which is generally problematic to remove from mine water. Possibly a future use, although the Committee on Intrinsic Remediation et al. (2000) acknowledge that both bio- and phytoremediation of PCBs is successful only in favourable cases, but mostly does not seem to proceed completely.

In the context of soil or stockpile remediation, as well as for tailings ponds, the term phytoremediation has been introduced since 1991 (Adams et al. 2000). However, Wilfried Ernst, who was one of the first to write about plants for soil remediation, was only partially optimistic about the future of phytoremediation: "There is still a long way to go from the potential small-scale […] to a realistic largescale approach" (Ernst 1996, p. 166).

### 1.2.17 pH Value

In each of my seminars or courses on mine water hydrochemistry, I ask the participants about the definition of pH and what is the range of values that pH can encompass. Never has a participant given the correct answer on the first try. If you search the Internet, you will certainly come across statements that will make you tear your hair out or stand on end.

When asked what numerical values pH can encompass (the non-existent "pH scale") , most of my course participants answer either 0–4, 1–13, or 3–10. Each of these pairs of numbers has its background, but as an answer to the question they are all incorrect because there is no established pH scale. Unless buffering minerals are present, the pH for mine water can take on negative pH values, such as −3.6, at the Iron Mountain Mine in California, USA (Nordstrom and Alpers 1995; Nordstrom et al. 2000; Stumm and Morgan 1996). At the other end of the range of "naturally" occurring pH values were Lake Velenje in Šaleška dolina, Slovenia, which had pH values of 11–12 (Stropnik et al. 1991, p. 217), and in the region around Lake Calumet in Chicago, Illinois, USA, with pH values as high as 12.8 (Roadcap et al. 2005). In both cases, the pH of the water is influenced by anthropogenic activities (Fig. 1.5).

Let us clarify any misconceptions at this point to create transparency:

The pH value is a function of the *hydrogen ion activity*, namely its negative decadic logarithm, as defined by Søren Peter Lauritz Sørensen in 1909 and called "hydrogen ion exponent" (Sørensen 1909, pp. 159–160; Tillmans 1919, p. 3, incidentally strongly contradicted this concept of pH). According to Cohen et al. (2008, p. 75), the correct definition today is shortened to:

$$\mathrm{pH} = f\left\{\mathrm{H}^+\right\} = -\log\left\{\mathrm{H}^+\right\} \quad (1.4)$$

Since activity is dimensionless (Cohen et al. 2008, p. 70), the pH value consequently also has no dimension. In many articles or reports, the bad habit seems to have established itself of giving the pH value the dimension "s.u." (*standard units* or *sine unitatis* [without unit]), but this must be rejected for the reasons mentioned above, because the logarithm of a dimensionless number cannot have a unit. Usually, pH is now written instead of $p_H$ (Buck et al. 2002; Jensen 2004). In highly dilute media, such as drinking water or groundwater, the hydrogen ion concentration $[H^+]$ largely corresponds to the hydrogen ion activity $\{H^+\}$, so it is usually irrelevant whether we speak of concentration or activity. In contrast, for more highly mineralised solutions such as mine water, this difference is substantial, as Nordstrom et al. (2000) demonstrate in detail using an extreme example in California, USA. Consequently, in the context of mine water, one should always speak of activity rather than concentration.

**Fig. 1.5** Stalactite of melanterite above the puddle where Alpers and Nordstrom measured the mine water to have a pH of −3.6. The mine water in the 2-L-beaker has a pH of −0.7 and a temperature of 35 °C. (Pers. comm. Kirk Nordstrom 2019; photo: Kirk Nordstrom & Charly Alpers). Scale: Plastic beaker

### 1.2.18 Acid Capacity ($k_A$; Alkalinity; p-Value)

The acid capacity ($k_A$) is the sum of all alkaline reacting substances in a mine water. This refers to all the ions in the mine water that are formed by the reaction of the water with carbonates and silicates. The acid capacity is determined by titrating the mine water with an acid (HCl or $H_2SO_4$) to the pH endpoints 4.3 or 8.2 (Fig. 1.2). It can only be determined mathematically from chemical analysis within very narrow limits. The term p-value comes from the equivalence point of the indicator phenolphthalein, and consequently, it has only historical significance and should be avoided (like titration with indicators).

Further guidance is given in the first paragraph of the base capacity section (Sect. 1.2.3).

### 1.2.19 Acid Mine Drainage

In this book, acid mine drainage is mine water whose base capacity is greater than its acid capacity. It is then referred to as net acidic water (Sect. 1.2.13). This definition is, first of all, independent of the exact value of the pH, which alone does not allow us to say whether a mine water is "acidic" or not. Wolfgang Helms therefore correctly wrote: "Acid mine drainage, i.e. water with a low pH and sometimes high concentrations of sulfate, iron and other metals, is formed when sulfide minerals come into contact with water and oxygen" (Helms 1995, p. 65).

Often, however, the term acid mine drainage is incorrectly referred to the pH value, considering mine water with pH values below 7 as acidic and those with pH values above 7 as alkaline. However, this definition is unsuitable regarding mine water treatment processes, as the type of mine water treatment is determined primarily by the sum of all alkaline and the sum of all acidic reacting components of a mine water. Internationally, the term "*circumneutral*" has become accepted for mine water with pH values between about 6 and 8 (Nordstrom 2011), first appearing in the literature on mine water in the mid-1980s (Cook et al. 1986; Edwards and Stoner 1990; Hammack and Hedin 1989). It generally encompasses mine water with pH values below 7 whose acid capacity is greater than its base capacity.

According to the definition in the GARD Guide (Sect. 1.2.1), acid mine drainage is present at pH values below 6, whereas mine water with a pH value above this – depending on the mineralisation – is referred to as neutral, circumneutral or saline mine water (brine). The exact pH value below which mine water is always acid mine drainage is 5.6, since below this pH value there are no longer any relevant concentrations of buffering carbon species present.

### 1.2.20 Sorption, Adsorption, Coprecipitation, Surface Complexation and Other Such Reactions

Most authors use the terms sorption, adsorption (surface sorption), absorption (matrix sorption), coprecipitation (co-precipitation) or (surface) complexation to describe reactions between different components in water, without specifically committing themselves to a mechanism (Table 1.3). As expected, utter confusion thus reigns in the literature regarding these terms as well (only very few authors in the field of mine water use the terms chemisorption or physisorption – I will therefore not introduce them here). Even within the same publication it can happen that the terms are used inconsistently or that the term *coprecipitation* without hyphen appears next to *co-precipitation* with hyphen (the former is used slightly more often than the latter, especially in US English). The reason for this inconsistent usage in many cases is that we often do not know with certainty the details of the reaction involved in sorption and (co-)precipitation of a water constituent (Langmuir et al. 2005, p. 26). Consequently, the coprecipitation of a potential contaminant

**Table 1.3** Definition of terms related to surface and precipitation reactions and selected literature

| Term | Meaning | Literature (selection) |
|---|---|---|
| **Sorption** | Generic term for all surface reactions | McBain (1909); Smith (1999) |
| Adsorption (surface sorption) | Two-dimensional | Stumm and Morgan (1996); Smith (1999) |
| Absorption (matrix sorption) | Three-dimensional | Stumm and Morgan (1996) |
| Complexation | Enclosing of an ion by molecules (cf. Fig. 2.6) | Smith (1999) |
| Polymerisation | Concatenation of molecules | Hem (1985) |
| **Coprecipitation** *s.s./s.l.* | Mostly incorporation of an ion into the crystal lattice (solid solution) | Langmuir et al. (2005); Plumlee and Logsdon (1999) |
| Precipitation | Formation of a mostly amorphous or crystalline precipitate | Blowes et al. (2014) |

The terms surface and matrix sorption are used by Merkel and Planer-Friedrich (2002)

with the iron hydroxide or other precipitate product may be one of several processes, or it may be that several processes occur simultaneously. Therefore, the authors try to remain "vague" and do not definitively define or are not aware of the differences between the various terms. This already caused James McBain to introduce the term "sorption" at the beginning of the twentieth century: "The non-binding name 'sorption' is intended to express the sum of phenomena, while 'absorption' and 'adsorption' should be restricted to proven cases of solution and surface condensation" (McBain 1909, p. 219). A neat separation of terms can be found in Langmuir et al. (2005), who use the term "coprecipitation" very narrowly (see below).

Consequently, there is no uniform usage of the terms in the mine water literature. All terms include reactions in which a water constituent (adsorptive) adheres to the surface of another water constituent (usually a solid: adsorbent) and thus becomes an adsorbate (Dörfler 2002; and an excellent account is given in U. S. Environmental Protection Agency 2007b, pp. 28–36), with reactions depending on pH (Fig. 3.5). Some modern authors, in the spirit of McBain (1909), consider "sorption" to be the generic term (Smith 1999), while others seem to use "adsorption" as the generic term (Stumm and Morgan 1996). The latter, however, write that adsorption is attachment to a two-dimensional surface, whereas absorption is incorporation into a three-dimensional matrix (Stumm and Morgan 1996, p. 520), though they again qualify this on page 764 by using *adsorption* once and then *absorption* in the caption to Fig. 13.2 – may we attribute this to a careless error. Merkel and Planer-Friedrich (2002) recognise the problem with the "small difference" in b and d and therefore suggest using the terms surface sorption and matrix sorption. Regardless of the term, different pH-dependent reactions occur at the surface of the adsorbent as well as at the solid–water interface, which may be, for example, electrochemical reactions, polymerisation, or surface complexation (Hem 1985; Stumm and Morgan 1996). The exact mechanism could only be established by elaborate chemical-physical investigations, such as those identified by Trivedi et al. (2001) for the sorption of zinc on iron oxides. For most practical applications of mine water treatment, it will be sufficient to speak of "sorption".

In the context of "coprecipitation", the term is most often used when the mechanism is unclear. Stumm and Morgan (1996) do not define the term and use it only four times in their standard work on aquatic chemistry: always in the context of molecular incorporation of an ion into the precipitant, which is consistent with the use of the term by Langmuir et al. (2005). There may be reasons why the two authors avoid the term. Let us refer to this incorporation of ions as ("coprecipitation *s.s.*") . The situation is quite different in Plumlee and Logsdon (1999), one of many publications on the geochemistry of ore deposits and mine water. Depending on the chapter author, the use of the term differs considerably: incorporation into the crystal lattice of the precipitant or sorption at the surface of the precipitant. The seventh edition of the *Oxford Dictionary of Chemistry* (Rennie and Law 2016, p. 145: coprecipitation) gives the following definition: "The removal of a substance from solution by its association with a precipitate of some other substance. For example, if A and B are present in solution and a reagent is added such that A forms an insoluble precipitate, then B may be carried down with the precipitate of A, even though it is soluble under the conditions. This can occur by occlusion or absorption."

Consequently, the definition includes both incorporation into the crystal lattice and sorption at the surface of the precipitant and coincides with that in the *Gold Book* of the International Union of Pure and Applied Chemistry (2014). Let us refer to this as "coprecipitation *s.l.*" . Langmuir et al. (2005, p. 26), on the other hand, refer to coprecipitation exclusively as incorporation into the crystal lattice to form a *solid solution* with, for example, metal hydroxides, carbonates, sulfates, or silicate minerals ("coprecipitation *s.s.*"). A good illustration explaining precipitation, coprecipitation and adsorption is provided by the U.S. Environmental Protection Agency (2007b, p. 29).

Let us summarise the above (cf. also Fig. 2.6): Sorption is a pH-dependent water-solids process in which a water constituent that is predominantly present in solution adheres to a solid in the water, the most important mechanisms being adsorption (two-dimensional surface sorption), absorption (three-dimensional matrix sorption), polymerisation (linking of molecules to form a polymer) or complexation. Solids are colloids such as iron hydroxides, clay minerals or organic substances (e.g. wood). Coprecipitation *s.s.* refers to the coprecipitation of a water constituent in the crystal lattice of metal hydroxides, carbonates, sulfate or silicate minerals and coprecipitation *s.l.* refers to the coprecipitation of a water constituent adhering to their surface. Many metals and metalloids are thus removed from mine water either by precipitation, coprecipitation or sorption on precipitated aluminium or iron phases (Blowes et al. 2014, p. 183).

### 1.2.21 Heavy Metal

There are at least 40 different definitions for the meaningless term "heavy metal", which sometimes differ greatly from one to the other (Sect. 1.2.1). The International Union of Pure and Applied Chemistry (discussion in Anonymous 2006; Duffus 2002) therefore recommends that the term should not be used at all. It therefore does not appear in the *Gold Book* (International Union of Pure and Applied Chemistry 2014), the "bible" in chemistry, and it is not allowed in recent articles in the journal *Mine Water and the Environment.* Therefore, the term will not be defined here, nor will it be used in this publication (except in the context of musical genres, as described in Martikainen et al. 2021).

### 1.2.22 Base Metal

With regard to the term "base metals", a distinction must be made as to whether it is used in a chemical or in a mining as well as economic context. While it is relatively well defined in the chemical sense, there is a lack of a clear definition in the mining context – similar to "heavy metals". A base metal is more likely to enter the ionic state than a noble metal and is therefore more reactive than a noble metal. In this process, the base metal gives up electrons (Hofmann 2013, p. 413). Chemically, however, this excludes copper, as it is a precious metal (Wiberg et al. 2016, p. 1686 f.). In a mining or economic context, however, the term "base metals" mostly includes copper and is used here in this sense. A better term for *base metals* would consequently be "non-ferrous metals", but even in this case there is no complete agreement. Giving a reference to a definition for this paragraph proves difficult – each one is different. Let us keep in mind: base metals in this book are all metals except the precious metals (silver, gold, platinum, iridium) and iron (Römpp online 2013, term "non-ferrous metals" by Piotr R. Scheller).

## 1.3 Formation of Mine Water and Buffer Mechanisms

There are hundreds of publications on the formation of mine water, especially acid mine drainage (Blowes et al. 2014; Jambor et al. 2003; Wolkersdorfer 2008; Younger et al. 2002), and almost every publication on mine water begins again with a description of the basic four reactions relevant to the formation of acid mine drainage by the oxidation of pyrite ($FeS_2$, cubic), marcasite ($FeS_2$, orthorhombic), or pyrrhotine ($Fe_{x-1}S_x$). Certainly, these equations are among the most frequently cited equations in publications on mine water. In doing so, we can save this space in the future by simply referring to the two fundamental publications on the subject: Singer and Stumm (1970, p. 1121) and repeated and summarised in Stumm and Morgan (1996, p. 690 ff.), or the entire process comprehensively presented in Wolkersdorfer (2008, pp. 10–13). So let us leave it at Fig. 1.6, which represents the abiotic part of the reactions and whose reaction rate can be increased

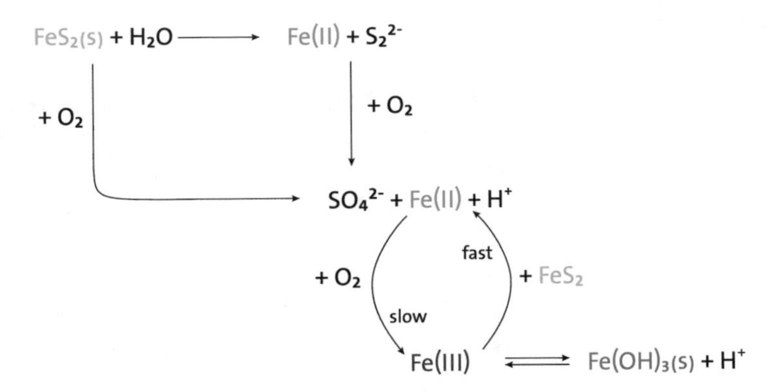


**Fig. 1.6** Schematic representation of abiotic pyrite and marcasite oxidation. (Based on information in Kester et al. 1975; Singer and Stumm 1970; Stumm and Morgan 1996)

by a factor of $10^6$ through microbial catalysis. For those interested in the role of microorganisms in this process, I can recommend the review article by Rawlings et al. (1999) (in which *Acidithiobacillus thiooxidans* is of course still called *Thiobacillus thiooxidans,* as this bacterium was only renamed in 2000). Unfortunately, as in so many cases, e.g. the development of the first constructed wetlands, basic German work on *Acidithiobacillus* or *Thiobacillus* has been neglected or even ignored in the literature. Marchlewitz (1959), Marchlewitz and Schwartz (1961), and Marchlewitz et al. (1961) studied 16 mine waters for their microbial composition and made the first systematic attempt to elucidate the role of fungi in mine water. In addition, they investigated the temperature behaviour of *Acidithiobacillus* and concluded that all the strains investigated survived temperatures up to 35 °C without any problems and that some of the strains even could exist at temperatures up to 45 °C.

Professional articles, bachelor theses, reports, export reports, dissertations and even post-doctoral theses continue to give *Thiobacillus thiooxidans* the crucial role in the oxidation of pyrite. However, I recommend reading Kelly and Wood, because they defined in 2000 that the species *Acidithiobacillus thiooxidans* (acid-loving iron-oxidising sulfur rod bacterium) is mainly responsible for the processes of pyrite oxidation: "The type species is *Acidithiobacillus thiooxidans* (formerly *Thiobacillus thiooxidans*)" (Kelly and Wood 2000, p. 513). The same applies to the oxidation of iron and *Thiobacillus ferrooxidans*. What I'm saying is that the small creature has had a new name since 2000, and you probably don't want to be misnamed either.

Microbially catalysed oxidation of disulfide sulfur produces acid (Fig. 1.7), which reacts with other metal and semimetal sulfides and enriches the mine water with a cocktail of elements. Some of these elements and their compounds are considered potentially toxic to organisms. It is not necessary to give a list of the elements found in mine water: A look at the periodic table of the elements may be enough, because there is hardly any element that is not found with an "increased" concentration in the mine water of one or the other mine (admittedly: $^{289}$Flerovium with a half-life of 30 seconds will hardly be found there –

**Fig. 1.7** Drops of acid mine drainage with pH 2 at a gallery roof in the former Glasebach mine in Straßberg/Harz, Germany. Second level: area of the pyrite passage. (Image width approx. 20 cm)

so let's say all elements up to uranium or plutonium can be detected in the mine water). As described above, the lowest pH measured to date in mine water is −3.6 (Nordstrom et al. 2000), where the total metal concentration was 200 g $L^{-1}$. Once buffering minerals such as carbonates, micas, or other silicates are present, the pH does not drop as much as without buffering minerals, but the mine water may still be highly mineralised and will require treatment. An example is the mine water of the former German Niederschlema/Alberoda uranium mine ("Shaft 371"), which has pH values in the circumneutral range, but a comparatively high mineralisation for that pH: pH 6.5–8.5, $\kappa$ 1–4 mS $cm^{-1}$ (Wolkersdorfer 1996, pp. 177–178). Even the smallest amounts of pyrite are capable of mineralising mine water to a relatively high degree. Wisotzky (2003, p. 31–35) found sulfate concentrations of up to 4000 mg $L^{-1}$ in the overburden of the Garzweiler open pit despite the low mass concentrations of 0.258% pyrite. This may be an extraordinarily high concentration compared to other mining regions. Nevertheless, relatively low pyrite concentrations can lead to substantial iron concentrations in mine water (Table 1.4).

Acid mine drainage can also result from groundwater flowing through soil containing disulfides and oxidising the disulfides or transporting the oxidation products already present (the problem of the term "soil" is discussed in Hoppe 1996). This causes a decrease in pH and consequently an increase in ion concentrations in the groundwater. The "soils" can be either natural soil, tailings, treatment residues or backfilled dumps. In the German-speaking world, this is often referred to as "pedogenic processes" (Neumann 1999; Schwertmann 1985), which includes all processes that occur in the soil or substrate. Internationally, discussions on this topic are usually conducted under the keyword *acid soils* (e.g. Hicks et al. 1999). In the context of mine water, the term "pedogenic" is deliberately avoided internationally, as this term suggests "naturally occurring", which, strictly speaking, is not the case with mining influenced water.

**Table 1.4** Prediction of maximum iron concentrations in mine water from newly flooded British deep mines as a function of the total sulfur concentration of the mined seams (Younger 2002a) and comparison with measured concentrations from Mpumalanga, South Africa. (Pers. comm. Altus Huisamen)

| Total disulfide concentration, mass % | Observed range of maximum concentrations in mine water, mg $L^{-1}$ | |
|---|---|---|
| <1 | 0.01…0.5 | |
| 1…2 | 0.5…100 | |
| 2…3 | 100…350 | |
| 3…4 | 350…1.200 | |
| 4…5 | 1.200…1.500 | |
| South African open pit coal mine | Pyrite concentration % | Total observed iron concentrations, mg $L^{-1}$ |
| Mine A | 2 | 0.0…0.07 |
| Mine C | 0.03 | 0.13 |
| Mine E | 6.5 | 22.570 |

For the processes in the partially renatured overburden, which take place for example in Lusatia, Germany, Neumann (1999, p. 110) has determined that after more than three decades no conditions have yet been established which correspond to natural conditions. Consequently, the author assumes that the acidification and the high sulfate concentrations in the soil do not lead to the expectation of conditions corresponding to the natural soils of Lusatia for a long time. These results can easily be transferred to other mining regions where pre-mining conditions do not exist even long after mining operations have ceased (e.g. Olyphant and Harper 1998; Yang et al. 2006).

An extreme example related to acid mine drainage is the Rio Tinto region in the Iberian pyrite belt of Spain and Portugal (Ariza 1998; Cánovas et al. 2005; Leblanc et al. 2000; Salkield 1987; Wolkersdorfer et al. 2021). Mining for non-ferrous and precious metals has taken place there since the middle of the second millennium BCE and has created an environment characterised by acidic waters through pyrite oxidation. Up to their confluence with the Rio de Huelva River, the Rio Tinto or Rio Odiel Rivers are strongly acidic with pH values ranging between 1.5 and 2.5 (Leblanc et al. 2000, p. 656; Olías et al. 2004; Olias et al. 2017; Sarmiento et al. 2005) and electrical conductivities ranging between 757 and 5500 μS $cm^{-1}$ (Torre et al. 2014, p. 220). The Cement Creek watershed near Silverton in Colorado (USA) has also been severely affected by historical mining and has low pH values of less than 4.5 (Kimball et al. 2002). In both regions, however, acid mine drainage generated due to anthropogenic activities is accompanied by natural acid mine drainage, as the comparatively high pyrite concentration of the source rock naturally causes pyrite oxidation (Fig. 1.8).

In individual cases, the extent to which an acidic and contaminated groundwater has natural or anthropogenic causes may be debated. It would be easy to assume in all cases that contamination originates from mining. Many NGOs (e.g. Earthworks) would probably welcome this immediately. Of course, a fascinating academic discussion could be

**Fig. 1.8** Red coloured water in a section of the Rio Tinto River in Spain. (Photo: Marta Sostre)

what proportion of the contamination is attributable to "pedogenic" or what to "anthropogenic" causes. To resolve this question in detail, elaborate isotope studies or tracer tests would be required. In addition, numerical modelling might be necessary, as in the case of the Kentucky-Utah Tunnel in the Wasatch Mountains of Utah, USA (Parry et al. 2000). The effort required for these studies is possibly disproportionate to the gain in knowledge.

In addition to acid-forming processes, nature provides processes that buffer the acid or otherwise lead to a reduction in the pollutant load. This process is collectively referred to as *natural attenuation* (e.g. Bekins et al. 2001; Committee on Intrinsic Remediation et al. 2000, p. 65 ff.) and the term first appeared in the mine water literature in 1994 (Kwong and Van Stempvoort 1994, p. 382; Webster et al. 1994, p. 244). More details on this topic are given in Sect. 5.2. Buffering involves the buffering of acid-forming protons by minerals such as carbonates, mica, or feldspars, and results in an increase in pH in mine water or contaminated groundwater. Other processes include precipitation of potential contaminants as their solubility is exceeded. This primarily involves iron and manganese compounds, with precipitation of iron or manganese oxides or hydroxides (often referred to as oxyhydrates).

Numerous publications are available on how to avoid the generation of contaminated mine water; however, there are too many to mention here. A good summary is provided by the GARD Guide (Chap. 6) or by Gusek and Figueroa (2009, pp. 15–80). Suffice it to say that it is important for the mine operator to be aware, even before mining begins, that measures taken at an early stage, with cost implications in the short term, can make mine water treatment less expensive in the long term.

Not all mine water is contaminated, as is commonly assumed. There are numerous examples where untreated mine water is even used for drinking water supply (e.g. Ypsilanta mine near Dillenburg, Germany), as medicinal water, driving water for turbines or for recreational purposes (Geisenheimer 1913; Razowska-Jaworek et al. 2008; Stengel-Rutkowski 1993; Szilagyl 1985; Wolkersdorfer 2008, pp. 270–275). However, even the lowest iron concentrations of 1 mg $L^{-1}$ or less can cause ochre deposits in receiving water courses (Glover 1975, p. 181). Not only do these look unsightly, but they can also contribute to the water body becoming biologically less fully functional. Ercker v. Schreckenfels (1565, pp. 11–12) already describes this problem and the environmental damage caused by acid mine drainage and natural attenuation as follows:

> "The two waters that flow out of the Rammelsberg through the adit, that, which is drawn out of the mountain by art, and that, which falls on the art wheel, are incidentally so great that it can drive a water wheel. There, where it flows through the adit, the wood becomes yellow and a yellow mud is attached to it by the thickness of a finger, and a yellow mud or sludge settles at the bottom of the water seams, almost hand-thick at some ends, which is called ochre yellow, which painters use for colours, and which is also sold in chemists' shops. The same water flows out of the adit into another water body, which flows together through the city of Goslar and is called the Abzucht. This then flows into the Oker River a quarter of a mile below the town of Goslar, and as the Abzucht flows into the Oker it poisons the water, so that the Oker carries no fish for two miles. So the wild ducks fall on it, and they become lame so that they cannot fly anymore and may be grabbed with the hands or caught. Afterwards again other freshwater rivers flow into it, and sweeten the water of the Oker so that it carries again fish of all kinds" (translated from 16th century German, trying to be as close as possible to the original meaning; one Prussian mile at that time was 7.9 km, the Rammelsberg mine was a lead-zinc mine in the German Harz Mountains, which operated for about a millennia until its closure in 1988).

At the end of the acidification process, the water is enriched with one or more of the following groups of substances or has the following properties (modified after Jacobs and Pulles 2007):

- Acidity and low pH
- Dissolved metals and semimetals (e.g. Fe, Mn, Al, As, Cr, Cu, Co, Ni, Zn, Hg, Pb)
- Radionuclides (e.g. U, Ra, Th, Po)
- Turbidity and suspended solids
- Total mineralisation (essentially from Na, K, Ca, Mg, Cl, $SO_4$)
- Nutrients (ammonium, nitrate, phosphate)
- Oxygen demand (organic impurities)
- Aesthetics (in the widest sense: scaling by ochre)

The aim of mine water treatment is to remove or alter the undesirable constituents or properties of the mine water to such an extent that the treated effluent is fit for its intended use or for discharge into the receiving water courses.

Numerous colleagues have developed methods to forecast the quantity and quality of mine water. These methods range from predictions before the start of mining to predictions when mining operations cease (Dold 2015, 2017; Renton et al. 1988; Wolkersdorfer 1996, 2008). However, we must accept that there is no universally accepted method to date and a large number of forecasts are wrong (Kuipers and Maest 2006). This is partly due to a lack of detailed investigations at the site, but also to the highly nonlinear behaviour of the chemical and physical processes – or to put it another way: The development of the chemical-thermodynamic-hydrodynamic processes is subject to chaotic behaviour and can therefore – like all chaotic systems – not be projected with sufficient accuracy into the future (Wolkersdorfer 2008, pp. 195f, 295). Renton et al. (1988, p. 76) have summarised this as follows: “There are simply too many random processes, variables, and interrelations associated with a coal mine site to allow the depiction of any environmental response within reasonable certainty using only basic scientific principles, equations, and empirical formulae.”

The easiest way to avoid the problem of acid mine drainage is to follow the recommendations of Glover (1975, p. 183):

> Prevention of the formation of an acid and ferruginous mine drainage is basically simple. Either the source of the drainage water can be cut off, contact between the water and the contaminating pyrite oxidation products can be cut, or production of the pyrite oxidation products can be prevented.

In the same vein, Earthworks and Oxfam America (2004, p. 30) state, “Refrain from projects that are expected to cause acid drainage.” If it were that simple, none of the thousands of publications on mine water would have a reason to exist. So let us move on to the classification of contaminated mine water.

## 1.4 Classification of Mine Water

Numerous classifications have already been proposed for mine water; these are based on a relatively easy-to-measure parameter such as the pH value, or a complete chemical analysis of the mine water is required. For the question of which treatment method a specific mine water requires, an exact classification – unlike a water analysis – plays a subordinate role, so that the existing classifications are rather of an academic nature for our question. However, when it comes to representing differences or changes of a parameter in a water analysis, such classifications are of indispensable value. Besides the classifications of water known from hydrogeology, such as the Piper diagram (Piper 1944, 1953) or Durov diagram (Chilingar 1956; Durov 1948), other representations are used for mine water, and different classifications are employed. Which classification or diagram you consider most suitable, you must decide on a case-by-case basis, or give in to your preferences.

Ficklin et al. first presented a diagram in 1992 and then extended it in 1999 (Ficklin and Mosier 1999; Ficklin et al. 1992), plotting pH on the abscissa and the sum of mass concentrations of characteristic base metals on the ordinate (Fig. 1.9). The diagram is suitable to depict differences in the water of different deposits or ore occurrences and to show the development or differentiation within a deposit (as you can see from the glossary, I am aware of the difference between deposit and ore occurrence – but always retain the term of

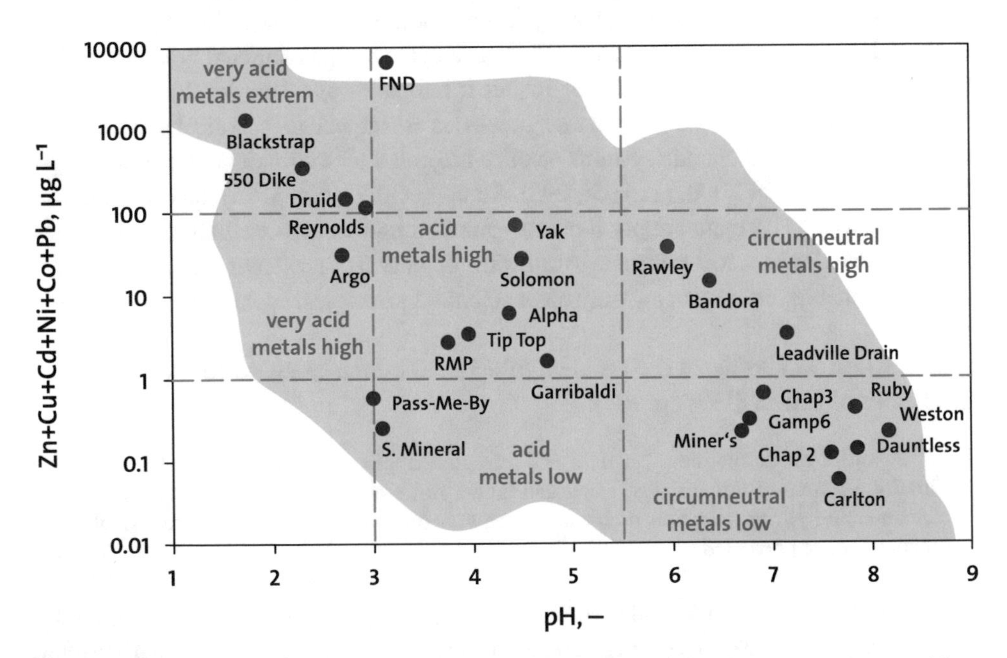


**Fig. 1.9** Ficklin diagram with mine waters from Colorado, USA. (Modified and supplemented from Ficklin et al. 1992). All waters investigated by Ficklin in the vicinity of mines are shown as a shaded background. "Metals" is a synonym for "metal concentrations" of the ordinate; FND: Fanie Nel Discharge, Witbank, South Africa

the respective author). The hypothesis behind the Ficklin diagram is that the paragenesis of the ore deposits causally determines the hydrogeochemical composition of the mine water. Based on a very large number of empirical studies with mine waters from a wide variety of deposits, the authors succeeded in identifying characteristic base metals that are suitable for describing the ore deposit: Zn, Cu, Cd, Pb, Co, and Ni. Based on these investigations, the metals Fe, Al, or Mn, which are characteristic for mine water, turned out to be less suitable to be used for the classification of ore deposits. Based on the results, if the genesis of the deposit is known, statements can be made within certain limits about the expected mine water chemistry.

Trace elements are not always analysed in mine water samples, whereas the major elements are usually present in their entirety. Therefore, the Younger-Rees diagram (Fig. 1.10) offers an advantage over the Ficklin diagram in the classification of mine water (Rees et al. 2002; Younger 1995, p. 106, 2007, p. 98): it requires only the equivalent concentrations of Cl and $SO_4$ in addition to the base and acid capacities. In the original version, the abscissa consisted of the equivalent sum of Cl and $SO_4$ and thus, for sulfate-rich mine waters, it had similarities to the NP–$SO_4$ representation discussed below. Rees et al. (2002) recognised the weakness of this method of representation and changed the abscissa to the relative proportion of $SO_4$ to Cl. This forms more distinguishable groupings of waters in the diagram. In terms of the Younger-Rees diagram (Younger 2007), these are divided into the following five classes:

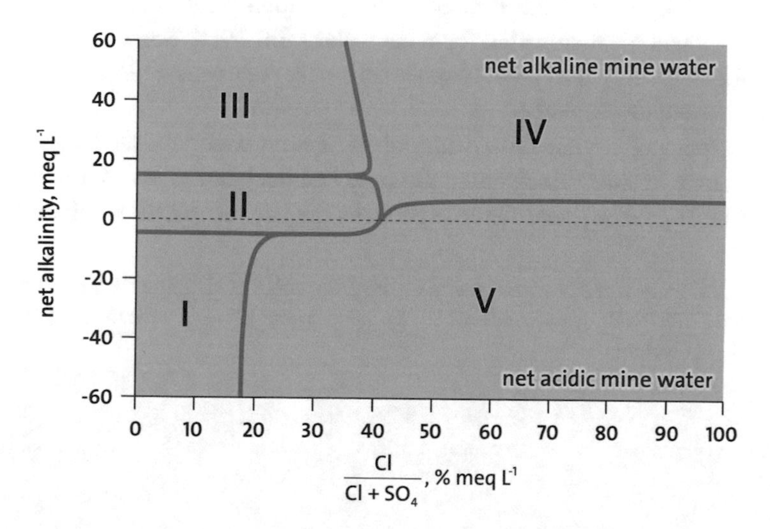


**Fig. 1.10** Expanded Younger-Rees diagram (Rees et al. 2002; Younger 1995, p. 106; 2007, p. 98). I. Acidic spoil leachates, tailings drainage, and shallow oxygenated workings in pyrite-rich strata; II. Majority of fresh, shallow, ferruginous mine waters; III. Previously acidic mine waters, since neutralised; IV. Deep-sourced pumped, saline, mine waters; and V. Field in which few mine waters plot

I. Acidic spoil leachates, tailings drainage, and shallow oxygenated workings in pyrite-rich strata
II. Majority of fresh, shallow, ferruginous mine waters
III. Previously acidic waters, since neutralised
IV. Deep-sourced pumped, saline, mine waters
V. Field in which few mine waters plot

Glover (1975, pp. 187–190) presented a classification for mine water based on acidity and the iron species $Fe^{2+}$ and $Fe^{3+}$. It can only be applied if the speciation of the iron is known. As will be described below, this is often not available, so mine water is rarely classified by this method. Its five groups include:

1. Acidic with low $Fe_{tot}$ concentration
2. Acidic with high $Fe^{3+}$ concentration
3. Acidic with high $Fe^{2+}$ concentration
4. Neutral with high $Fe^{2+}$ concentration
5. Iron hydroxide in suspension (combined with dissolved $Fe^{2+}$ and $Fe^{3+}$)

Because the classification contains a subjective component and allows little to be said about the genesis of the water or the potential hydrochemical relationships, this classification has not gained acceptance. However, Glover (1975) suggests one or more treatment options for each of the five mine water types, which makes his work interesting.

Based essentially on pH, the Federal Water Pollution Control Administration in the USA has developed a classification for mine water (Hill 1968; Scott and Hays 1975). It comprises four classes, which are differentiated by six parameters: pH, acidity, $Fe^{2+}$, $Fe^{3+}$, aluminium and sulfate (Table 1.5).

A simple (but not trivial) mine water classification was published by Hedin et al. (1994a). In terms of this classification, the base and acid capacities of a mine water are determined, and by taking the difference between the two, the net alkalinity or net acidity

**Table 1.5** Classification of mine water based on the proposal of the Federal Water Pollution Control Administration (Hill 1968; Scott and Hays 1975); units in mg $L^{-1}$; pH without unit; and acidity in mg $L^{-1}$ $CaCO_3$ equivalents

| | Class 1 | Class 2 | Class 3 | Class 4 |
|---|---|---|---|---|
| Parameter | Acid mine drainage | Partially oxidised and/or neutralised | Oxidised and neutralised and/or alkaline | Neutralised and not oxidised |
| pH | 2…4.5 | 3.5…6.6 | 6.5…8.5 | 6.5…8.5 |
| Acidity | 1000…15.000 | 0…1000 | 0 | 0 |
| $Fe^{2+}$ | 500…10.000 | <500 | 0 | 50…1000 |
| $Fe^{3+}$ | 0 | <1000 | 0 | 0 |
| Aluminium | <2000 | <20 | 0 | 0 |
| Sulfate | 1000…20.000 | 500…10.000 | 500…10.000 | 500…10.000 |

of the water is obtained (Sect. 1.2.13). This in turn gives the two classes, namely net acidic or net alkaline mine water. Despite – or perhaps because of – its simplicity, this is the most important of all classifications regarding passive mine water treatment.

Puura and D'Alessandro (2005, p. 48) classify mine water based on the total discharge and the exceedance of a given standard. They introduced the *pressure factor,* which is the logarithm of the exceedance of the standard and the discharge. This in turn can be used to group mine water into environmental pressure classes. In essence, this approach represents the relationship between the load and a standard, which is advantageous in that mine waters with high loads can be clearly distinguished from those with lower loads.

Schöpke has since 1999 repeatedly presented an NP–$SO_4$ diagram (Schöpke 1999, p. 31). As this is very specifically tailored to the sulfate-rich waters of the open pit mining lakes in Lusatia, Germany, it has hardly been applied to mine waters in other regions. In addition, NP (neutralisation potential) is often very well correlated with $SO_4$, so that linear dependencies inevitably arise in the NP–$SO_4$ representation, which are only of limited use for classifying mine water. A detailed discussion of how the NP–$SO_4$ representation is used to allow statements about the development of mine water is given by Schöpke and Preuß (2012).

Based on mine water titration curves, Ott (1988) and later Totsche et al. (2006) presented a method to classify mine water. They were able to show that metal concentrations of the mine water can be estimated from the titration curves and the respective buffer ranges that are passed through during the titration. Since the titration curves can be reproduced very well and their curve progression is characteristic for the water chemistry, the latter were able to design a classification for open pit mining lakes in Lusatia. This is essentially limited to the groups "hydrogen sulfide buffered", "iron buffered" and "aluminium buffered". In my view, this classification has considerable potential, and it should be explored with a view to general application at a wide variety of sites around the world. In particular, a tool would need to be developed to describe the curve numerically and to allow automated comparison and classification. A first attempt to do this was made by van der Walt and Wolkersdorfer (2018), but it turned out that an automated approach is more complicated than originally anticipated.

To be able to compare the water quality of water from waterworks with that of mine water, Rottmann (1969) introduced a classification according to the quality index. His investigation was triggered by the possible "replacement of drinking water from public supplies by mine water" (p. 166) in connection with emergencies (Wassersicherstellungsgesetz [Water Security Act] of 24th August 1965 as amended on 19th June 2020). Rottmann's figure of quality represents a sum of the parameter values calculated from the mass concentrations of chloride, sulfate, and total hardness (°dH) and the deviation of the pH value from 7. Based on this, he classifies the mine water of the Ruhr, Aachen and Saar mining districts and compares their quality figures with those of published drinking waters of the year 1959, without taking into account the limit values for the parameters used at that time. Of the approximately 1800 drinking waters of the waterworks, he only disregards those with quality numbers above 500 and assumes that mine waters with a maximum

**Table 1.6** Visual pH meter scale. Colour indications are based on the Munsell scale. (After Younger 2010)

| pH | Description/explanation of appearance | Color |
|---|---|---|
| 2 | Clear, intensely red colored water (10R 3/6) due to high concentrations (> 500 mg $L^{-1}$) of dissolved $Fe^{3+}$ | |
| 2.5 | Mainly clear water (occasionally with a faint blue-green tint due to high concentrations of dissolved $Fe^{2+}$) with tendrils of green to dark greenish-yellow (10Y 6/6) acidophilic microorganisms adhering to the stream bottom | |
| 2.5 | Clear water with high dissolved $Fe^{3+}$ (> 500 mg $L^{-1}$), which has an intense red color (10R 4/8) | |
| 3 | Yellowish-white (10YR 8/2) precipitates, usually from iron hydroxysulfates or from mixtures of orange iron hydroxides with white Au precipitates from aluminum hydroxide | |
| 4 | White (N) patina and foam of aluminum hydroxide/hydroxysulfate. These aluminum precipitates are usually insoluble from a pH value around 4.5 | |
| 5 | Clear water above thick precipitates of yellow (2.5Y 8/8) ochre (mainly X-ray amorphous $Fe^{3+}$ hydroxides) formed by surface-catalyzed oxidation at high concentrations of dissolved $Fe^{2+}$ (≈ 400 mg $L^{-1}$). Above the water level, the ochre dries dark red (10R 3/4) | |
| 6 | Clear water above moderate precipitation of light red (2.5YR 6/8) ochre formed by surface catalyzed oxidation at moderate concentrations of dissolved $Fe^{2+}$ (≈ 40 mg $L^{-1}$). Above the water surface, the ochre dries dark reddish brown (2.5YR 3/4) | |
| 7 | Pale reddish yellow (7.5YR 7/5) ochre patina on stones formed by surface-catalyzed oxidation at low concentrations of dissolved $Fe^{2+}$ (≈ 5 mg $L^{-1}$) | |

quality index of 500 are also suitable as drinking water after treatment. The introduction of this figure represents a redundancy as far as the same statement can also be obtained with the electrical conductivity or the TDS. My comparative calculation based on mine water and groundwater analyses showed a linear dependence of the figure of merit on these two parameters with a correlation coefficient of 0.9…1.0.

Younger (2010) presented a short-range pH determination that is at least suitable for a rapid check of measured values in the field. He describes how the colour of mine water, together with knowledge of local geological conditions, can be used to estimate the pH of mine water (Table 1.6). This approach is not new, as you and I likely have developed our own internal pH scale over time.

# Preliminary Investigations 2

## 2.1 Introductory Remarks

To design an optimal mine water treatment plant, several preliminary investigations are necessary. Without these, a plant, whatever its design, will fail in most cases. As a rule, the goal will be a treatment plant that is precisely required for the given mine water, but in the rarest cases what would be feasible with the most sophisticated technology – much as I would occasionally wish for the latter. Furthermore, the process selection should be the result of a multi-stage planning process, in which, moreover, tests for alternatives take place. In the end, the individual style and marks of the planner also play a role in the final concept, as well as the person ultimately responsible for the functionality of the completed system.

Regardless of the method selected, the following principles must be observed when planning a system:

- Effectiveness in achieving the remediation objective, in particular the effluent values
- Cost
- Operational and occupational safety
- References of the application under comparable boundary conditions
- Secondary environmental effects
- Socio-economic aspects
- Sustainability aspects
- Availability of operating resources
- Legal requirements



C. Wolkersdorfer, *Mine Water Treatment – Active and Passive Methods*,
https://doi.org/10.1007/978-3-662-65770-6_2

A frequently selected approach, even if it is not explicitly named as such, is the best available technology economically achievable (BATEA) principle, also abbreviated BAT or BATA (Train et al. 1976), which was already introduced in the USA in the 1970s. The European IPPC (Integrated Pollution Prevention and Control) office has also written numerous BAT documents, e.g. for tailings and overburden (European Commission 2009), but not for mine water *per se*, as the listing in Vidal-Legaz (2017) shows. For Finland, Kauppila et al. (2013) or Punkkinen et al. (2016) have summarised what should be considered best technologies in the mining sector. A listing of the methods currently considered BATEA in Canada and the MEND partner companies shows that membrane processes, ion exchange and classical neutralisation processes can remove most potential pollutants from mine water (Table 2.1). As expected, sorption plays only a minor role, and passive technologies tend to receive a poorer assessment, as is usual with MEND reports, since MEND reports generally target active mining.

In the preliminary investigation for a treatment plant, it is not necessary to arrange for a complete mine water analysis. Instead, one should rather concentrate on a few key parameters, which are described below (Sect. 2.3.1). Only then a more comprehensive

**Table 2.1** Mine water treatment technologies depending on the parameter to be treated, which can be considered BAT/BATEA (Hatch 2014; Pouw et al. 2015)

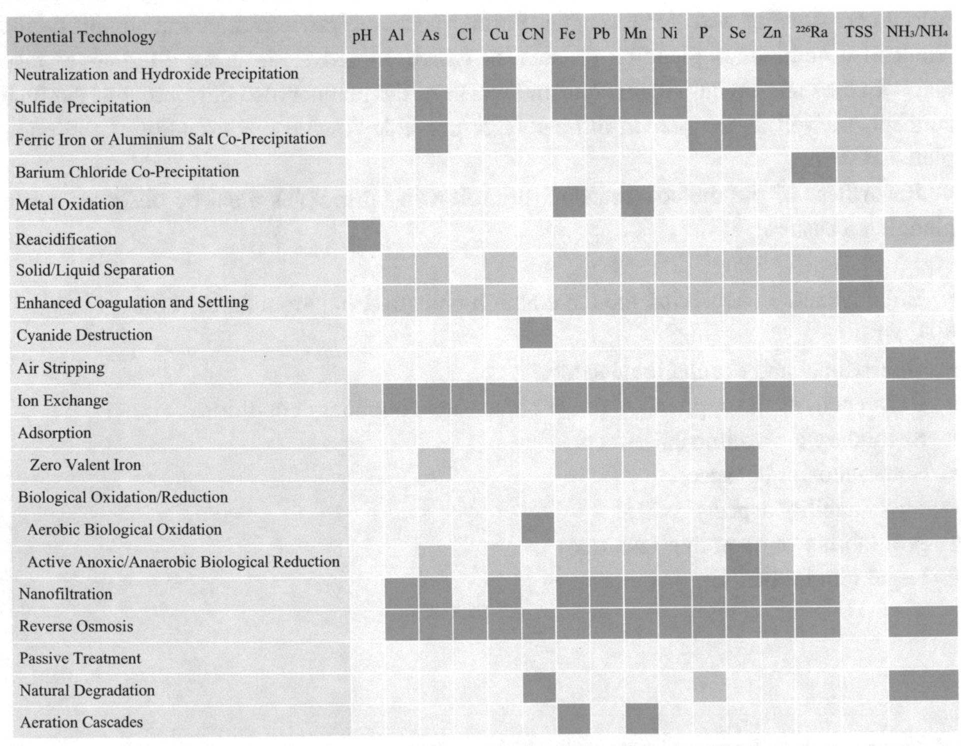

| Potential Technology | pH | Al | As | Cl | Cu | CN | Fe | Pb | Mn | Ni | P | Se | Zn | $^{226}$Ra | TSS | $NH_3/NH_4$ |
|---|---|---|---|---|---|---|---|---|---|---|---|---|---|---|---|---|
| Neutralization and Hydroxide Precipitation | | | | | | | | | | | | | | | | |
| Sulfide Precipitation | | | | | | | | | | | | | | | | |
| Ferric Iron or Aluminium Salt Co-Precipitation | | | | | | | | | | | | | | | | |
| Barium Chloride Co-Precipitation | | | | | | | | | | | | | | | | |
| Metal Oxidation | | | | | | | | | | | | | | | | |
| Reacidification | | | | | | | | | | | | | | | | |
| Solid/Liquid Separation | | | | | | | | | | | | | | | | |
| Enhanced Coagulation and Settling | | | | | | | | | | | | | | | | |
| Cyanide Destruction | | | | | | | | | | | | | | | | |
| Air Stripping | | | | | | | | | | | | | | | | |
| Ion Exchange | | | | | | | | | | | | | | | | |
| Adsorption | | | | | | | | | | | | | | | | |
| Zero Valent Iron | | | | | | | | | | | | | | | | |
| Biological Oxidation/Reduction | | | | | | | | | | | | | | | | |
| Aerobic Biological Oxidation | | | | | | | | | | | | | | | | |
| Active Anoxic/Anaerobic Biological Reduction | | | | | | | | | | | | | | | | |
| Nanofiltration | | | | | | | | | | | | | | | | |
| Reverse Osmosis | | | | | | | | | | | | | | | | |
| Passive Treatment | | | | | | | | | | | | | | | | |
| Natural Degradation | | | | | | | | | | | | | | | | |
| Aeration Cascades | | | | | | | | | | | | | | | | |

Passive technologies tend to receive lower ratings in MEND reports, as MEND methods often target active mining picture of the chemistry can be obtained by a full mine water analysis. There is a large body of literature on the detailed procedure for water sampling, and some of the references should be mentioned here (e.g. American Public Health Association (APHA) et al. 2012; Lloyd and Heathcote 1985; Sächsisches Landesamt für Umwelt und Geologie 1997; Sächsisches Landesamt für Umwelt, Landwirtschaft und Geologie 2003; U.S. Geological Survey (USGS) 2015). For mine water sampling, on the other hand, there is little generally available literature (e.g., Ficklin and Mosier 1999; PIRAMID Consortium 2003), so commonly used sampling procedures for mine water will be addressed first. A more comprehensive description of mine water sampling can be found in Wolkersdorfer (2008, pp. 171–194) and McLemore et al. (2014). In any case, keep in mind that analytical results from a water sample improperly collected in the field are questionable at best, but useless at worst (Ficklin and Mosier 1999, p. 249).

Essential to mine water analysis is the fact that mine water is subject to strong seasonal and sometimes diurnal variability (e.g., Duren and McKnight 2013; Frau and Cidu 2010; Younger and Banwart 2002; Younger et al. 2002). Therefore, it is not sufficient to take only a random sample, but measurements should be taken over a hydrological year, especially when planning a treatment plant. Unlike groundwater, however, changes in mine water chemistry occur because of hydrological changes and because the mine water comes into contact with different parts of the mine. As a result, water quality can change drastically during the mining and post-mining phases. This effect is usually described as first flush (see Sect. 1.2.8) (Younger 2000a, p. A211), and it means that both sampling and considerations of a potential treatment facility must take this effect into account. The varying rates of weathering of the minerals that the mine water is in contact with may also be responsible for a change in pH. For example, if buffering carbonates have weathered while pyrite, marcasite or pyrrhotine are still present in the mine workings, the formerly well-buffered mine water will become acidic (Fig. 2.1).

In addition, it is essential to measure the flow rate together with the water sample (Sect. 2.3.5) and to record it regularly to obtain its maxima and minima. PIRAMID Consortium (2003, p. 14) states: "The chemical data can be considered practically useless for the planning of a treatment plant if a flow measurement is not carried out at the same time" (Fig. 2.9).

At the end of the field survey, the data are evaluated, and the original conceptual model is verified using statistical methods, chemical-thermodynamic codes (Nordstrom et al. 2017b) and, if necessary, numerical hydrogeological models. The tools you decide to use will vary, depending on your personal preferences, your employer's practices, or else your client's specifications. For chemical-thermodynamic modelling alone, there are nearly a dozen codes available today (Table 2.2), and for the characterisation of residues from mines even more, still (Nordstrom et al. 2017b, p. 89 ff.).

Part of the following sections will follow the recommendations given in PIRAMID Consortium (2003, pp. 8–41). These recommendations have been developed to ensure the best possible planning basis for mine water treatment plants and should always serve as a resource for you. Many treatment plants, whether passive or active, could be planned more quickly and easily if every person working on them were aware of this guideline. The

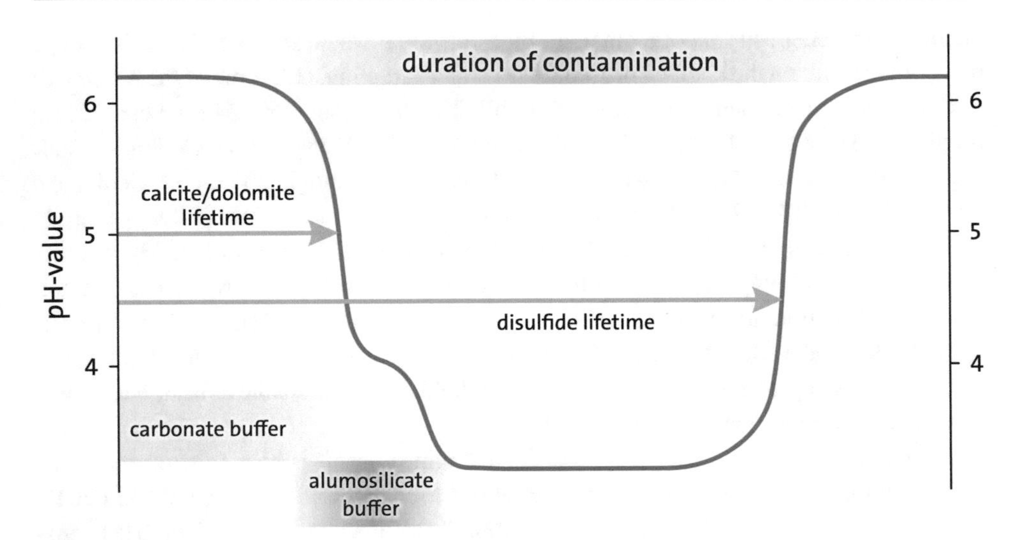


**Fig. 2.1** Possible development of mine water quality in a mine where the pyrite concentration exceeds that of the carbonates. The second pH increase is due to different buffer processes and the fact that the pyrite is largely weathered. (Modified after Younger et al. 2002)

**Table 2.2** Selection of numerical, chemical-thermodynamic codes, arranged according to the actuality of the version

| Name | Version I Date | Developer |
|---|---|---|
| Software based on law of conservation of mass | | |
| CHEAQS Next (64 bit version) | 0.2.1.4 I May 2022 | Wilko Verweij[a] |
| The Geochemist's Workbench | 16.0 I March 2022 | Aqueous Solutions LLC[b] |
| HSC Chemistry | 10.1 I March 2022 | Metso Outotec (Finland) Oy[c] |
| PHREEQC | 3.7.3 I December 2021 | U.S. Geological Survey[d] |
| DataBase and Spana | 0.2.1 I June 2020 | Ignasi Puigdomenech[e] |
| JESS | 8.7 I May 2019 | Murdoch University[f] |
| HYDRA and MEDUSA | r201 I August 2016 | Ignasi Puigdomenech[e] |
| MINEQL$^{+}$ | 5.0 I September 2015 | Environmental Research Software[g] |
| Visual MINTEQ | 3.1 I December 2013 | KTH Royal Institute of Technology[h] |
| EQ3/6 | 8.0a I June 2013 | Lawrence Livermore National Laboratory[i] |
| WATEQ4F | 3.00 I March 2011 | U.S. Geological Survey[j] |
| CHEPROO | 1.0 I October 2008 | Technical University of Catalonia[k] |
| MINTEQA2 | 4.03 I May 2006 | U.S. Environmental Protection Agency[l] |
| CHESS | 3.0 I April 2002 | École des Mines de Paris[m] |

(continued)

**Table 2.2** (continued)

| Name | Version \| Date | Developer |
|---|---|---|
| Software based on Gibbs Free Energy | | |
| FactSage | 8.2 \| July 2022 | École Polytechnique de Montréal, GTT-Technologies[n] |
| GEMS3K | 3.9.5 \| June 2022 | Paul Scherrer Institute[o] |
| Thermo-Calc | 2022b \| June 2022 | Thermo-Calc Software AB[p] |
| Thermodynamic reference database | | |
| TDB | TDB-6 \| May 2022 | Thermochemical database[q] |
| ThermoChimie | 10d \| October 2021 | ANDRA thermodynamic database[r] |
| THEREDA | 2021 \| 2021 | Chemical-thermodynamic Reference Database[s] |
| Software with Transport Modelling | | |
| TOUGHREACT | 4.13 \| September 2021 | Berkeley Lab[t] |

The range of functions of the individual codes and their handling as well as the type of licence differs considerably in some cases. Details can be found on the specified websites (current as of 30th June 2022)

[a]https://www.cheaqs.eu
[b]https://www.gwb.com
[c]https://www.mogroup.com/portfolio/hsc-chemistry
[d]https://www.usgs.gov/software/phreeqc-version-3
[e]https://sites.google.com/site/chemdiagr
[f]http://jess.murdoch.edu.au
[g]https://www.mineql.com
[h]https://vminteq.lwr.kth.se
[i]Seems to be unavailable at the moment
[j]https://wwwbrr.cr.usgs.gov/projects/GWC_chemtherm/software.htm
[k]https://h2ogeo.upc.edu/en/investigation-hydrogeology/software/158-cheproo-en
[l]https://www.epa.gov/ceam/minteqa2-equilibrium-speciation-model
[m]https://chess.geosciences.mines-paristech.fr
[n]https://www.factsage.com
[o]https://gems.web.psi.ch
[p]https://www.thermocalc.com/products-services/software/thermo-calc
[q]https://www.oecd-nea.org/dbtdb
[r]https://www.thermochimie-tdb.com
[s]https://www.thereda.de
[t]https://tough.lbl.gov/software/toughreact_v4-13-omp

chapter is not intended as a cookbook describing exactly how measurements should be made. Rather, it provides guidelines on what to look for when mine water samples are taken or when on-site measurements are to be made. Some of the pitfalls that can arise when taking mine water samples are discussed. The standard procedures can be found in the relevant literature.

There are hardly any examples described in the literature where an incorrect investigation led to the failure of a plant (e.g. Johnson and Hallberg 2002; Rose et al. 2004a;

Watzlaf et al. 2000, p. 268); rather, operators and the consultants often overlook the problems that occurred. However, it is precisely these that could contribute to the optimisation of novel procedures or help to learn from the mistakes.

► An example of what can happen if a parameter is either not measured or overlooked during final planning is the former water treatment plant in Lehesten/Thuringia, Germany. There, a highly acidic mine water containing aluminium was to be treated by means of an anoxic limestone drain. However, none of the published articles or project reports provided any detailed discussion of the importance of the oxygen concentration of the water in an anoxic limestone drain, and even the final report makes no reference to oxygen saturation. Yet, especially for anoxic limestone drains, the international literature consistently points out the importance of oxygen concentration: "For an anoxic limestone drain to function successfully, the unaerated mine water must come into contact with the overburden limestone" (Brodie et al. 1991, p. 5). However, since the mine water in Lehesten first entered the Rehbach stream and then flowed for 350 m before it was routed into the limestone drain, the oxygen saturation is 100%, as evidenced by my students' measurements in May 2008 (Table 2.3). Nevertheless, an anoxic limestone drain was constructed in 1997, and as expected, the system failed after the usual 1 year. Since the end of the project's funding period, the plant has not been functioning. In 2008, the plant was already completely overgrown and the pH values in the inlet and outlet did not differ substantially.

Although the criticism of certain procedures or projects is, respectively, based on a specific example, this should not be understood as criticism of precisely this company or person. Rather, the examples have been selected in such a way that they show by way of example which mistakes have been made and are still being made. This applies to smaller engineering firms with one or two employees as well as to larger, globally active companies with several thousand employees, and research institutions alike. No one is immune to mistakes, so the mistakes shown should be understood as a reflection of one's own behaviour and raise awareness about one or the other problem.

**Table 2.3** On-site parameters in the Rehbach stream at the former passive mine water treatment plant Lehesten, Germany in May 2008

| Measurement | pH value, – | EC, μS $cm^{-1}$ | $O_2$-concentration, % ǀ mg $L^{-1}$ | Redox potential, mV | Temperature, °C |
|---|---|---|---|---|---|
| Base of waste rock pile | 4.26 | 646 | 101.0 ǀ 12.13 | 552 | 5.6 |
| Inflow ALD | 4.71 | 661 | 106.8 ǀ 12.08 | 581 | 7.5 |
| Outflow ALD | 4.82 | 627 | 118.5 ǀ 12.03 | 546 | 7.9 |

EC Electrical conductivity, ALD Anoxic limestone drain

## 2.2 Mine Water Sampling

### 2.2.1 Checklists and Notes

You should familiarise yourself with the sampling regulations relevant to you before hydrogeological sampling. I am aware that there are many regulations in which almost all relevant procedures are described. Therefore, allow me to touch on a few "carelessnesses" here that I have observed in the past.

It may sound trivial, but one of the most common causes of measurement errors in the field are incorrectly calibrated electrodes, probes that have reached the end of their service life, dead batteries, broken electrodes, or "forgotten" calibration solutions. Occasionally, water-level measurement in boreholes are neglected, so that the water level correlations cannot be determined correctly, or plugs of measuring instruments are corroded and do not work. In addition, it occurs that storage containers for pH or redox probes tend to move parallel to the direction of the gravitational field – to put it simply: they fall down. Possible mistakes are as numerous as the number of day-holes (or adits) in a historic mining area, and therefore, you should do everything possible to avoid falling into one of the pitfalls of daily business. Or, in other words, anything that could go wrong will go wrong. Therefore, the U.S. Geological Survey (2015) also recommends thinking about back-up sampling equipment.

► The following happened to me during a drilling campaign downstream of a contaminated site: At the end of the drilling day I asked the drilling foreman whether he had also logged water-bearing layers in the drilling log and whether he had finally recorded the groundwater level. He had forgotten that, he explained to me, and called into the nearby borehole: "Hello water", turned his ear in the direction of the borehole, listened and said: "It's dry, you can put it in the log book like that." For understandable reasons, I explained to the drilling foreman that this had been the last job for me. Don't think this is an isolated incident; I could list other stories of drilling contractors at this point. This must not be seen as a general criticism of drilling contractors, just a reminder to be careful in your choice of contractors and to check occasionally. However, you don't need to take such extreme measures as Saudi Arabia did in the case of the SSSP project, where numerous international engineering firms were entrusted with third-party inspection and mutual monitoring in addition to the extensive in-house inspection.

Therefore, I strongly advise you to keep checklists that you should work through before going into the field. These checklists should include all the equipment and items that are essential for sampling and taking measurements. Examples included are gloves, field books, pencils, buckets, stopwatches or your camera, as well as bug spray, sunscreen and a first aid kit. I have been using a small portable box we've christened the "Yellow ToY" for nearly 20 years that contains a majority of these items. Every time I need a new small item, it goes into the box – even if it sometimes takes another few years before I use it again.

**Fig. 2.2** Sampling waste from a less environmentally conscious colleague. (Cape Breton Island, Nova Scotia, Canada – but not only found there)

Leave your sampling location as you would like to find it. This means that any waste is removed again. At numerous sampling locations, especially underground, strips of paper, filter papers or syringes are found lying around; this is evidence of carelessness (Fig. 2.2). This kind of behaviour does little to promote public acceptance of mining. Even in this context, the exemplary behaviour of a consulting company that is committed to the environment should be obvious.

Another aspect is smoking during hydrogeological sampling. Under no circumstances should smoking be allowed during sampling, as the constituents of the smoke dissolve in the water sample and thus increase the inorganic and organic constituents of the sample. This applies, for example, to cadmium, which is particularly highly enriched in smoke (Lewis et al. 1972), or to other metals that may dissolve in water. Heavy smokers should be sure to wear gloves when taking samples, and no smokers should be present in the immediate vicinity of the sampling location either.

At the end of this introduction, I would like to introduce a technology that will certainly revolutionise the sampling of mine water. Up to now, our possibilities to sample mine water have been limited to taking water samples at the adit portals or, at best, in boreholes or flooded shafts. This may change in the future. If camera drones (UAVs – Unoccupied Aerial Vehicles) have already helped change the way we look at our world, autonomous robots capable of diving, photographing or taking water samples in flooded mines will do the same. To this end, the EU project UNEXMIN (Underwater Explorer for Flooded Mines) started at the beginning of 2016, developing the Robotic Explorer (UX-1) robot for autonomous 3D mapping of flooded mines (Milošević et al. 2019; Žibret and Žebre 2018). The robot utilises various sensors to gather data, such as electrical conductivity, pH, gamma activity, magnetic field, sonar, fluorescence (365 nm), and is also equipped with a multispectral camera (14 wavelengths between 400 and 850 nm) and a device for water sampling. The mines studied included the former Ecton metal mine (North Derbyshire) in the United Kingdom, as well as the Urgeiriça mine in Portugal, Idrija in Slovenia and Kaatiala in Finland (all information about the project is available on the UNEXMIN website). Since 2017, the research project ARIDUA (Autonomous Robots and Internet of Things in Underground Facilities) has been established at the TU Bergakademie Freiberg, Germany, which is also working on the development of autonomous robots (Grehl et al. 2018). Hydrogeological sampling does not seem to be the focus of the research group for now. Such a device for flooded mines is currently under development at the University of Bochum, Germany, and underwater cameras developed specifically for underground mines are already being used successfully (Stemke et al. 2017).

A new development, besides camera, thermal or multispectral (Isgró et al. 2022) scanner drones, are drone based sampling procedures (Castendyk et al. 2018, 2020). This research group equipped a Matrice 600 drone with bailers used in hydrogeological and limnological studies. Besides single samples close to the lakes' surfaces, samples for limnological depth profiles were also obtained using this method.

### 2.2.2 Note on Occupational Health and Safety

First, of relevance here is that occupational safety is obligatory when taking samples in the mining environment. It is the responsibility of each sampler or analyst to ensure that his or her life or health is not endangered during sampling and that the applicable occupational health and safety regulations are observed. A good sample in exchange for a health hazard is a bad exchange! Occupational health and safety measures must be given special consideration, particularly when taking samples in underground mines, mine shafts, on slopes of open pit lakes, with boats, helicopters or during dives. Sometimes, official permits must be obtained. Wearing a safety helmet or safety shoes should be the least of the personal protective equipment (PPE) you have to use! In addition, in mines, the rule is always that you do not move forward until you have verified that there are no obstructions or openings in front of you, above you, or next to you. Under no circumstances do you take steps backwards – you always walk forwards! And never trust a wooden beam, plank or ladder in an underground mine – you may not realise it, but they are most likely to be rotten. By the way, a very good presentation on safety in hydrogeological sampling can be found in Chapter A9 of the *National Field Manual for the Collection of Water-Quality Data* (U.S. Geological Survey 2015) – but in my opinion you can safely disregard Sects. 9.9.3 through 9.9.5 there. *If* it happens that you do encounter an alligator, bear, or mountain lion, please remember: "Do not approach an alligator on land or water", and in the case of a bear or mountain lion, "Do not run".

However, mine gases can occur in abandoned underground mines (Hall et al. 2005; Plotnikov et al. 1989), which – detected too late – inevitably result in death! But even too low oxygen concentrations in the mine air can mean that a working place is no longer safe. In coal mines in particular, flammable methane is also to be expected (e.g. British Coal

Corporation 1997), so that only explosion-proof equipment must be used for sampling there. In a mine near Bad Ems, Germany, for example, there is outgassing of $CO_2$ from a downcast mine shaft (Ofner and Wieber 2008). Working without a gas detector in places such as these must be avoided at all costs! It endangers one's own safety and that of students or employees. The Lausitzer und Mitteldeutsche Bergbau-Verwaltungsgesellschaft mbH (2019, p. 21) also point out the following: "If there is a risk of slope slides and flow slides that do not permit entering the slope areas and driving on the lakes (i.e. passenger traffic) under occupational safety law, helicopters may be used" – yet, as we read above, there are now also unoccupied aerial vehicles (UAV) available for these cases.

Although the following tip cannot in principle be regarded as a warning, if headaches, a slight feeling of malaise, a distinct smell of hydrogen sulfide ("rotten egg smell") or, after a while, tiredness should develop underground, the sampling should be stopped immediately. All these signs can be an indication of dull, toxic or nasty mine air, commonly referred to as "bad air" (Weyer 2010); this is the air in a mine that contains either too little oxygen (dull) or toxic gases (toxic or nasty, e.g. carbon monoxide). One of the Tyrolean state geologists once felt a severe malaise while investigating an old exploratory tunnel for a power plant project through the Tauferberg near Umhausen/Ötztal, Austria, and turned back. His companion wanted to continue into the tunnel and was later found unconscious after he had not returned to the tunnel portal for some time (pers. comm. Gunther Heißl 2018).

Please also stay away from restricted areas. One of my students was detained for several hours near Logatec in Slovenia a few years after the Ten-Day-War *(desetdnevnavojna)*. He had been mapping during a hydrogeological field course along the border to a military area and was then taken in for interrogation by friendly soldiers (and of course our Slovenian speaking contact person was not available just that day).

Finally, it should be noted that mine water may contain bacteria, germs (Double and Bissonnette 1980), fungi (Marchlewitz 1959), algae, archaea or toxic organic as well as inorganic compounds. Disposable powder-free latex or nitrile gloves are therefore recommended when taking samples. These protect both the sampler from contamination by the sample and the sample from contamination by the sampler itself (Ficklin and Mosier 1999, p. 261; U. S. Geological Survey 2015, Section 2.0.2).

### 2.2.3 Sampling Methods

There are numerous procedures and specifications for sampling water (e.g. ERMITE Consortium et al. 2004; Ficklin and Mosier 1999; Wolkersdorfer 2008; Younger et al. 2004). However, general guidelines are of limited use when sampling mine water. McLemore et al. (2014) published a comprehensive account of mine water sampling, Cravotta and Hilgar (2000) produced a guide for coal mines, and generally applicable guidance can be found in MEND (2001). Yet, the approach presented here essentially follows the requirements for *practical* sampling of mine water. It may not always meet the requirements for an exact scientific question or may deviate in detail from operational regulations (e.g. Lausitzer and Mitteldeutsche Bergbau-Verwaltungsgesellschaft mbH

**Table 2.4** Recommended water quantities and pre-treatments for standardised mine water sampling

| Parameter | Pre-treatment | Water quantity |
|---|---|---|
| On-site parameters (e.g. temp., pH, redox) | Probe calibration | In situ measurements |
| Main ions | None | 250…500 mL, better 1…2 L |
| Filtered trace elements | Filter, $HNO_3$ (ultrapure!) | 50…150 mL |
| Unfiltered trace elements | Unfiltered | 50…150 mL |

2019). Consequently, it must be left to each sampler or operation itself to decide on the amount of effort to put into sampling. However, model-based monitoring is not necessarily better than expert knowledge of professionals or monitoring relying on multi-parameter statistics. Irrespective of the method used, there are some quality criteria that must not be neglected under any circumstances – these are briefly outlined below.

In addition to the on-site parameters described below, a water analysis is essential to plan mine water treatment (Table 2.4). Usually, two or three mine water samples are sufficient for the preliminary analysis, namely one for the major ions and the other two for the trace elements. For the final design, it is necessary to sample the mine water regularly over the course of a year and to measure the flow. Care must be taken to ensure that the mine water comes into as little contact as possible with atmospheric oxygen during sampling, as oxidation of the metals should be avoided at all costs. In general, it should be noted that mine water behaves fundamentally differently from groundwater or surface water that is not affected by mining, and its chemical composition changes constantly after sampling – even if it is cooled or frozen and thawed again.

For the main ions (Na, K, Ca, Mg, $SO_4^{2-}$, $Cl^-$, $NO_3^-$, $HCO_3^-$), an unfiltered 250–500 mL sample (PE or glass bottle) is usually sufficient, which – like all water samples – must be stored at or below 4 °C until analysis (in the 22nd edition of the *Standard Methods for the Examination of Water and Wastewater* [2012] and in *USEPA Region 9 Management and Technical Services Division* [2012/2015], a minimum temperature of 6 °C is now prescribed). Refrigerated boxes or bags, which come in a wide variety of sizes and are handy enough for an underground crawl space, can be used for transport. Cooling elements, ice or electrical energy are usually used for cooling. Since cold air is heavier than warm air, the cooling elements should be placed on top of the sample vessels, not vice versa. If the filter residue is to be determined at the same time, it is advisable to use pre-weighed filters or to determine the dry residue immediately in the field. Because of the rapid onset of hydrolysis of the iron, determination of the filter residue in the laboratory is not sufficiently reliable. Whether the specified amount of water is sufficient should be agreed with the laboratory. To obtain reliable data, the analysis in the laboratory should be carried out immediately, because the composition of mine water usually changes rapidly. Accordingly, sampling should be planned so that samples are not stored in the refrigerator over the weekend or on a holiday. Details of individual parameters can be found in American Public Health Association et al. (2012) or Wasserchemische Gesellschaft – Fachgruppe in der Gesellschaft Deutscher Chemiker in Gemeinschaft mit dem Normenausschuss Wasserwesen (NAW) im DIN Deutsches Institut für Normung e.V. (2021).

For the trace elements, which may occasionally have concentrations in the range of the major ions, two further samples of about 50–150 mL are necessary. One of these should be filtered and the other unfiltered (filter sizes see below). For the filtered sample, add a few drops of $HNO_3$ (ultrapure!) until the pH is at least pH 2 (this should be checked individually, as some mine waters are very well buffered and the usual 2–3 drops are occasionally not sufficient). This acidification prevents reactions taking place in the sample during storage or transport (Lloyd and Heathcote 1985).

Regarding organic water constituents (e.g. PAHs, PCP, PCBs, BTEX, VOCs), considerably less experience or treatment methods are available for mine water than for groundwater or surface water. Let me put it this way: the number of publications on this subject is extremely limited. On the one hand, this is since the concentrations of potential organic contaminants in mine water are often low (if they are measured), or the problems caused by acid mine drainage and increased metal concentrations are considered more relevant than the organic contaminants. In Germany, however, PCBs in mine water from coal mining have gained attention in recent years (AG "PCB-Monitoring" 2018; Landesamt für Natur, Umwelt und Verbraucherschutz NRW et al. 2018; Merkel et al. 2016).

With regard to filters (hydrophilic and made of polycarbonate or cellulose acetate), there is a debate in academic circles as to whether 0.45 μm or 0.2 μm pore size filters should be used in hydrogeological sampling (e.g., McLemore et al. 2014; Ranville and Schmiermund 1999; Shiller 2003). It is often argued that filtered samples are free of microorganisms (bacteria, viruses, fungi) or that all colloids are removed. However, this is by no means the case (Fig. 2.3, also compare Fig. 2.6). Neither the 0.45 μm nor the 0.2 μm filter completely fulfils this purpose. Rather, the 0.45 μm pore size is a purely practical aspect: a water sample can be filtered in the field in an acceptable time using this particular pore size filter. Already at

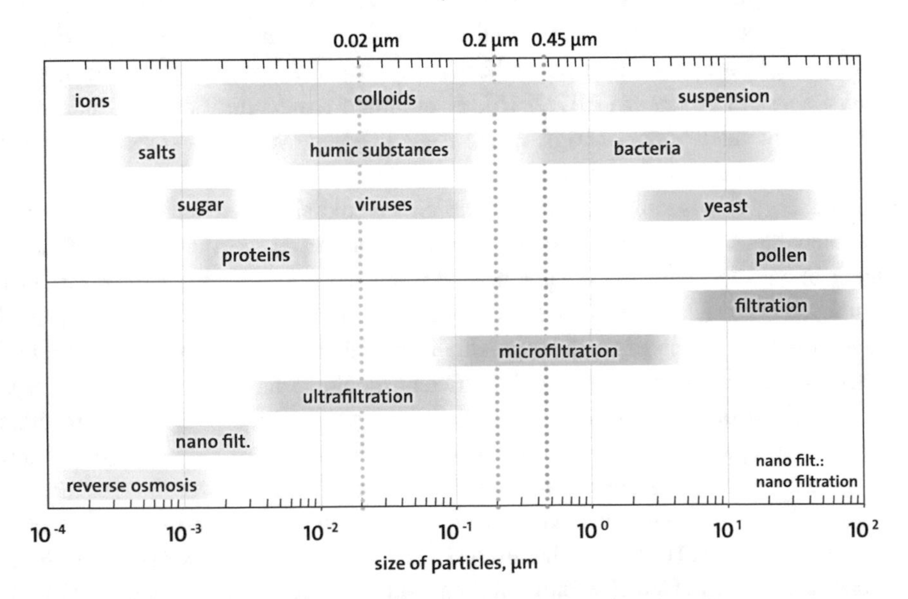


**Fig. 2.3** Particles in water and filtering techniques. The limits of 0.02, 0.20 and 0.45 μm refer to membrane filters. (Adapted from Ranville and Schmiermund 1999, p. 185; supplemented and modified from Stumm and Morgan 1996, p. 821)

0.2 µm or even at 0.1 µm pore size, the time required to filter 50 mL of mine water occasionally increases to several tens of minutes, or even up to 1 h (therefore a 0.45 µm filter should be installed ahead of the smaller pore size filter). Ranville and Schmiermund (1999), on the other hand, suggest historical reasons for selecting the 0.45 µm pore size, as early bacteriologists believed bacteria were not smaller than this. However, in the case of aluminium the influence of the correct choice of filter pore size on the concentration in the mine water becomes apparent (Fig. 2.15). In any case, the filtration of the samples should already be done in the field (McLemore et al. 2014, p. 41) and not in the laboratory.

It is also important that the sampling vessels (glass, amber glass or HDPE, depending on the parameter) are completely filled with water and that as far as possible no air bubble remains, as otherwise hydrolysis or gas–water reactions may occur. Unless the sampling vessels are already filled with acid, they must always be rinsed three times with the mine water to be sampled before the final sample is taken (Ficklin and Mosier 1999, p. 261). In addition, during mine water sampling, agitation of the channel or stream bed should be avoided, as the agitated sediment can lead to misidentification of ions. Furthermore, a sampling site should always be approached from downstream of the direction of flow, the first sampling site should be the furthest downstream, and one should then move upstream to avoid mutual interference between sampling sites.

In the context of a main investigation, it may be necessary to investigate additional parameters that depend on the particular issue or on regulatory requirements (e.g. organic compounds). This may include speciation of arsenic or other constituents, or sampling for isotope studies (Wolkersdorfer et al. 2020). In this case, the water sample will need to be conditioned or collected in a different manner than described above. In any case, this must be agreed with the laboratory in advance.

The Arbeitsausschuss Markscheidewesen im Normenausschuss Bergbau (FABERG) im DIN (2003) [German Working Committee on Mine Surveying in the Mining Standards Committee] has compiled the following parameters that are usually determined underground (the 20 °C for pH is an old temperature compensation reference, which nowadays must be 25 °C):

- Temperature of the inflow in °C
- Pressure (with closed bores) in Pa
- Inflow rate (e.g. in L $min^{-1}$ or drops $min^{-1}$), including indication of the determination procedure
- Conductivity at 25 °C in µS $cm^{-1}$
- Density in g $cm^{-3}$
- In coal mining: pH value, chloride (Cl), sulfate ($SO_4$), hydrogencarbonate ($HCO_3$), sodium (Na), iron (Fe), manganese (Mn), barium (Ba), strontium (Sr)
- In potash and rock salt mining: pH at 20 °C, sodium (Na), potassium (K), magnesium (Mg), calcium (Ca), chloride (Cl), sulfate ($SO_4$) and hydrogencarbonate ($HCO_3$). The data are given in g $L^{-1}$ or mol $L^{-1}$.
- Furthermore, trace elements (e.g. Br, Rb and Li) as well as isotopes (e.g. tritium, $^{13}C$, $^{18}O$) can be determined for the safety assessment

Since there was no relevant metal ore mining in Germany at that time (but its legacies number in the thousands), there is no information on the parameters usually measured there. However, they write that "depending on the mining sector and the need … further chemical parameters [can] be recorded". This last sentence is important, because the parameters listed above alone cannot be used to calculate an ion balance.

ERMITE Consortium et al. (2004, p. S49), based on many years of experience and from literature studies, have produced a more comprehensive table with parameters to be analysed for different stages of exploration (Table 2.5).

**Table 2.5** Recommended scope of analysis for different types of mine water investigations; parameters in bold should be determined on-site; for parameters in italics, total concentrations as well as those of individual species should be determined; additional parameters should be analysed depending on geological conditions or presumed history of the mine

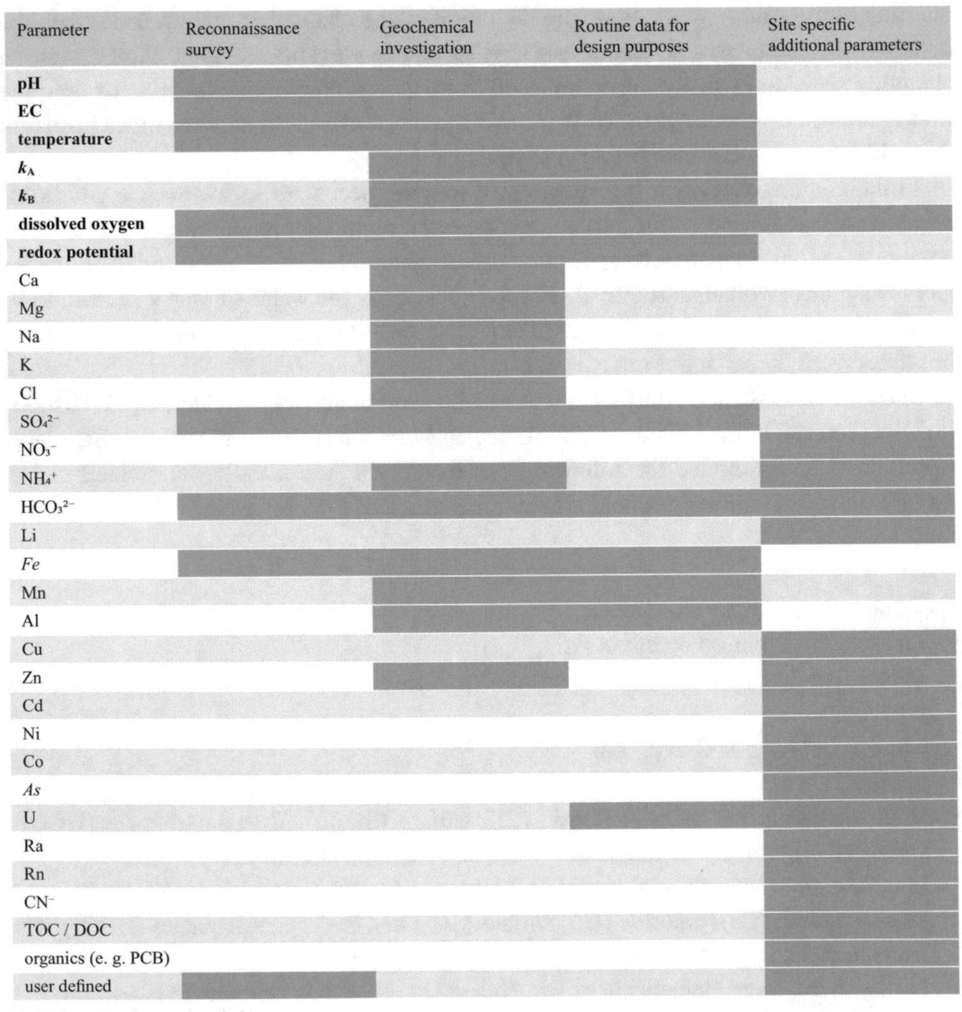

| Parameter | Reconnaissance survey | Geochemical investigation | Routine data for design purposes | Site specific additional parameters |
|---|---|---|---|---|
| **pH** | ■ | ■ | ■ | |
| **EC** | ■ | ■ | ■ | |
| **temperature** | ■ | ■ | ■ | |
| $\boldsymbol{k_A}$ | | ■ | ■ | |
| $\boldsymbol{k_B}$ | | ■ | ■ | |
| **dissolved oxygen** | ■ | ■ | ■ | ■ |
| **redox potential** | ■ | ■ | ■ | |
| Ca | | ■ | | |
| Mg | | ■ | | |
| Na | | ■ | | |
| K | | ■ | | |
| Cl | | ■ | | |
| $SO_4^{2-}$ | ■ | ■ | ■ | |
| $NO_3^-$ | | | | ■ |
| $NH_4^+$ | | ■ | | ■ |
| $HCO_3^{2-}$ | ■ | ■ | ■ | ■ |
| Li | | | | ■ |
| *Fe* | ■ | ■ | ■ | |
| Mn | | ■ | ■ | |
| Al | | ■ | ■ | |
| Cu | | | | ■ |
| Zn | | ■ | | ■ |
| Cd | | | | ■ |
| Ni | | | | ■ |
| Co | | | | ■ |
| *As* | | | | ■ |
| U | | | ■ | ■ |
| Ra | | | | ■ |
| Rn | | | | ■ |
| $CN^-$ | | | | ■ |
| TOC / DOC | | | | ■ |
| organics (e. g. PCB) | | | | ■ |
| user defined | ■ | | | ■ |

EC Electrical conductivity
Modified from ERMITE Consortium et al. (2004, p. S49)

### 2.2.4 Quality Control

Without an ion balance (also ionic balance or ion[ic] balance error), the simplest way of checking the reliability of a chemical analysis is lost. This has also been pointed out by Nordstrom et al. (2010), who, in addition to the ion balance, recommend the conductivity balance as a measure of whether a chemical analysis is complete: "Calculated conductance can improve QA/QC by determining whether a cation or anion is in error" (Nordstrom et al. 2010, p. 382). Consequently, instead of merely using the ion balance as a quality criterion, the conductivity balance should also be used. Wolkersdorfer (2008, pp. 421–423, 177 f.) describes how the latter should be carried out, reduced to the bare minimum. Alternatively, and this is possibly the easier way, it can be calculated using the phreeqc.dat database of PHREEQC (Parkhurst and Appelo 2013). It is reported there as Specific Conductance (μS cm$^{-1}$) in the "Description of solution", where the electrical conductivity temperature is that of its input value in the pqi file. The Geochemist's Workbench® (Bethke 2008) also calculates the electrical conductivity from your mine water analysis and is reported there as `Elect. conductivity = x uS/cm` (or umho/cm).

When establishing an ion balance for acid mine drainage, it is important to consider the proton activity as soon as the pH falls below 4 (Nordstrom et al. 2009). Because of the logarithmic nature of pH, an activity greater than $1{\cdot}10^{-4}$ already is relevant for the ion balance (Fig. 2.4). To calculate the ion balance ΔIB according to DIN 38402-62 or the algorithm in WATEQ4F (Ball and Nordstrom 1991, p. 4), the following equation should be used (DIN Deutsches Institut für Normung e. V. 2013, where $k$ refers to the cations and $j$ to the anions):

$$IB = \frac{\sum_k c_{eq,k} - \sum_j c_{eq,j}}{\left(\sum_k c_{eq,k} + \sum_j c_{eq,j}\right) \times 0.5} \times 100 \qquad (2.1)$$

On the aqion website (www.aqion.de) the use of ΔIB is described as a "naïve approach" – naïve or not: this is what is stated in DIN and DVWK and is consequently the relevant criterion for companies that work in accordance with DIN. However, in the case of low pH values, the notes in the next paragraph should not be disregarded.

The equation used in PHREEQC (Fig. 2.8), in Langguth and Voigt (2004, p. 320) or Wisotzky et al. (2018, p. 14) to estimate the error of a water analysis is not the ion balance ΔIB, as assumed elsewhere (Lausitzer und Mitteldeutsche Bergbau-Verwaltungsgesellschaft mbH 2019), but more correctly the charge balance ΔCB or the "percent charge-balance error" (Parkhurst and Appelo 2013):

$$CB = \frac{\sum_k c_{eq,k} - \sum_j c_{eq,j}}{\sum_k c_{eq,k} + \sum_j c_{eq,j}} \times 100 \qquad (2.2)$$

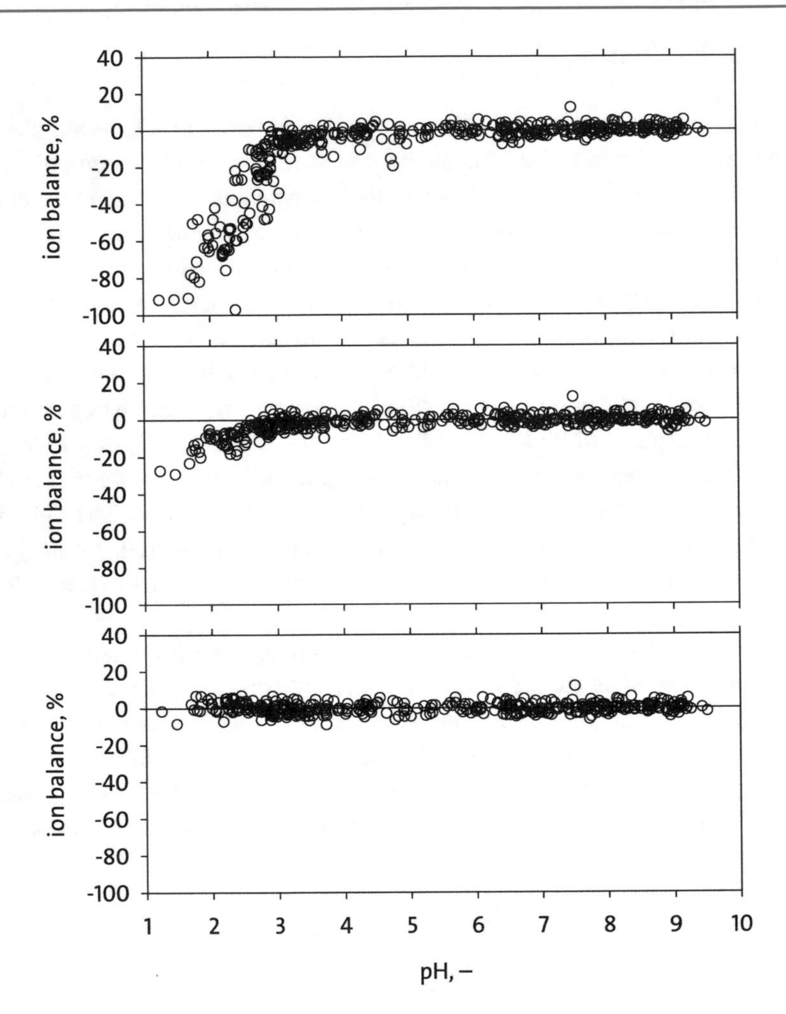


**Fig. 2.4** Ion balance calculated by different methods. Top: without speciation; middle: without speciation, protons taken into account; bottom: with speciation and protons. (After Nordstrom et al. 2009)

In the PHREEQC geochemical code and in other chemical-thermodynamic codes, the speciation of the water constituents is used when calculating the charge balance. As already pointed out by Nordstrom et al. (2009, Fig. 7) in three graphs (Fig. 2.4), speciation is relevant for acidic mine water. Without speciation, the charge balances drift further and further into the negative with increasing proton activities – and this is what aqion are referring to: Considering the charge balance without taking into account speciation at high proton activities could indeed be called naïve.

Since the two equations differ from each other, it must consequently always be stated which of the two ion balances have been used. For example, the ΔIB of the water sample listed in the DIN 38402–62 is −3.45% and according to the equation for ΔCB in PHREEQC it is −1.89% (at pH 7 and a temperature of 25 °C). These differences are also relevant to the question of whether a water analysis is still correct or not.

In relation to quality assurance or quality control, it goes without saying that blind samples and duplicates must also be prepared for correct mine water sampling. These should include at least three blank samples and a few duplicates adapted to the total number of samples. In addition, standards for mine water should always be included in the scope of analysis:

- Acid blank
- Process blind test
- Blank test with the distilled water used to clean the equipment and vessels
- Duplicates after about every 10th sample
- Standard for mine water

All blank or duplicate samples should follow the standard sample naming scheme and should not be identifiable to the laboratory as duplicates or blanks. A sample designation of “FDR-3101-PN1-duplicate” or “process blank” should be omitted. For example, “FDR-3101-BSW” for a blank sample of distilled water or “FDR-3101-MWS” for a mine water standard would be more appropriate.

### 2.2.5 Measuring Instruments and Sampling

Various measuring devices and electrodes have proven useful for laboratory and field applications. However, these have predominantly been developed for “normal” water and not specifically for mine water. Although there are probes for wastewater and industrial water with higher ion concentrations, these are often bulky or expensive to purchase, so that not every consulting firm has them on hand. Several electrodes should be used, especially if mine water of higher ionic strength and lower ionic strength is to be measured alternately or if the pH values change *considerably* from measurement to measurement. It is also essential to calibrate the meters and electrodes before each daily measurement campaign or to have them calibrated at the calibration office. When measuring oxygen using a Clark electrode, calibration must even be carried out before *each* individual measurement. It is advisable to use the modern luminescent optical electrodes, which are self-calibrating, and, what’s more, they do not consume oxygen during the measurement (Jackson and Hach 2004; Klimant et al. 1995). Another problem with measurements in mine water is the clogging of the electrodes or the formation of a biofilm (“biofouling”). This can already occur with several successive measurements in mine water, or only after a considerable time (Fig. 2.5). It is therefore essential to clean the electrodes regularly as described in the respective manuals.

There has been much discussion about whether a mine water sample is representative or not, whether the sample was taken correctly or whether this or that parameter accurately

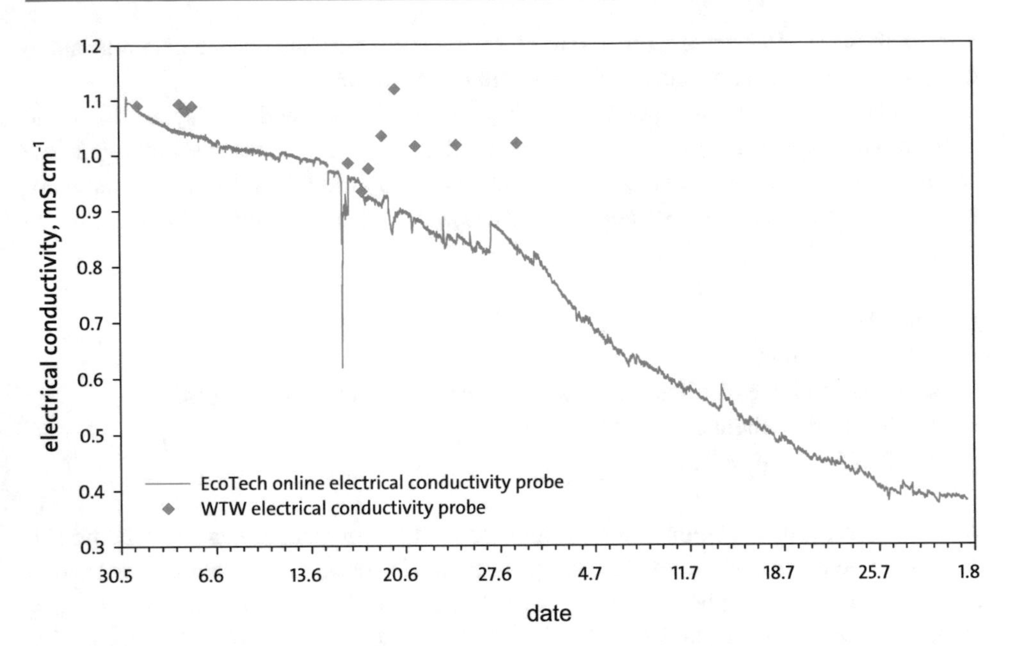


**Fig. 2.5** Decrease in electrical conductivity due to scaling and biofouling on an online probe. Over the course of 9 weeks, the electrical conductivity decreases by 0.7 mS $cm^{-1}$. For comparison, the initial individual measurements taken with a hand-held meter. An evaluation of the measurement in the context of a salt tracer test was not possible

reflects the actual conditions. Of course, one is free to pursue hydrogeological sampling in the mine environment to the point of excess by taking samples under a nitrogen atmosphere or by shock freezing them directly at the time of collection. From a practical point of view, however, this is not very effective, considering how rapidly the composition of mine water can change in the mine – this is not intended to encourage a sloppy approach to sampling, but merely to raise the question of the extent to which mine water sampling should be carried out.

Herbert and Sander (1989) provide detailed guidance for hydrogeological sampling in high-salinity mine water (brines) and indicate that these waters require exceptional attention during sampling. To ensure representative sampling, the following precautions must be observed in this environment:

- Avoid evaporation during sampling
- Avoid contamination and alteration of the sample by the mine air
- Avoid suspended solids in the sample
- Take the sample at the original temperature and pressure conditions corresponding to those of the solutions in the rock matrix

During the flooding experiment at the Hope Mine, Germany, Herbert and Sander (1989) developed and patented a sampling device for taking samples and on-site parameters under in situ conditions in a flooded mine shaft. This device made it possible to measure all relevant parameters without interference and to take an undisturbed water sample.

### 2.2.6 Sample Names

In a lot of laboratories there are many water samples in the refrigerators, all of which are called P1, P2 or P3. In some cases it is no longer possible to assign these incomplete and insufficiently labelled samples to a student or a location. This common practice should therefore be abandoned in favour of an individual sample name. Many of the above-mentioned publications deal with the correct labelling of specimens, but the basics will be repeated here, and a generally valid proposal for sample names will be presented.

Each sample should be labelled with at least the following information: project, sampling location, sampling point, date, and sampler name. Barcodes or RFIDs can only supplement this information, but not replace them. Further details could be dispensed with only if it is guaranteed that there is a continuous chain of custody (CoC), i.e. a complete documentation, which is also suitable for electronic documentation. Experience tells us that there is always a mix-up of samples – and unfortunately also loss. Recently I received an e-mail from a laboratory telling me that they had dropped one of my bottles. In 32 years, this has never happened to me. I guess we can't be prepared for every eventuality – so a backup sample for every sampling point is not something I would want to consider because of this rare event. If the electrical conductivity was measured in the field and then compared to the electrical conductivity determined in the laboratory, errors can be traced in many cases, as it is exceedingly rare for two different water samples to have the same electrical conductivity.

▶ Sample names that consist almost exclusively of numbers are more error-prone in handling than those with number-letter combinations. A colleague had once sent about four dozen samples A1.2–A2.9 continuously up to H1.0–H2.9 to a laboratory. She later realised that there had been quite a few mix-ups in the laboratory. There was no way to trace the errors. A less monotonous numbering of the samples might have prevented this.

Over the past decades, the following approach to sample naming has proven useful in my projects: Project abbreviation-Short date-Sample location. This results, for example, in a sample name such as RZS-1208-RSA (Reiche Zeche Shaft – August 12 – Rothschönberger-Stollen-Adit). The date can be extended by the year to avoid duplicate sample names; however, to date, in my experience this has never occurred. Further, I repeat the last three letters on the bottom of the sampling bottle. Even if the labelling (sticker or permanent pen) should partially come off, the complete information mentioned above, and the double labelling will always keep a sample uniquely assignable. If a naming convention for samples already exists for a project, it is of course preferable to the one proposed here. Many projects keep going for years, if not decades, and then it is essential to describe the measuring points precisely in a master file. There, sampling must be documented on an ongoing basis, structural changes must be recorded, or the relocation of monitoring sites must be outlined. In addition, a procedure must be developed for taking duplicate or blank samples and how these should be named so that the laboratory does not have to guess what type of sample it might be.

### 2.2.7 Dissolved and Total Concentrations

Regarding the terms "dissolved" and "total" concentrations of water constituents, there are unfortunately inconsistent conceptual uses. If one were to follow the diagram in Stumm and Morgan (1996, Fig. 6.1), the situation would be relatively simple (Fig. 2.6 and also Fig. 2.3). However, the diagram shows that the range of membrane filtration reaches into the range of particulate water constituents. Particularly with iron concentrations, but also with other water constituents, there is a confusion of terms used by different researchers, which makes a technically sound discussion difficult. So the question is: what does "dissolved iron" or "total iron" mean? The terms are defined in EN ISO 11885, but in doing so it does not conform to other DINs and ISOs (e.g. DIN 38406-E1-1 May 1983; EN ISO 11885 November 1997; DIN 38406-32-1 May 2000). According to EN ISO 11885, the dissolved metals are those that can be analysed in the filtrate of a 0.45 μm filter; thus, in the case of iron, the sum of $Fe^{2+}$ and $Fe^{3+}$. Other researchers refer to only the $Fe^{2+}$ as dissolved iron and the sum of the iron species as total iron – all of which seems ill-suited to objective discussion. The Oberflächengewässerverordnung (OGewV, German Surface Water Ordinance) also describes, "With the exception of cadmium, lead, mercury and

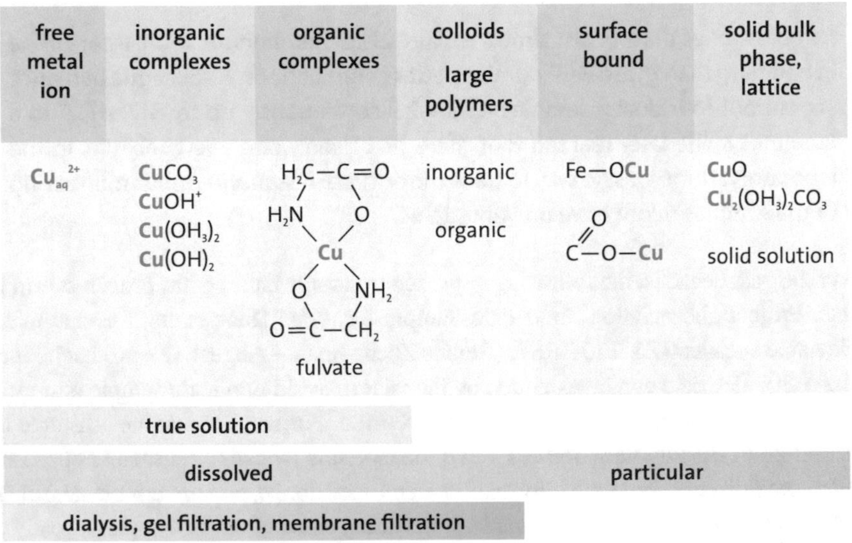

| free metal ion | inorganic complexes | organic complexes | colloids large polymers | surface bound | solid bulk phase, lattice |
|---|---|---|---|---|---|
| $Cu_{aq}^{2+}$ | $CuCO_3$<br>$CuOH^+$<br>$Cu(OH_3)_2$<br>$Cu(OH)_2$ | $H_2C-C=O$ / $H_2N$ O / Cu / O $NH_2$ / $O=C-CH_2$<br>fulvate | inorganic<br>organic | Fe–OCu<br>O=C–O–Cu | CuO<br>$Cu_2(OH_3)_2CO_3$<br>solid solution |
| true solution | | | | | |
| dissolved | | | particular | | |
| dialysis, gel filtration, membrane filtration | | | | | |

**Fig. 2.6** Forms in which metal species occur, using copper species as an example and potential filtration methods. (Modified after Stumm and Morgan 1996)

**Table 2.6** Expressions that should be used for analysis in mine water, using iron as an example

| Ingredient | Filter method | Recommended designation | DIN/ISO application | APHA method 3010 A |
|---|---|---|---|---|
| $Fe^{2+}$ | Filtered | $Fe^{2+}$ filtered | – | – |
| $Fe^{3+}$ | Filtered | $Fe^{3+}$ filtered | – | – |
| $Fe^{2+} + Fe^{3+}$ | Filtered | $Fe_{total}$ filtered | Dissolved iron | Dissolved iron |
| $Fe^{2+}$ | Unfiltered | $Fe^{2+}$ unfiltered | – | – |
| $Fe^{3+}$ | Unfiltered | $Fe^{3+}$ unfiltered | – | – |
| $Fe^{2+} + Fe^{3+}$ | Unfiltered | $Fe_{total}$ unfiltered | Total iron | Suspended iron |
| $Fe^{2+} + Fe^{3+}$ | Unfiltered, acidified | $Fe_{total}$ unfiltered, acidified | – | Acid-extractable iron |
| $Fe^{2+} + Fe^{3+}$ | Unfiltered, digestion | $Fe_{total}$ unfiltered, digested | – | Total iron |

These terms should be used analogously for other constituents. In addition, the filter pore size (e.g. 0.02 μm; 0.2 μm; 0.45 μm) and the filter material should be indicated

nickel (metals), the environmental quality standards are expressed as total concentrations in the whole water sample. For metals, the environmental quality standard refers to the dissolved concentration, i.e. the dissolved phase of a water sample obtained by filtration through a 0.45 μm filter or equivalent pre-treatment" (Bundesministerium der Justiz 2016).

Only DIN ISO 10566 April 1999 seems to provide clarity with regard to the analytes but refers exclusively to aluminium. There, the term "filterable aluminium" is to be used for the aluminium concentration in the filtrate. The other analytical standards should also ensure more correct terminology in future versions. Until that time, the terminology in Table 2.6 should also be used *mutatis mutandis* for other water constituents, as they rule out any confusion.

In the USA, the term *potentially dissolved metal* is used in some states (e.g. Colorado). This refers to the fraction of (semi)metals in the filtered sample that is detectable in the filtrate (0.4–0.45 μm membrane filter) after acidification with nitric acid to below pH 2.0 and an exposure time of 8–96 h. Consequently, it is assumed that coprecipitation *s.l.* of the (semi)-metals occurs after sampling and initial filtration, that they are re-dissolved after the standing time of up to 96 h and are then analytically detectable (Colorado Department of Public Health and Environment – Water Quality Control Commission 2013; Weiner 2010, p. 120). Details about filtration and the terms used can also be found in American Public Health Association et al. (2012, method 3010 A): "*Acid-extractable metals* – The concentration of metals in solution after treatment of an unfiltered sample with hot dilute mineral acid. To determine either dissolved or suspended metals, filter sample immediately after collection. Do not preserve with acid until after filtration; *Dissolved metals* – Those metals in an unacidified sample that pass through a 0.45-μm membrane filter; *Suspended metals* – Those metals in an unacidified sample that are retained by a 0.45-μm membrane filter; *Total metals* – The concentration of metals determined in an unfiltered sample after vigorous digestion, or the sum of the concentrations of metals in the dissolved and suspended fractions. Note that total metals are defined operationally by the digestion procedure".

Yet it could be so simple: In water chemistry we are concerned with constituents that are present as atoms or molecules (i.e. “dissolved”). Separation of the non-dissolved elements, can operationally only be achieved by filtration with 0.001 μm filters or centrifugation (Fig. 2.3). Since filtration with 0.001 μm is very tedious in the field or laboratory, 0.1 μm or 0.2 μm is a reasonable compromise to capture the dissolved water constituents. The differences between filtration fractions can be substantial (Fig. 2.15), as also shown by Shiller (2003), who even used 0.02 μm filters with a special apparatus. However, care should be taken: The last filtration step with a 0.1 μm filter and an apparatus like the one used by Shiller (2003, p. 3954) took more than a day with 50 mL of a mine water sample from Cape Breton Island, Canada.

Please also bear in mind that not all types of membrane filters are suitable for all water parameters. For example, you cannot use filters made of cellulose nitrate, also called mixed cellulose ester (MCE), when analysing PCBs, since up to 100% will be sorbed by the filter, depending on the congener. Well-suited filter materials for almost all types of mine water are PC (polycarbonate), PES (polyethersulfone), and NY (nylon). There are also hydrophobic (PTFE: polytetrafluoroethylene) and hydrophilic membrane filters – it almost goes without saying that only hydrophilic filters are used for mine water hydrogeological sampling. Before filling your sample bottles, discard a small amount of the filtered water sample so that potential contaminants from the filter do not get into your sample.

► Here is a current example in relation to the PCB discussion in German coal mine water. The issue there is whether and how the highly persistent PCBs can be removed from mine water. It is known that most hydrophobic PCB congeners are colloidally bound in water (IARC Working Group on the Evaluation of Carcinogenic Risks to Humans 2016; Karickhoff et al. 1979), and consequently they are essentially transported with the “suspended load”. In the water phase, the less soluble PCBs are generally difficult to detect analytically. During the ongoing discussion, it became apparent that some believed PCBs never occur in the dissolved phase as a matter of principle (PCBs 28 and 52 have solubilities of 134 and 34 μg $L^{-1}$, calculated from Li et al. 2003a), while others tended to see PCBs in the dissolved fraction as well. In reviewing the published and unpublished reports, it became apparent that two different concepts were being referred to. Some use the term “dissolved” and mean the colloidally bound PCB that passes through a 0.45 μm filter, while others refer to the PCBs that are actually present in dissolved form. In order not to complicate the discussion further, it would be appropriate to define which term means what, and always use that definition at the beginning of a discussion, article or expert opinion. Table 2.6 can provide an indication of this. In contrast, the IARC Working Group on the Evaluation of Carcinogenic Risks to Humans (2016, p. 64), provide the following definition:

- Unfiltered water includes dissolved components and those bound to colloids and particulate matter

- Filtered water gives PCB concentration on particulate matter (residue on the filter) and dissolved or bound to colloids (filtrate)
- Passive sampling targets the dissolved fraction

This definition is chemically and physically correct because it distinguishes between the actually dissolved and colloidally bound fraction (which passes through the filter) on the one hand and the suspended particulate matter (which remains on the filter) on the other. Unfortunately, it does not agree with the discussion and definition in Germany, according to which everything that passes the filter is – incorrectly – described as dissolved.

### 2.2.8 Documentation

As with all hydrogeological investigations, complete documentation of results and sampling is essential. Only when complete and reliable data are available, they can be transformed into meaningful information through evaluation and analysis (McLemore et al. 2014, p. 169). To this end, the first and most important tool is your field book, which should have numbered pages to have indubitable probative value in the event of a legal dispute. In the field book, all observations and measurements should be recorded with the date, name of the person who made them, and description of the on-site conditions, as well as a sketch if necessary. When giving site coordinates, it is important to give the full geodetic datum, which must consist of the projection and the underlying geoid (e.g. UTM WGS84). However, underground you will struggle with a satellite-based navigation system; in this case, ask your trusted miner for the name of the sampling site. Simply giving coordinates is not sufficient, as there are several thousand projection and geoid combinations. Experience shows that reliable notes in the field not only help to avoid misunderstandings but, in case of ambiguity, provide immediate information about the actual conditions. Jumbles of bits of paper or even sticky notes (Post-it™) are to be avoided at all costs for the sake of good working practice. In addition to field book documentation, field computers, handheld devices, tablet computers, or even smart phones can now be used to record data. Furthermore, it should be determined where the field books or copies of them are to be deposited in case an employee from a company or research institution decides to "reorient" him- or herself personally. The same applies to data or photo documentation.

In the laboratory or office, it is essential to feed the data into a data processing system. If possible, a professional database system should be used, by means of which the recorded data can also be accessed by future colleagues. Data recording in a word processor or spreadsheet without a database connection should be avoided for the sake of good working practice. If you are looking for advice on this topic, you should contact one of the World Data Centres.

At this point it should be added for the sake of completeness that measured values also have units. These must always be noted in the field book together with the measured value. There are numerous field books that say, for example, "Temperature 15.3" – without giving a unit behind it. This is not good scientific or engineering practice and must be avoided at all costs. How can your colleague know if you had a Mr. Kelvin, Celsius, Fahrenheit, Rankine, Newton, or Rømer in mind (you may object at this point: How can water have 15.3 K, 15.3 °F, or 15.3 °R – you're right about that, but 15.3 °C, 15.3 °N, or 15.3 °Rø are all within the realm of possibility)? Also, there are numerous field books that only have some values recorded without saying what they are – it may well be that you can keep something like that in your head for a few hours, but by the following week the memory has faded, and a month later the readings can't be reconstructed at all. And what about a few years later, when you are in court because of an incorrectly calculated parameter and the judge confronts you with incomplete records?

In addition to the usual information, details of the weather, the equipment used, and, above all, any special features observed or deviations from the usual conditions should be noted. Often, such details can help to identify why an analysis shows different results than usual.

▶ During a tracer test, the sampling person observed that after an intense rainfall event another water leakage occurred next to the usual measuring point. She recorded this correctly in the field book, and thus it could be explained that the tracer concentration on this day was slightly lower than usual – dilution by the precipitation, which had also been observed by the breakthrough curve of the tracer within the flooded mine.

## 2.3 Essential On-Site Parameters

### 2.3.1 Introductory Note

The following sections provide some guidance on the on-site parameters that need to be measured in the field to reliably design a mine water treatment plant (Table 2.7, Fig. 2.7). As a rule, the measurement methods themselves will not be discussed, as these have been described in detail in the literature on hydrogeology (e.g. Brassington 2017; Hölting and Coldewey 2019).

For all measurements with electrodes, it has proven useful to use flow-through cells so that the electrodes are completely covered by water. This is particularly important for oxygen measurements with a Clark electrode (I hope you don't use one of those anymore,

**Table 2.7** Essential parameters to be recorded when sampling mine water (details in Sect. 2.3 and Table 2.5)

| Parameter | Parameter |
|---|---|
| Electrical conductivity | Flow rate |
| pH value | Base capacity |
| Oxygen concentration | Acid capacity |
| Redox potential | Iron concentration |
| Temperature | Iron species |

**Fig. 2.7** Paul Younger† during on-site determination of iron concentrations (August 2000)

but an optical oxygen sensor instead), because it consumes oxygen in an electrochemical reaction. It may sometimes be problematic to carry a flow-through cell with you, especially underground. In this case, the electrodes should be completely covered by the flowing mine water, or the electrodes should be slowly agitated in a beaker or bucket or stirred with an inert plastic stirrer. However, my personal experience is that the variability of the on-site parameters of mine water is greater than the difference between measurements with or without a flow-through cell.

Another problem, which is particularly relevant for underground sampling, is the contamination of the devices and especially the cables. To date, there are hardly any manufacturers who take this circumstance into account and offer untethered devices (powered without cables). One example is the Californian company, Myron L® Company, which has been producing a measuring system for most field parameters for years that does not require cables (ULTRAMETER II™; in the meantime, it is also marketed by the company Hach under the name MP-6). It would be desirable if such a system were also available for oxygen. Since 2012, the company YSI, a Xylem Brand, has the EXO1 multiparameter probe on offer and since the beginning of 2016, the company WTW, A Xylem Brand, markets the IDS measuring system; these devices establish a connection between the sensor and the measuring device by means of Bluetooth Wireless Technology. Hopefully, these devices will allow future underground sampling to be as simple as possible (in Eliot's sense, so to say).

In acid-base titration, it is important that no colour indicator is used, as older manuals and textbooks still recommend, but only a pH meter that continuously measures the pH value. The inherent colour of the mine water, the incidence of light, the colour of the clothing, even a colour weakness of the colleague sampling the mine water can have a substantial influence on the detection of the inflection point of the titration. Then titrate to pH 4.3 (the exact end point can vary between pH 3.7 and 5.1) or pH 8.2 (occasionally one reads 8.3) with the base depending on the initial pH of the mine water and record the amount of base used. When using an indicator, it has been shown that different persons independently of each other always overestimated the base capacity and estimated the acid capacity very differently (Wolkersdorfer 2008, p. 185).

▶ You might have asked yourself at some time, which of the temperatures provided by your numerous measuring probes you should use. The pH sensor, conductivity sensor, redox sensor and oxygen sensor – each one shows a slightly different temperature because each of the PT100 or PT1000 temperature sensors installed in them provides a slightly different resistance, thus temperature. So at best you can choose between four values or note all four and average them. DIN Deutsches Institut für Normung e.V. (2009) recommends: "Deviation of the temperature measuring device is to be calibrated with a calibrated thermometer." However, only in the rarest of cases, you might have this thermometer on you. I have gotten used to the following procedure: since oxygen saturation and oxygen concentration are directly dependent on temperature, I always take the temperature of the oxygen probe. If I don't have one with me, I take the temperature of the conductivity probe as an alternative, because this also has a strong dependence on the temperature. This saves me having to constantly think about which value is the correct one. However, if you want to be very precise, then you cannot avoid using a temperature measuring instrument calibrated with a calibrator or simulator, often equipped with a negative temperature coefficient (NTC) sensor (Lipták 2003, pp. 594–598).

### 2.3.2 Electrical Conductivity (Specific Conductance)

Electrical conductivity or specific conductance is the property of a liquid to conduct electricity. It is usually measured with a conductivity electrode (Kohlrausch measuring cell), whereby both two-electrode and four-electrode measuring cells are used. Because of the property of mine water to coat electrodes with a layer of "ochre" within a relatively short time, four-electrode measuring cells are preferable to two-electrode measuring cells. In addition to electrical measurements, electrical conductivity can also be calculated from analysis by calculation (see Wolkersdorfer 2008, pp. 421–423) or using PHREEQC (Parkhurst and Appelo 2013) (Fig. 2.8). However, if you use pitzer.dat for mine water of lower ionic strength, there may be substantial discrepancies between the calculated and actual electrical conductivity (pers. comm. Kirk Nordstrom 2019).

As the conductivity of water is temperature-dependent, you must ensure that a distinction is made between conductivity and electrical conductivity. Conductivity is a parameter without reference to a standardised temperature, whereas electrical conductivity or specific conductance is the conductivity at 25 °C (in some countries still 20 °C). The values can vary considerably in individual cases, so care should be taken when reporting "conductivities". It is important to calibrate the probe regularly with a suitable standard, and when measuring mine water, a "natural water" standard is best. It is also important to ensure, when making the settings on the meter, that the correct temperature compensation is used – otherwise fatal consequences, i.e. misinterpretation, could result. DIN Deutsches Institut für Normung e. V. (1993) state: "The values given for the

```
Example 1.--Add uranium and speciate seawater.

-------------Description of solution----------------------------

                                       pH  =   8.220
                                       pe  =   8.451
      Specific Conductance (µS/cm,  25°C)  = 52630
                          Density (g/cm³)  =   1.02323
                               Volume (L)  =   1.01282
                        Activity of water  =   0.981
                 Ionic strength (mol/kgw)  =   6.747e-01
                       Mass of water (kg)  =   1.000e+00
                    Total carbon (mol/kg)  =   2.182e-03
                       Total CO2 (mol/kg)  =   2.182e-03
                         Temperature (°C)  =  25.00
                  Electrical balance (eq)  =   7.936e-04
 Percent error, 100*(Cat-|An|)/(Cat+|An|)  =   0.07
                               Iterations  =   7
                                  Total H  = 1.110147e+02
                                  Total O  = 5.563054e+01
```

**Fig. 2.8** Result of a chemical-thermodynamic calculation with PHREEQC (PHREEQC 3.5.0–1400). Result of the electrical conductivity calculation highlighted in colour

temperature correction factors are mean values from measurements of various natural waters. It should be noted that they may only be used for measurements of such waters which have electrical conductivities for $\gamma_{25}$ of about 6–100 mS/m and a composition corresponding to that of natural groundwaters, spring waters and surface waters."

Electrical conductivity is a measure of the total dissolved solids in water. Its symbol is usually the $\kappa$ (occasionally also $\gamma$ or $K$), and the SI unit of electrical conductivity is the Siemens per metre (S $m^{-1}$) although the units S $cm^{-1}$, µS $cm^{-1}$ or mS $cm^{-1}$ are commonly used for water. In some countries (e.g. South Africa or Finland), and in DIN EN 27888, mS $m^{-1}$ is also common (the ban on prefixes in the denominator of units has since been lifted by the Bureau international des poids et mesures). Occasionally, in Anglo-Saxon countries, "mho" instead of "S" is still used, although ASTM D1125 (2014) has clarified since at least 1995 that "the unit of electrical conductivity is siemens [*sic!*] per centimetre". Distilled water usually has electrical conductivities in the range of a few µS $cm^{-1}$, whereas high-salinity mine water can have electrical conductivities up to several mS $cm^{-1}$. You can use the electrical conductivity for first statements about the mine water quality, because contaminated mine water often has increased electrical conductivities compared to the background. By systematically measuring electrical conductivities along a receiving watercourse, it is possible, for example, to locate mine water seeps in a targeted manner.

It is imperative to always measure the electrical conductivity of a water sample in the field. If samples are mixed up in the laboratory, they could be identified by the electrical conductivity measured in the laboratory, provided you have not acidified your sample.

Particularly in mine water with high electrical conductivity and high iron concentrations, the formation of a biofilm and clogging occurs extremely rapidly, as is also known from borehole water (Cullimore 1999). This effect must not be neglected, as it causes the electrodes to age more quickly, and measured values may differ from each other only because the electrode has aged.

It is often stated in the literature that the total mineralisation *R* (total dissolved solids, TDS, in mg $L^{-1}$) or the density can simply be calculated from the electrical conductivity $\kappa$ (in µS $cm^{-1}$). This may be a good sales argument for the manufacturers of measuring instruments, but in most cases the result is wrong, because very few users read the operating instructions describing the exact procedure for the correct use of these parameters. It is then encouraging to see a publication discussing why there is a correlation coefficient of approximately 1 between $\kappa$ and *R*. However, especially for mine water, no simple relationship exists between the two. Let me be specific: While the relationship between the two parameters is simple, finding the correct coefficient *f* (in dag $m^{-2}$ $S^{-1}$) is rather less so:

$$R = f \cdot \kappa \tag{2.3}$$

In Wolkersdorfer (2008) I had written that *f* ranges between 0.6 and 0.725. However, this assumption is not correct: rather, it is a rough estimate (Scofield and Wilcox 1931). For example, for saline water the factor is 0.5, and for mine water it appears to range from 0.25 to 1.34 ($n = 45$; Hubert and Wolkersdorfer 2015, p. 495). Consequently, it is essential to determine the conversion factor gravimetrically for each individual monitoring well, e.g. according to the American Public Health Association et al. (2012), or to calculate it from the overall analysis (consider $HCO_3$ properly). Details can be found in the aforementioned publication.

### 2.3.3 Base Capacity ($k_B$; Acidity)

The base capacity ($k_B$) is a concentration that represents the sum for all acids present in the liquid (see also Sect. 1.2.3). In the Anglo-Saxon literature this is almost predominantly referred to as acidity, and in the past, p- and m-values have also been used (Fig. 1.2). There are numerous methods of determining base capacity, and a discussion of which one is correct would go too far at this point. Therefore, you are referred to the fundamental work of Kirby and Cravotta (2005a, b). Regardless of which method is selected, the following statement should be borne in mind: The correct base capacity of mine water can only be determined by on-site measurement or in the laboratory immediately after sampling. As soon as more than approximately 6 h have elapsed between sampling and laboratory analysis, the measured value must be considered incorrect. As the base capacity is essential for correctly dimensioning a mine water treatment plant, special attention must be paid to an accurate measurement. The acids present in mine water are mineral acids (e.g. Fe, Mn, Al), organic acids (e.g. tannin – I am aware of the current academic discussion about this term and concept, but since these "humic substances" contribute to the acidity of a mine water, I decided to keep the older term "organic acid") and inorganic acids (e.g. carbonic acid, sulfuric acid). As Hedin et al. (1994a) have shown in a publication that, unfortunately, is often misunderstood in this respect, the base capacity can be calculated very well from the mine water analysis with sufficient accuracy. However, it should be noted that the equation only gives correct results for acid mine drainages with a pH below 4.5. The following equation, modified from the original publication (Hedin et al. 1994a, p. 5), can be used for this purpose (Wolkersdorfer 2008, p. 185):

$$\text{Acidity} = \gamma_{\text{H}_2\text{O}} \times \left( 10^{(3-\text{pH})} + \sum_{n}^{1} \frac{z_i \times [Me_i]}{M_i} \right) \tag{2.4}$$

where:

$z_i$ Charge of species $i$, no unit
pH pH value, no unit
$[Me_i]$ concentration of species $i$ (e.g. Fe, Mn, Zn, Al, Co, Cu, Cd), g $L^{-1}$
$M_i$ Molar mass of species $i$, g $mol^{-1}$
$\gamma H_2O$ Activity of the mine water, no unit.

The activity is not given in the original work, but this has proved helpful, as can also be seen from the figure in Hedin et al. (1994a, p. 6), where there is a slight deviation from the ideal values at higher concentrations. This calculated acidity is largely identical to the potential acidity $ACI_{pot}$ used by Lausitzer und Mitteldeutsche Bergbau-Verwaltungsgesellschaft mbH (2019, p. 14, Annex VI-a). There, too, their Fig. 2.2 shows a deviation from the ideal line at higher acidities, which could possibly be corrected by taking activity into account. However, to obtain a quick estimate, titration in the field is preferable.

In practice, cold titration in the field with NaOH has proven to be practicable for dimensioning a mine water treatment system, whereby a pH electrode should be used. Indicators should be avoided because each person perceives the colour change differently (Lieber 2003). At all times, the concentration of NaOH should be adjusted to the base capacity of the mine water so that the dilution does not become too large. If the pH electrode does not respond quickly enough during the titration, it is recommended to use different electrodes for the determination of $k_B$ and $k_A$.

### 2.3.4 Acid Capacity ($k_A$; Alkalinity)

The acid capacity ($k_A$) is a concentration that represents the sum of all the bases present in the liquid (cf. the summary at the beginning of the book). In the Anglo-Saxon literature, this is almost predominantly referred to as alkalinity, and in the past also the terms p- and m-values were used (Fig. 1.2). Strictly speaking, only the term acid capacity would be chemically correct. In the context of mine water, however, the two terms have become established side by side, with “alkalinity” being used far more frequently than “acid capacity”.

In practice, the cold method with sulfuric acid ($H_2SO_4$) has proven useful for dimensioning a mine water treatment system. However, there are also methods with hydrochloric acid (HCl) or with hot acid. In any case – as described above – the concentration of the $H_2SO_4$ must be adapted to the acid capacity of the mine water so that the dilution does not become too large.

U.S. Geological Survey (2015; 6.6 Alkalinity, version 4.0 [9/2012]) recommends using reversal points rather than endpoints because of the difficulty in accurately determining the pH 4.5 endpoint of a titration. Although this titration requires a slightly greater effort than for endpoint titration, this method should be used in the future to determine acid and base capacity. Unfortunately, the prescriptions in American Public Health Association et al. (2012) and U.S. Geological Survey (2015; 6.6 Alkalinity, version 4.0 [9/2012]) differ. While the latter require filtration of samples, the former do not. It seems more sensible to me to abstain from filtration, because filtration would disturb the carbon balance, leading to changes in base or acid capacity. The terminology in both publications is also different. American Public Health Association et al. (2012) does not distinguish between alkalinity and acid neutralizing capacity, whereas U.S. Geological Survey (2015; 6.6 Alkalinity, version 4.0 [9/2012]) does. This refers to alkalinity as the acid capacity of the filtered sample, whereas acid neutralisation capacity refers to the unfiltered sample. If we now also consider DIN 38409-7:2005-12, the chaos seems perfect, as it so often does: no filtration and an endpoint titration to pH 4.3 or 8.2 is required.

For a correct determination of the acid capacity, the total inorganic carbon (TIC) concentration should be determined. From this, the acid capacity can be easily calculated by means of the pH value of the sample or by chemical-thermodynamic modelling.

### 2.3.5 Flow and Loads

As already explained above, the flow rate is of extraordinary importance in the treatment of mine water. From the flow rate $Q$ (formula symbol also $\dot{V}$) and the concentration $c$, the load $M$ can be calculated, and this is necessary for mine water treatment plants in order to design its size:

$$M = Q \cdot c \tag{2.5}$$

However, it turns out that flow measurement is more complicated than often assumed, and from experience it can be said that most flow measurements are wrong. This is mainly due to the fact that the measuring devices or measuring instruments for flow measurement look comparatively simple, whereas the physical principles are decidedly complicated. To determine the flow rate, there is a whole range of procedures, some of which differ fundamentally in their methodological bases:

- Bucket and stopwatch method (up to 500 L min$^{-1}$ in larger vessels; in confined underground conditions, a water-impermeable plastic bag can be used instead of a hard vessel)
- Impeller flow meter
- Weirs (V-notch weir, rectangular weir, trapezoidal weir)

- H-flumes, Parshall flumes
- Venturi flume
- Pump run time and rate
- Water meters
- Pressure measurements in boreholes
- Ultrasonic flow meters, acoustic flow meters (e.g. ADCP, Fig. 2.9),
- Tracer tests (e.g. salt dilution method).

In many instances, whether using measuring weirs or flow meters, the respective measuring method is incorrectly applied at the chosen location, or the conditions under which the measurement can be carried out are not met. Wolkersdorfer (2008, pp. 142–172) has summarised on 31 pages what is important for flow measurements in the mine water context. In addition, an eight-page appendix reproduces recalculated weir equations based on the original published data. In a few cases, the equations published in the usual literature prove to be incorrect or too inaccurate to be used in weir measurements – such as the three equations and associated explanations in Younger et al. (2002, p. 158) or in PIRAMID Consortium (2003, p. 11).

To use a V-notch weir, 12 conditions must be met (U.S. Department of the Interior – Bureau of Reclamation 2001). As soon as one of these conditions is not met, the measurement is not only "a little" wrong but must be considered *fundamentally* wrong. The reason for this is that the flow through a weir depends on eight parameters, but usually only one

**Fig. 2.9** Flow measurement with an acoustic digital flow meter in a mine water influenced stream (Cape Breton Island, Nova Scotia, Canada). The operator correctly stands downstream of the meter and holds the meter far away from the body to avoid interfering with the flow as far as possible

of them – namely the water height above the weir – is measured. Thus, as soon as the base conditions are no longer met, the values of the other seven parameters change and the equations are no longer valid.

In the case of incorrectly installed measuring weirs, the flow measurements can deviate from the actual flow by up to 64%, although the usual deviations in the case of incorrect installations are between 5 and 20% (Thomas 1959). An academic discussion of errors in flow measurements with weirs is given by de Vries (1989), who found computational errors between 0.1 and 30%, depending on the source of the error. The most common installation or measurement errors I have observed are:

- Incorrect reading of the water height above the weir (measured too close to the weir: Reading must be taken at a distance equal to four times the weir height)
- Overflow must be fully aerated (free overflow)
- Weir crest on air side instead of water side or no weir crest present at all (Fig. 2.10)
- Wrong construction material (wood or plastic, material too thick; Fig. 2.11)
- Incoming flow not perpendicular to weir plate
- Weir dimensions not considered
- Inflow channel not free of debris, vegetation or sedimentation

To be sure that the weir measures correctly, the discharge should be verified by means of another measuring method (e.g. bucket and stopwatch method, with impeller flow meters or other velocity-area methods) if the water level differs, especially since each weir is slightly different due to its construction.

This is illustrated by an example for a small ½ 90° V-notch weir (Thomson type weir) that can pass one-half as much flow as the 90° V-notch weir. According to the Kindsvater-

**Fig. 2.10** Incorrectly installed measuring weir at the former IMPI pilot plant in Middleburg, Gauteng Province, South Africa. The sharp edge on the water side of the weir plate is missing

**Fig. 2.11** Incorrectly constructed triangular weir for a mine water outlet of the 1 B Mine Pool on Cape Breton Island, Nova Scotia, Canada. The weir plate is too thick (2 cm), is made of non-stainless material, and lacks the sharp edge on the water side. The weir was in use for a period of about 1 year to measure flows with "high accuracy"

Shen equation, the flow $Q$ through a V-notch weir with water height $h$ can be calculated as follows (Eq. (2.6)):

$$Q_{1/2}^{90^\circ} = f \times h^{2.5} \tag{2.6}$$

where, according to Wolkersdorfer (2008, p. 161), $f = 0.684$.

A V-notch weir was installed at the abandoned Finnish Metsämonttu mine to determine the flow through a passive treatment system. For this purpose, 13 individual measurements were made at different water levels. By means of a correlation analysis, the calibrated factor $f = 0.627$ was found to reflect the flow more accurately (Fig. 2.12).

The same applies when measuring flow using the impeller flow meter or ultrasonic method. As soon as the parameters mentioned in the manual or those of the respective method are not observed, the measurement cannot be used. When measuring flow using the surface velocity method, often too few lamellae (measuring sections in the watercourse's cross-section) are selected, the selected measuring depths are not exactly correct, or the marginal parameters of the channel are unsuitable for the measuring method. As a result, measurement errors of up to 20% are possible.

With the impeller method, the velocity of the water is measured with an impeller flow meter at one or more depths of the water body and in three to several water body sections (lamellae). At least one single measurement per lamella should be used, but it is better to

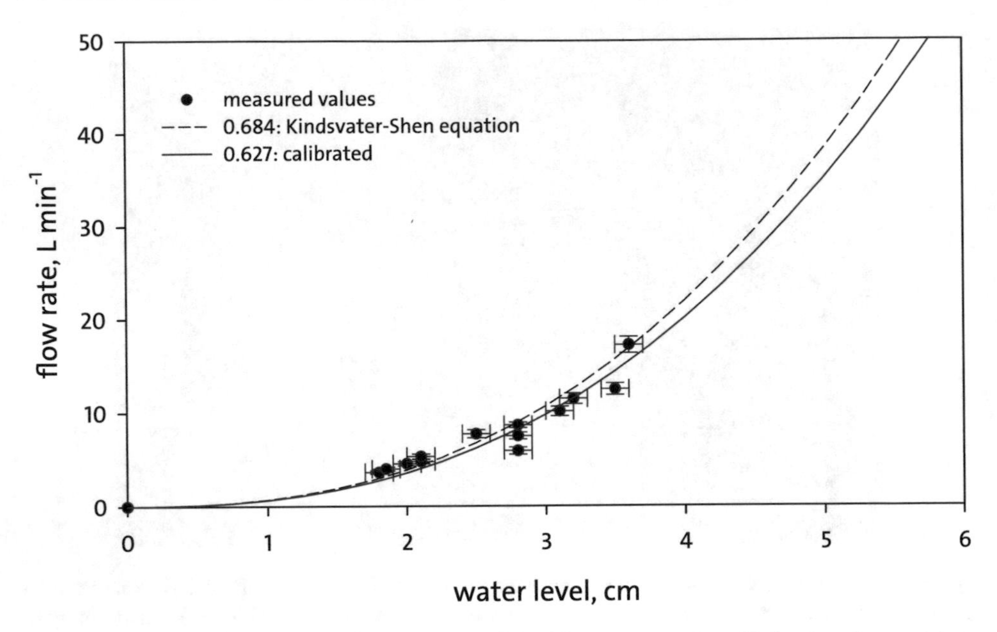


**Fig. 2.12** Calibrated flow at a measuring weir of the Metsämonttu mine, Finland, and comparison with the flow from the Kindsvater-Shen equation

choose a multi-lamella 3-point method. Flow measurements using the ultrasonic method are identical to the method using an impeller flow meter. Details of the measurement method relating to mine water can be found in Wolkersdorfer (2008, pp. 152–155).

Gees (1990) compared three different methods of flow measurement with the results of the stage-discharge curves on eight Swiss mountain streams. He found an exact measurement with the impeller flow meter (20-lamellae-5-point method) and optimum conditions for the salt dilution method; the latter showed slightly smaller deviations from the stage-discharge curve results (1.1%) than that with the impeller meter (1.3%). Since hardly anyone in the mine water sector uses the 20-lamellae-5-point method, but at best a 4-lamellae-3-point method, it follows that the salt dilution method is the most accurate method for flow measurements. This also works for high electrical conductivities because it is not the absolute electrical conductivity that is important for the measurement method, but the difference in the background conductivity during the measurement.

In the case of the electromagnetic flow measurement method, some colleagues sometimes say that it is not applicable if the iron concentrations in the mine water are too high, because the manual states that the device cannot be used if there is iron or metal in the vicinity. However, this guideline in the manuals refers to metallic iron and not to ferrous or ferric iron dissolved in the mine water or to iron that is present in colloidal form. Thus, the electromagnetic flow measurement method is not precluded from being used in mine water that is rich in iron.

A measurement method that has hardly been used in the mine water sector to date is the Acoustic Doppler Current Profiler (ADCP). These devices allow a continuous velocity–

depth profile of a water body to be established and are increasingly found instead of the more complex impeller flow meter measurements (Herschy 1995, pp. 244–267; Muste et al. 2007), using two to four transmitters and receivers operating at the same frequency. However, the minimum water depth is 30 cm, with frequencies of 1200 Hz and 3000 Hz applied, and turbulence from the device or the operator must be avoided at all costs. Mayes et al. (2008) used an ADCP to measure flows in the River Gaunless catchment in England, where they quantified diffuse inflows of mine water.

▶ An example from the German Harz Mountains demonstrates the errors caused by incorrect flow measurements. There, a consultant, who had been familiar with mine water issues for a long time, carried out flow measurements in an open channel. This channel started on the right side at an opening in a sealed adit portal, seen in the direction of flow. Since the channel had a rectangular cross-section, the consultant used a 1-lamella-1-point measurement with an impeller flow meter and the area-velocity method. At the same time, my group made measurements there using the salt dilution and the impeller meter method. It turned out that the measurements of the consultant were regularly 20% lower than those with the salt dilution method. The consultant could not be convinced that the salt dilution method was the most accurate of all indirect measurement methods for flows, so that a reference date measurement was initiated, in which the salt dilution method and various lamella methods were carried out by my working group and the consultant. As could be shown, neither I nor the consultant was able to obtain reliable results with the lamella method in the channel, since there was no homogeneous flow within the measurement cross-section. Rather, there was a preferential flow in the lower right-hand corner of the channel that was not correctly detected by any of the lamella methods used (Fig. 2.13). Only the flow rates measured with the salt dilution method were resilient.

As the consultant still did not trust our measurements, he carried out measurements with the "bucket and stopwatch method", to verify his measurements. Although this is usually the most accurate method for measuring discharges, it finds its limits at about 500 L min$^{-1}$ – however, the flow at said mine was 840–1080 L min$^{-1}$, so the bucket-and-stopwatch measurement is no longer sufficiently accurate. After the consultant involved had seen our salt tracer results, I received the following letter referring to the salt-tracer-method:

If the salt tracer tests give different results, the calculation mode and the accuracy of the conductivity measurement should be checked – just yesterday, during joint investigations with A-student at the B-well in C-city, I was able to identify again that the two conductivity meters we used showed substantial deviations and different behaviour at higher salt concentrations. In this respect,

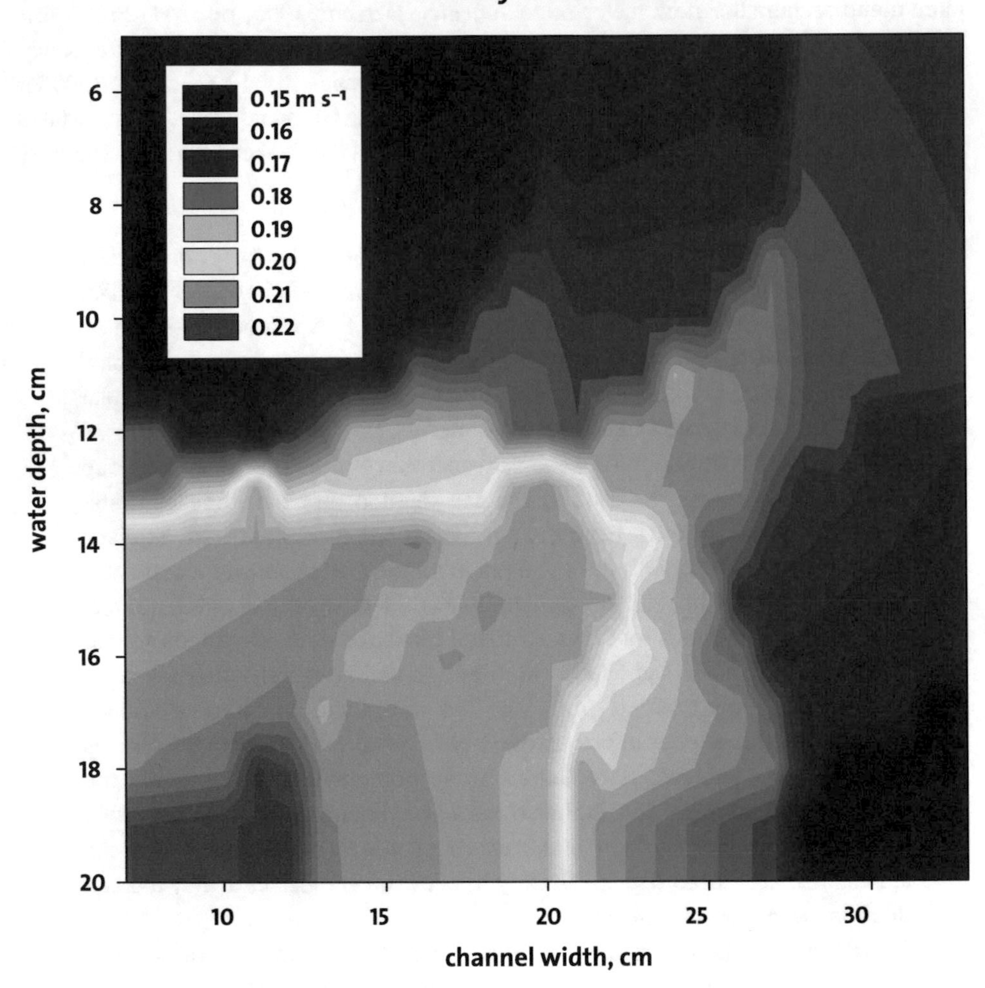


**Fig. 2.13** Flow measurements in an open channel using the 4-lamella 3-point method and preferential flow in the channel's lower right corner (the figure is a section against the direction of flow; velocity in m $s^{-1}$). The single-point lamella method in the middle of the channel gave readings that were 20% too low compared to the salt dilution method. The water level in the channel on the measurement day was 15 cm (Gernrode/Harz, Germany; outlet Hagenbachtal adit, 15th October 2003)

I fundamentally doubt the accuracy of conductivity measurements at higher salt concentrations.

I do not want to discuss the missing calibration of the consultant's meters here, but due to the necessary "calibration" in the salt dilution method (e.g. Gees 1990; Landesanstalt für Umweltschutz Baden-Württemberg 2002; Spence and McPhie 1997), it is irrelevant whether the measuring instrument measures "correctly" or not, as long as the instrument is calibrated with the water to be measured and the salt to be used. In fact, the method does not use the *value* of the conductivity, but only the integral under the curve – and what offset the conductivity measurement has in the process is completely irrelevant. I had a similar discussion on Cape Breton Island, Canada, where the local remediation company rejected not only the salt dilution method but my measurement results and commissioned a local engineering company to carry out impeller meter measurements – any questions?

Care should also be taken when using meters that are manufactured in the UK or in the US. These meters usually measure in gallons – and the UK (imperial system) gallon (1 imp gal = 4.54609 L) and American gallon (1 US liq. gal = 3.785411784 L) differ by 17–20% depending on the conversion direction! At a mine water discharge on Cape Breton Island, Canada, where the UK and US gallon are still being used side by side, a flow rate of about 1 $m^3$ $min^{-1}$ was determined by Cape Breton University using an acoustic digital current meter (OTT ADC). The remediation company, on the other hand, determined a flow rate of approximately 317 gpm (1.2 $m^3$ $min^{-1}$) using a Greyline ultrasonic AVFM-II Area Velocity Flow Meter (Wolkersdorfer 2008, p. 108). From the 20% difference in measurement, it could be concluded that the company must somewhere have made a conversion error from UK gallons to US gallons (or vice versa). Finally, it turned out that the transmitter was calibrated to US liq. gal ("264" gpm), but the programmer of the data logger assumed that the voltage supplied by the transmitter came from a calibration to imp gal and converted this reading to US liq. gal units ("317" gpm). Therefore, it is imperative for instruments displaying gpm (gallons per minute) to verify exactly what gallons are involved and that the transmitter calibration matches the conversion in the data logger. You don't want your mine water treatment plant to have a hard landing on Mars.

Before I come to the last section of this chapter, a warning that Georg Wieber, Director of the German Rhineland-Palatinate State Office for Geology and Mining until April 2023, pointed out to me: The metal edges of weirs can sometimes be very sharp, and therefore caution is advised during installation and measurements to avoid injuring yourself. I can agree with this observation from my own experience!

Let me end the flow measurement section here – I could write a whole book about it. Maybe it will be my next project – you can already send me your personal experiences.

### 2.3.6 pH Value

The pH value is a measure of how acid or alkaline a mine water is (for definition see Sect. 1.2.17). It is erroneously assumed by many users that the pH value alone can be used to determine whether a mine water is acid mine drainage. However, this is not correct in relation to mine water. Rather, the base and acid capacities must be used, as explained in

Sects. 2.3.3 and 2.3.4. When mine water has a pH value of between about 6 and 8, it is usually referred to as circumneutral (Nordstrom 2011). As illustrated above, it is often assumed that pH values can only range between either 0 or 1 and 13 or 14. However, this assumption is incorrect, and even Evangelou (1995, p. 8) states that incorrectly. Rather, the range between pH 0/1 and 13/14 is where most electrodes exhibit a linear voltage–proton activity dependence. With proper calibration, electrodes can measure lower or higher pH values. How to proceed in this case, especially for negative pH values, has been described by Nordstrom et al. (2000).

Today, pH is usually measured with an electrode that must be calibrated with standards before each daily measurement campaign (for standards, see Buck et al. 2002, pp. 2177–2187). This can be a 1-point, 2-point or multi-point calibration. By carefully stirring the buffer solution, avoiding the introduction of air into the buffer, you need to wait until a constant pH or voltage has been established. Modern measuring instruments calculate the calibration curve independently, so that no further calculations are necessary if one remains within the linear range of the measuring probe. In heavily mineralised mine waters, the electrode may age rapidly, requiring more frequent cleaning or replacement. In addition, a memory effect (hysteresis) may develop over time. In these cases, it may take a long time for a constant pH value to appear on the meter. If you need to measure pH at depths greater than 100 m, see Schindler et al. (1995) for options up to 10 MPa and 100 °C.

Since the activity coefficient is temperature-dependent, the pH value is converted to a reference temperature to obtain comparable values. Modern measuring instruments automatically compensate the temperature to 25 °C, recognise the buffer solution and are battery-operated (as you can see, this is in contradiction to the specifications of the Arbeitsausschuss Markscheidewesen im Normenausschuss Bergbau [Working Committee for Mine Surveying in the Mining Standards Committee], which I quote in Sect. 2.2.3. There, 20 °C is specified for the pH value. Why? There must be historical reasons). If the individual measured values fluctuate greatly during the day, it may be necessary to recalibrate the electrode. However, the optimal solution for such conditions would be to use two electrodes: one for the acidic water and the other one for the alkaline. It is also important that the electrodes are always kept in KCl solution between measurements, with the KCl concentration equal to that in the electrode. Always carry a second vial of KCl solution, as the storage containers for the probes have a life of their own that is always directed toward the centre of the earth. In an emergency, use normal water for a short time – under no circumstances should you store your probes dry!

In addition to the pH probes, there are pH test strips or indicator liquids, which, by the way, you should not use for mine water, because this often results in incorrect pH values due to the many interfering factors (pers. comm. Arthur Rose 2006). Let me add that there are other methods for determining pH. This involves using the colour intensity of one or more indicators along with an algorithm to determine pH by optical calculation (Martinez-Olmos et al. 2011). Currently, this method may seem unsuitable or a gimmick for mine water, but in a future mine water lab-on-a-chip, I can certainly envision such applications.

First such experiments are shown by Am et al. (2011), and Yang et al. (2020) used microfluidic screening with a lab-on-a-chip for the study of pyrite oxidation.

As shown above, mine water pH values range widely. Nordstrom et al. (2000) measured pH values of −3.6 in the Iron Mountain, California, USA, copper and pyrite mine, and the mine-affected Lake Velenje in Slovenia at times had pH values of 12 and more (Stropnik et al. 1991). For this reason, the term acid mine drainage (AMD) is no longer used internationally when acid mine drainage is meant, but rather acid mine leachate (AMD/ML) or mining influenced water, since even alkaline or circumneutral mine water can cause considerable environmental damage.

Without doubt, acid mine drainage is always present when the pH value is below 5.6. Below this pH value, buffers are no longer relevant for the mine water treatment plant, so that the mine water is always "net acidic". This is considered, for example, in the Ficklin diagram (Fig. 1.9).

A pH-neutral water exists when the activity of the OH ions corresponds to that of the $H^+$ ions (protons). At 25 °C, the pH value is then exactly 7.00. Since the self-dissociation of water increases with temperature, a neutral water has a pH value of 7.08 at 20 °C and a pH value of 6.92 at 30 °C (pH = pOH). Natural waters typically have pH values ranging between 4.3 and 5.3 (at the current partial pressure of $CO_2$, and without the input of acids and bases, the pH in equilibrium with $CO_2$ should be 5.6). However, there is theoretically no limit to the activity of hydrogen ions in a solution, either downward or upward, so far more negative and positive pH values exist than mentioned in the previous paragraph.

Finally, a note on calculating pH averages (you can find an online calculator here: www.wolkersdorfer.info/ph). Since the pH value is a function of the hydrogen ion activity, the inverse function must first be applied before the average can be calculated (Table 2.8):

**Table 2.8** Calculation of pH averages. The correct average pH value is 3.41 and not 3.63 (pH values of a mine water near Carolina in Mpumalanga Province, South Africa; online calculator: www.wolkersdorfer.info/ph)

| | Measured pH value | Calculated proton activity $\{H^+\}$ | pH |
|---|---|---|---|
| | 2.82 | $1.51 \cdot 10^{-3}$ | |
| | 3.30 | $5.01 \cdot 10^{-4}$ | |
| | 4.16 | $6.92 \cdot 10^{-5}$ | |
| | 3.90 | $1.26 \cdot 10^{-4}$ | |
| | 3.25 | $5.62 \cdot 10^{-4}$ | |
| | 4.17 | $6.76 \cdot 10^{-5}$ | |
| | 4.01 | $9.77 \cdot 10^{-5}$ | |
| | 4.02 | $9.55 \cdot 10^{-5}$ | |
| | 3.40 | $3.98 \cdot 10^{-4}$ | |
| | 3.30 | $5.01 \cdot 10^{-4}$ | |
| **Mean value** | ***(3.63)*** | $\mathbf{5.90 \cdot 10^{-8}}$ | |
| **pH average** | | | **3.41** |

$$\overline{\mathrm{pH}} = -\log\left(\frac{1}{n}\cdot\sum_{i=1}^{n}10^{-\mathrm{pH}_i}\right) \quad (2.7)$$

The program AMDTreat+ includes a module that easily calculates the pH averages (unfortunately with inaccuracies in the last digit of the result). But I think you can calculate it yourself quite easily in a spreadsheet, with your calculator, in an app or the aforementioned web page. Though the differences between the two calculation methods are mostly small they should not be considered trivial, because the pH is a logarithmic number.

### 2.3.7 Iron Concentration

In most cases, the iron concentration is an essential decision criterion for the dimensioning of a mine water treatment plant. In addition to the concentration of total iron, the two redox species of iron, $Fe^{2+}$ (*ferrous*, reduced species) and $Fe^{3+}$ (*ferric*, oxidised species) are important. Often the species are determined in the laboratory, where usually only one of the two species and the total iron are measured and, in many cases, considerable analytical inconsistencies occur.

Many laboratories measure $Fe_{tot}$ and $Fe^{2+}$ using different analytical methods. Therefore, $Fe^{2+}$ concentrations may occasionally exceed those of $Fe_{tot}$. This seems to be particularly the case when almost all iron is in the reduced form. At a mine water discharge in Nova Scotia, Canada, an accredited laboratory regularly detected $Fe^{2+}$ concentrations up to 15% higher than the $Fe_{tot}$ concentrations. This is not an isolated case, but a systematic error that must be sought in the measurement method and is also known from other laboratories. To obtain correct results, it is therefore essential to determine $Fe^{2+}$ and $Fe_{tot}$ concentrations by photometric methods in the field, taking care that the iron does not oxidise too rapidly during sampling (Fig. 2.14). High Fe concentrations can be detrimental because the mine water must first be diluted prior to photometric analysis and oxidation of the iron can occur during this process. If care is taken, the readings will be identical to those measured in an accredited laboratory. It is important to always measure standards to verify the accuracy of the photometric measurement. The commercially available, preconditioned powder packs or cuvettes have been shown to be adequate for this purpose and are in no way inferior to elaborate measuring solutions prepared in the laboratory (Stookey 1970; To et al. 1999; Viollier et al. 2000).

Both the different iron concentrations in mine water and the terminology used often causes discussion, because they are handled differently by different colleagues. It sometimes happens that someone refers to the $Fe^{2+}$ as dissolved iron and the sum of $Fe^{2+}$ and $Fe^{3+}$ as total iron. This is, of course, incorrect. Table 2.6 therefore provides an overview of the different terms and how they should be used correctly to avoid future ambiguities.

Particularly, the iron concentration in a mine water sample is subject to rapid change as soon as the sample "stands" for a while. This also applies if the sample is cooled. Acidifying the sample with $HNO_3$ only helps if the total iron concentration is of interest; speciation will only be maintained in a few cases. Instead, you need to acidify with HCl (American Public Health Association et al. 2012; Ficklin and Mosier 1999).

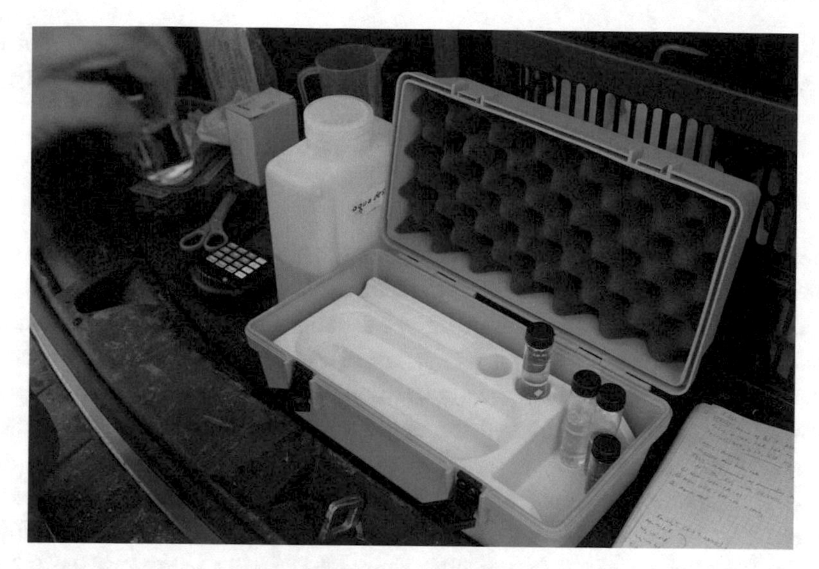

**Fig. 2.14** Photometric on-site analysis of total iron and $Fe^{2+}$

The reason why the iron concentration in the water decreases between sampling and analysis is the hydrolysis of the iron. During this process the dissolved iron reacts with the oxygen in the water and precipitates as colloidal iron oxyhydrates, turning the mine water brown or causing a brown precipitate at the bottom of the vessel. The hydrolysis of the iron is particularly detrimental when laboratory tests are to be carried out with "real" mine water. Consequently, it is always problematic to carry out laboratory experiments on mine water treatment; therefore, synthetic mine water is often used, in which the iron concentration can be adjusted by means of suitable salts. In a mine water from the Schwefelstollen near Alexisbad/Harz, Germany, hydrolysis lowered the total iron concentration from 0.25 mg $L^{-1}$ to 0.15 mg $L^{-1}$ (Wolkersdorfer and Kubiak 2008). Since the hydrolysis of iron generates protons, the pH value of the sample also decreases between sampling and the laboratory.

A colleague has had very good experience with the following procedure (pers. comm. Broder Merkel 2017): Filter the sample with 0.2 µm in the field, acidify and determine the total iron with ICP-MS. In the field, determine $Fe^{2+}$ photometrically in the filtered, non-acidified sample. Repeating the $Fe^{2+}$ determination in the laboratory often reveals a lower concentration, which can be interpreted as an indication of oxidation in the sampling bottle. If $Fe^{2+}$ is to be measured in the laboratory, then $Fe^{2+}$ must be stabilised separately with HCl.

### 2.3.8 Manganese Concentration

Manganese is another metal that should be removed from mine water during mine water treatment. Although it can be determined photometrically in the field, it is not necessary to perform an analysis in the field. It is sufficient to have manganese measured in the laboratory. In addition, it is not necessary to treat the sample in any special way; the general sampling procedures described earlier are sufficient. Although manganese oxidises and hydrolyses in a similar way to iron, the pH value of the sample drops only slightly.

### 2.3.9 Aluminium Concentration

Substantial concentrations of aluminium are found in acid mine drainage. These concentrations can reach the milligram per litre (mg $L^{-1}$) range and aluminium must therefore be removed from the mine water. Consequently, a reliable analysis of dissolved aluminium is essential. However, unlike iron or manganese, aluminium forms colloids even below 0.2 μm and therefore is often analytically overdetermined, and what is often reported as "dissolved" aluminium when filtered through 0.45 μm is not "dissolved" in the mine water sample and can account for an over-determination of aluminium by up to eight times the actual aluminium concentration (Fig. 2.15). It is therefore recommended to use 0.2 μm or better 0.1 μm filters for a correct aluminium analysis (Wolkersdorfer 2008). To reduce the filtration time, a 0.45 μm filter should be installed ahead of the 0.2 μm or 0.1 μm filters.

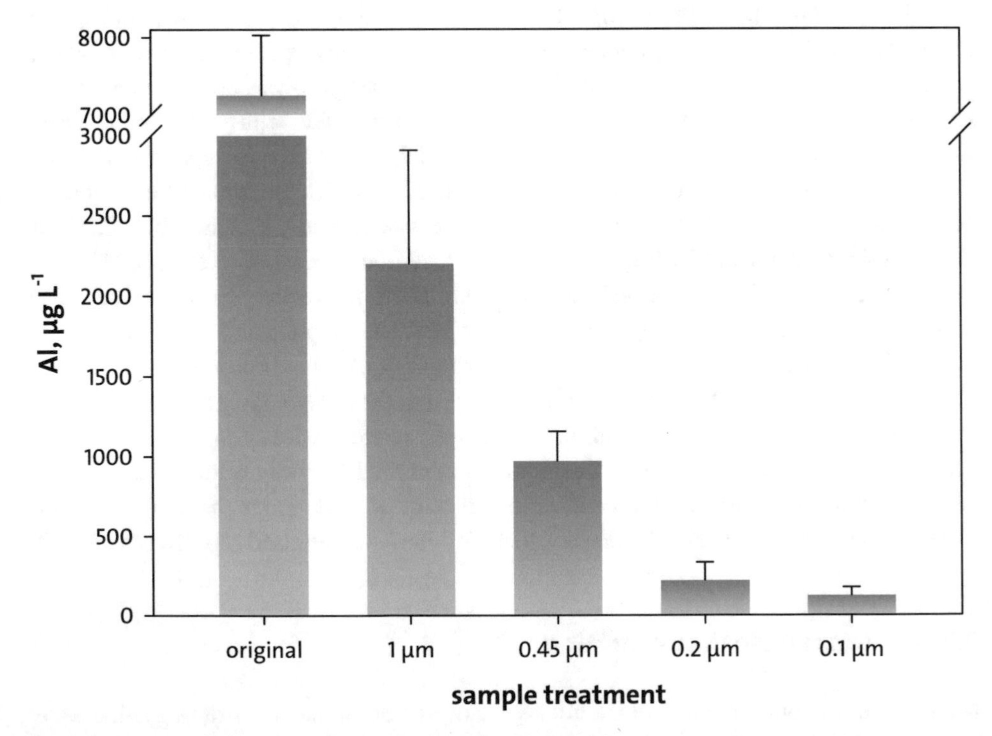


**Fig. 2.15** Aluminium concentrations in sequentially filtered mine water from the 1 B Mine Pool in Canada. Of the 970 ± 184 μg $L^{-1}$ Al filtered with a 0.45 μm pore size filter, 750 ± 148 μg $L^{-1}$ is still present in colloidal form (B183 well, Cape Breton Island, Canada)

### 2.3.10 Redox Potential ($E_h$, ORP)

No other parameter causes more problems in measurement and interpretation than the redox potential. Ficklin and Mosier (1999, p. 255) even write that $E_h$ measurements in the field are of doubtful meaning and value. This is partly because it can sometimes take a very long time for a constant redox potential to be established (Schüring et al. 2000, p. 10), but also because some researchers might have unresolved questions regarding the redox potential. Lloyd and Heathcote (1985, pp. 113–142) observed times of up to 30 min until a constant redox potential was reached, which is consistent with the data in Deutsches Institut für Normung e. V. (DIN) (1984). As shown by Gezahegne et al. (2007), this appears to occur when the groundwater or mine water is not in equilibrium and there is little buffering capacity for the redox potential. The most important redox couple in mine water is the $Fe^{2+}/Fe^{3+}$ iron species (Stumm and Morgan 1996), but numerous other redox couples also contribute to the redox potential. While the iron species reach redox equilibrium relatively quickly, this is only conditionally the case for other redox pairs.

Operators often forget that the ORP value they read on their meter is that of the cell they are using. However, the redox potential ($E_h$) value is given in relation to the standard hydrogen electrode (SHE). Only a few electrode manufacturers and measuring instruments allow the reading on the instrument to be the $E_h$ value and not the electrode potential. However, for hydrogeological issues, the $E_h$ value is relevant and not the cell potential. Thus, the value provided by the cell usually must first be corrected to the standard hydrogen electrode at 25 °C. Equation 2.8 and the corresponding Table 2.9 can be used for this purpose (accuracy of the equation ±5 mV at temperatures ranging between 5 and 65 °C).

$$E_{0(25°C)} = E_t - 0.198 \times (T - 25) + \sqrt{a - b \times T} \tag{2.8}$$

where:

$E_0$ Redox potential of the standard hydrogen electrode, mV
$E_t$ measured potential of the electrode at temperature $T$, mV
$T$ Measured temperature, °C.

The data given in Table 2.9 can be explained with an example: At the Metsämonttu mine in Finland, the Hach measuring device showed an ORP of −152.4 mV at 13.4 °C. Since

**Table 2.9** Coefficients for the calculation of the redox potential according to Eq. 2.8. Further coefficients and an online calculator can be found at www.wolkersdorfer.info/orp

| Sensor type | Coefficient *a* | Coefficient *b* |
|---|---|---|
| Mercury Calomel KCl, saturated | 67,798 | 324 |
| Ag/AgCl KCl 1 mol $L^{-1}$ | 62,775 | 284 |
| Ag/AgCl KCl 3 mol $L^{-1}$ | 50,301 | 297 |

the measuring probe was a 3 molar Ag/AgCl probe, the values of the coefficients *a* and *b* are 50,301 and 297, respectively. Thus, a redox potential $E_h$ of 65 mV is calculated (www.Wolkersdorfer.info/orp). Decimal places can be omitted, because the measurements are only accurate to about ±10 mV anyway.

For each redox measurement, the type of probe must be noted to allow calculating the correct redox potential. Redox electrodes are subject to rapid surface poisoning because redox reactions take place on the surface of the cell. They should therefore be cleaned regularly to provide correct results.

▶ It should be noted at this point that pH electrodes also provide a value in mV, but these values are in no way suitable for calculating the redox potential. "Beautiful" examples of such incorrect measurements can be found in numerous publications (e.g. Özcan et al. 2007, Table 3). In the specific case, the redox potentials are given to two decimal places, which contradicts DIN Deutsches Institut für Normung e. V. (1984) or American Public Health Association et al. (2012), according to which an accuracy of about ±10 mV can be achieved. Besides, the values reproduced there are extremely low, ranging from 48 to 84 mV, and correlate with pH ($R^2 = 0.3$; since pH is temperature compensated, the $R^2$ is not close to 0.99). This is a clear indication that a pH electrode was used, and the meter was switched from the pH display to the mV display. However, the natural waters found in this region should have redox potentials ranging between 200 and 500 mV (Wolkersdorfer et al. 2004, $R^2$ between redox potential and pH: 0.0003). Unfortunately, such incorrect measurements are found again and again in expert reports and the literature. Only a few manufacturers have probes in their portfolio by means of which both the pH value and the redox potential can be measured simultaneously (e.g. YSI).

Another problem arises when no redox potential has been measured at all because it is considered inaccurate or unnecessary by the researcher or consultant. This may be irrelevant in some cases, but when it comes to chemical-thermodynamic modelling of mine water, reliable redox measurements should be available. In another unpublished example from an open pit mine in Ghana, elevated ammonium concentrations were present in the sump located across from the active mining area where blasting was used. The question was whether the ammonium concentrations, ranging from $<1$ mg $L^{-1}$ to $>40$ mg $L^{-1}$, could be a natural source or whether it was an anthropogenic source entering the sump water because of the mining operations. The consulting company with an internationally well-known name discusses nitrification and denitrification of ammonium in the report; at no point, however, is the redox potential discussed, nor are measurements of it found in the report. Instead, the measured ammonium concentrations obtained over a period of 2 years were fed into a numerical simulation software which this company likes to use. They failed in assessing the cause of the strong fluctuations or the source, which would certainly have been possible with redox measurements.

In another case involving monitoring in an experimental reducing and alkalinity producing system (RAPS), the researchers noted a redox potential of −81 mV. All iron was present as the oxidised species ($Fe^{3+}$) according to the analysis, and the $Fe^{2+}$ concentration was <0.006 mg $L^{-1.}$ This is evidence that the redox potential cannot be −81 mV as reported, but rather must be around 110 mV. It was clear that the researchers forgot to convert the probe's potential to the potential of the standard hydrogen electrode.

Jang et al. (2005) have developed miniaturised redox probes that allow detailed observation of redox potential changes within a medium, e.g. a biofilm. Meyer and Dick (2010), in turn, report novel redox probes based on the green fluorescent protein GFP. I consider both developments to be exciting. However, until market-ready, redox-sensitive fluorescence sensors for mine water are available or the miniaturised redox probes become widely used, a lot of water will still run down the Nile River. Rather, the usual glass electrodes will still have to be used for quite a long time.

### 2.3.11 Oxygen Saturation

For some passive mine water treatment systems it is essential to know the oxygen saturation of the water. In principle, of the numerous methods for oxygen determination (e.g. iodine titration according to Winkler, electrochemical, optical), only two are still used today: electrochemical with the obsolete Clark electrode (Clark et al. 1953) or optical with the modern LDO sensors (luminescent dissolved oxygen, Klimant et al. 1995). In contrast to the Clark electrode, the LDO sensors do not consume the oxygen, so the measurement is easier and more reliable to perform than the one with the Clark electrode. Therefore, I do not recommend using the latter electrodes anymore. When measuring oxygen with the Clark electrode, the electrode must be calibrated in the white storage container before *each* individual measurement, as the electrode is temperature and pressure dependent (Schwedt 1995). Oxygen measurements with values below 0.5 mg $L^{-1}$ are meaningless since the detection limit of the Clark electrode is 0.5 mg $L^{-1}$. This must be considered in the evaluation. It is also recommended that the sensor be used in a flow-through cell since oxygen is consumed during the measurement (Lloyd and Heathcote 1985). When measuring, the saturation in percent and the mass concentration in milligram per litre (mg $L^{-1}$) should always be reported together with the temperature.

In warm water containing organic material, the oxygen saturation can rise to values above 100%. Such readings sometimes occur, especially in stagnant water. This is not a measurement error, but the water is oversaturated with oxygen. In rare cases it can also be observed that the electrode does not reach constant values at all. These effects occur (particularly with the Clark electrode) when the flow of water towards the electrode is too fast or if dissolved gases from the water reach the electrode.

## 2.4 Water Analysis

To analyse mine water chemically or physico-chemically, several analytical regulations exist. Most of these are based on regulations for groundwater or surface water and are generally suitable for analysing mine water. Unfortunately, the common regulations differ depending on the country, which sometimes makes the interpretation of a parameter problematic, as the examples of the alkalinity/acidity exhaustively show. It is therefore necessary to familiarise oneself with the applicable analytical regulations at the beginning of a sampling campaign to carry out the analysis correctly. Depending on the country, various regulations must therefore be considered. I will refrain from a detailed presentation and listing here, since the regulations are subject to changes adapted to the scientific state of knowledge.

Consequently, there is no uniform sampling and analysis procedure for mine water, as is available in the case of hydrogeological sampling (e.g. Deutscher Verband für Wasserwirtschaft and Kulturbau e. V. 1992; Sächsisches Landesamt für Umwelt und Geologie 1997; Sächsisches Landesamt für Umwelt, Landwirtschaft und Geologie 2003; Selent and Grupe 2018). However, if the "usual" procedures of hydrogeological sampling for surface waters are followed, sampling and analysis of mine water can also largely be considered representative.

This means that mine water analysis should broadly follow local standards or *Standard Methods for the Examination of Water and Wastewater* (American Public Health Association et al. 2012). In individual cases, mining companies may have issued their own company standards that deviate from these. Where this is the case, these procedures should be applied, whether they follow national or international practice.

In the USA and Canada, the hardness of a water in mg $L^{-1}$ $CaCO_3$ eq. is now increasingly used (I will spare you the equations) to define limits for several parameters, or limits are set for acute or chronic toxicities. In addition, the use to which the particular water body is put is taken into account when setting a discharge limit (e.g., Colorado Department of Public Health and Environment – Water Quality Control Commission 2013). Unfortunately, they do not explicitly state the "hardness" they are referring to, namely whether it is total hardness, carbonate hardness, or non-carbonate hardness (I'll spare you a colleague's comment on this here as well). Let us assume that they mean the hardness defined in the *Standard Methods for the Examination of Water and Wastewater* according to Method 2340 A: "Hardness [mg eq. $CaCO_3$ $L^{-1}$] = 2.497 × [Ca, mg $L^{-1}$] + 4.118 × [Mg, mg $L^{-1}$]."

## 2.5 Lime Addition or Column Tests

Quite often, laboratory or field tests are undertaken to determine how the pH of a mine water is affected by the addition of limestone or "lime" (often used colloquially for calcium oxide or calcium hydroxide). Let's note at this point: positive. This attempt has been made thousands of times, and the results are almost always identical: to obtain

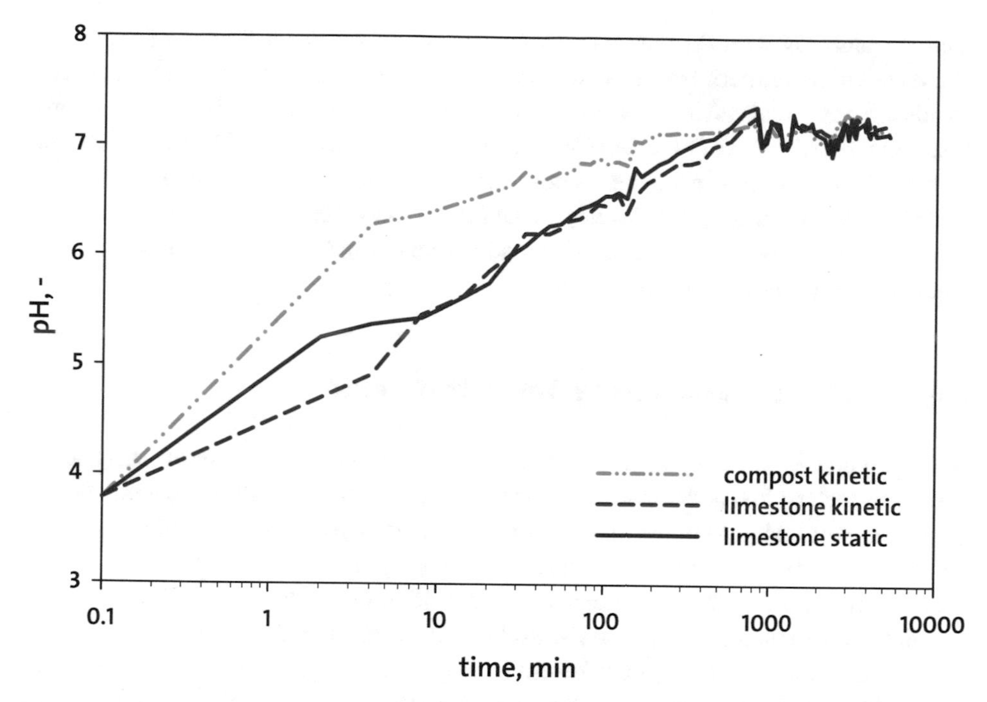


**Fig. 2.16** Limestone and compost experiment (horse manure) with acid mine drainage (100 L tank, approx. 5 kg limestone or manure). A stable pH of 7 is reached after only 2–8 h, as much of the acidity comes from the $CO_2$ in the mine water. Kinetic: water was stirred regularly; static: water was not stirred. Error in pH measurement 0.01

stable pH values of around pH 7, the contact time between limestone and mine water must be about 10–20 h, and that between "lime" and mine water about 1–2 h (e.g., Cravotta et al. 2008; Hedin et al. 1994b; Kostenbader and Haines 1970; Maree et al. 1996; Schipek and Merkel 2012; Singh and Rawat 1985; Uhlmann et al. 2001; Watzlaf et al. 2000; Zinck et al. 1997; Fig. 2.16). The results of elaborate pilot tests have shown that mine water can be treated with "lime", namely, that any mine water can be treated by adding "lime" or limestone. You can control the amount to be added in each case by monitoring the pH of the treated water of the output of the mine water treatment system. Let me point out just one of the results of the aforementioned experiments: The processes are influenced considerably by a high $CaCO_3$ percentage in the limestone, a limestone grain size optimally matched to the respective method (not too small, not too large) and potential inhibitors in the limestone such as Mn, $SO_4$, Fe or Cd (Schipek 2011).

Elaborate column tests are regularly planned and carried out to study a wide variety of physicochemical reactions between overburden or tailings ponds or ore. The number of these laboratory tests also runs into the thousands, and only in a few cases can the results be transferred from the laboratory to the field. However, the systematic experiments conducted by Schipek on the addition of lime to residual lakes in Lusatia are interesting, and in my opinion novel (Schipek 2011; Schipek and Merkel 2012).

Unfortunately, to date, there are no reliable and standardised methods to interpret the results of column experiments (Van der Sloot and Van Zomeren 2012). Each method has its advantages and disadvantages, and before planning a column experiment, the descriptions given in the GARD Guide (Chap. 5) should be considered. Germer (2001) has shown on some German ore samples how difficult it is to deduce from one method how the mine water chemistry will develop in the future. These tests, which, incidentally, can take years, are, however, internationally and nationally indispensable in many cases to obtain a mining permit for coal or ores.

## 2.6 Active or Passive Mine Water Treatment?

Whether you need to set up an active or passive mine water treatment system depends on numerous factors: first and foremost, however, it depends on the quality of the mine water, the quantity of mine water, the financial expenditure that you or your client can provide, and the required discharge limits. Although low-cost systems are often referred to in connection with passive mine water treatment, this is not necessarily true in individual cases. The initial investment costs of a passive system can be considerable, and after a few years the sludge – as in an active system – must be disposed of or the substrate renewed. In an active plant, on the other hand, the investment costs can be relatively low, whereas the running costs are considerable – especially if a large amount of sludge must be dried and disposed of in an environmentally sound manner, or if large amounts of leachate are produced. Morin and Hutt (2006) have attempted to establish a relationship between the capital and operating costs of an active mine water treatment plant and the volume of water to be treated. Indeed, the cost of a plant increases with the volume of mine water to be treated; however, the relationships are less clear for operating costs (Fig. 2.17). A somewhat older but much more detailed listing of the costs of various passive treatment systems is given by Tremblay and Hogan in the MEND Handbook (MEND 2000, pp. 5-96–5-99).

In simple terms, an active treatment plant should be constructed whenever large loads are to be removed from the water, and a passive treatment plant whenever the loads are relatively small (Fig. 2.18). There are several exceptions to this, as the example of the Wheal Jane pilot plant in Cornwall shows – albeit with all the initial problems of a pilot plant (e.g. Younger et al. 2005). Despite these difficulties, through optimisation of the system and intensive scientific support, it was possible to passively clean a partial stream of mine water. At this abandoned mine, three separate passive treatment systems were designed to clean mine water using different operating procedures (Hamilton et al. 1999). In addition, a preliminary stage was installed at two of them to raise the pH of the mine water by neutralisation. However, the main stream of mine water at Wheal Jane is actively treated (Coulton et al. 2003a).

To help you decide on a suitable treatment method, you can use freely available software. Earth Systems, in cooperation with INAP, has developed the relatively simple, MS Excel™-based solution ABATES (Acid-Base Accounting Tool), which helps to character-

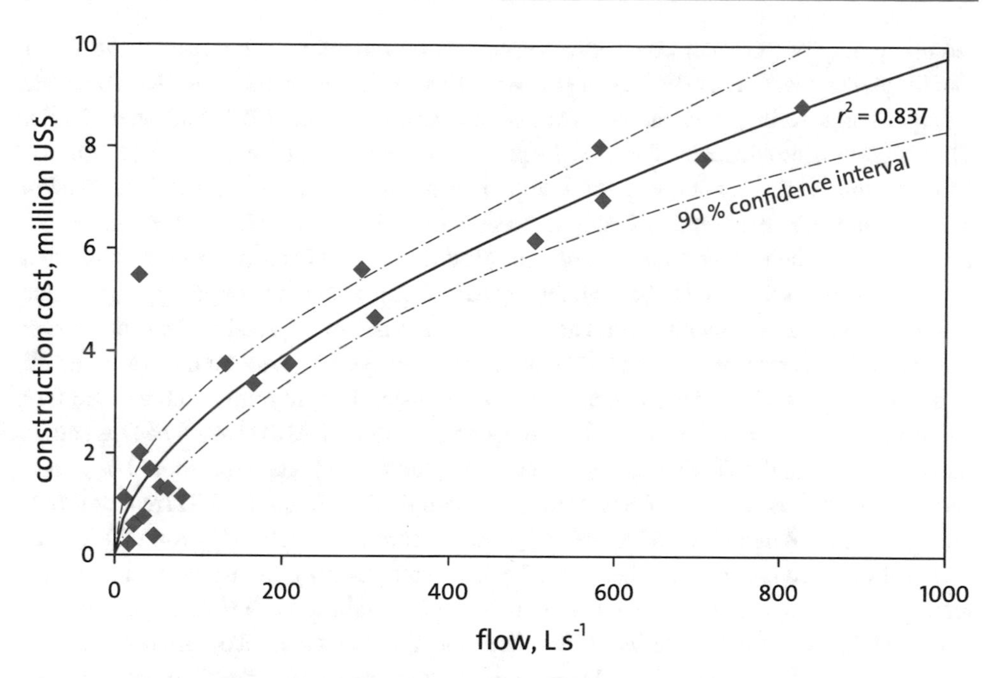


**Fig. 2.17** Relationship between the investment costs of an active mine water treatment plant and the volume of mine water to be treated. (After Morin and Hutt 2006)

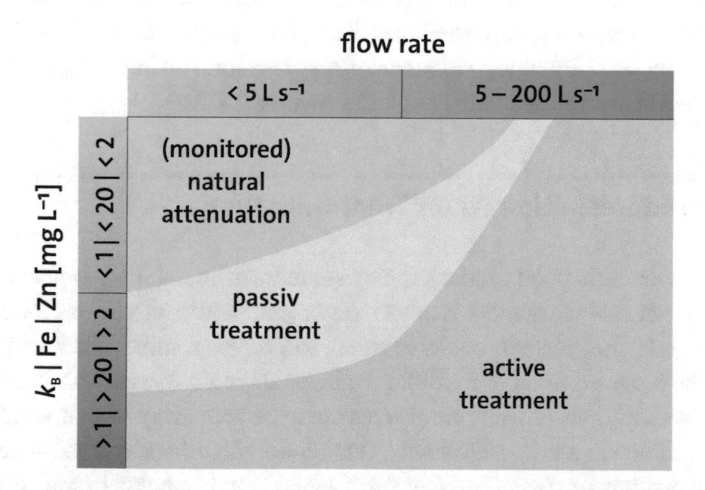


**Fig. 2.18** Graphical decision aid for and against active or passive water treatment. The boundaries between the processes must be considered blurred. On the ordinate there is an upper area with less contaminated mine water (lower concentration of iron, lower base capacity and flow) and a lower area with more heavily contaminated mine water. $k_B$ in mmol $L^{-1}$, Fe and Zn in mg $L^{-1}$ (ERMITE Consortium et al. 2004; Younger 2002b)

ize the quality of the mine water or to forecast its development. It consists of the Water Quality Assessment Tools (WQA) and the Acid Base Accounting Tools (ABA). To decide on a particular facility, you can use AMDTreat+ (Cravotta et al. 2010; McKenzie 2005). This software, provided by the U.S. Dept. of the Interior, Office of Surface Mining, Appalachian Region, is a decidedly comprehensive tool that was available in version 5.0.2 in January 2013 and includes a version of PHREEQC (Cravotta et al. 2015). Unfortunately, the installer does not run on Windows 10 or 11 machines, and although a new version was scheduled to be available by early 2021, at the time of going to print, this was not the case. Meanwhile, you can download a customised version from my server (www.Wolkersdorfer.info/docs/AMDTreat52.exe); yet, you should familiarise yourself with the specifics of Anglo-American units before using the program. Such a cubic foot has it in itself – to be exact: 28.32 L (the conversion tool of AMDTreat+ has numerous inaccuracies from the 7th decimal place on – use a more exact one – if it should be necessary). In the newest version of AMDTreat, the module PHREEQ-N-AMDTreat will provide geochemical modelling of active and passive treatment systems (Cravotta 2021).

I will not hide the fact that there have been several passive mine water treatment systems in the past that did not meet the design criteria (e.g. Johnson and Hallberg 2002; Rose et al. 2004a; Rose and Dietz 2002; Skousen and Ziemkiewicz 2005). Errors occurred either because of insufficient knowledge of the technology or because scientific and technical principles were not implemented correctly or not at all. However, there are also sometimes considerable problems with active mine water treatment plants (e.g. Horenburg 2008), which occasionally lead to regular plant operation after a considerable time. "Problems" are usually not published (see Sect. 1.1), only discussed in the hallways of meetings or circulated by non-governmental organisations or interest groups, and often only mentioned behind closed doors (e.g., Earthworks 2012).

## 2.7 The Endless Mine Water Treatment Plant

From a socio-economic point of view, it may sometimes be required to plan a mine water treatment plant in such a way that it must treat or pump out water for as long as possible (in perpetuity). In this context, one also refers to perpetual mine water treatment costs, which must be provided in order to fulfil certain legal requirements. This is a good idea in structurally weak regions or when employees are to be kept away from unemployment for as long as possible via a hive-off vehicle. This section describes how to proceed. Aspects of environmental law or reasons arising from general environmental protection are deliberately disregarded in the discussion here.

Firstly, the mine water must be monitored for a sufficiently long period to identify the critical water constituents. These parameters may also be defined by the authorities, or legal requirements, for example by the Water Framework Directive, a Groundwater Ordinance, Surface Water Ordinance or locally defined discharge parameters for the receiving water course. The evaluation should be accompanied by an extensive investigation of the background values. Then, by means of multivariate statistical investigations, a characteristic value must be extracted from the available data sets, which is suitable for determining the discharge limits for the mine water treatment plant. The characteristic values, and based on them the discharge limits, must be chosen in such a way that always a part of the water that corresponds to the background concentration of the chosen parameter is treated (for the determination of background values and the definition of terms see Hobiger et al. 2004; Mast et al. 2007; Matschullat et al. 2000; Schneider 2016; Wagner et al. 2003; Walter 2006). If this is ensured, a mine water treatment plant can operate ad infinitum.

How to proceed is described by an example where more than 300 water analyses of a former uranium mine were available. Empirical methods were used to isolate parameters that were suitable for classifying the water in the mine as infiltration water (largely geogenic background value) and mine water (largely contaminated). Infiltration water was groundwater with a fairly short residence time in the mine, and mine water was groundwater that was in contact with the primary and secondary minerals. As could be shown, the parameters As, U and Ca were suitable for a classification at this mine. For this purpose, two indices were formed from the molar concentrations of the three elements and plotted graphically (Fig. 2.19). The figure shows that infiltration waters have an index $q_1$ greater than 0.1 and $q_2$ greater than 2.6 (these figures are based on a detailed

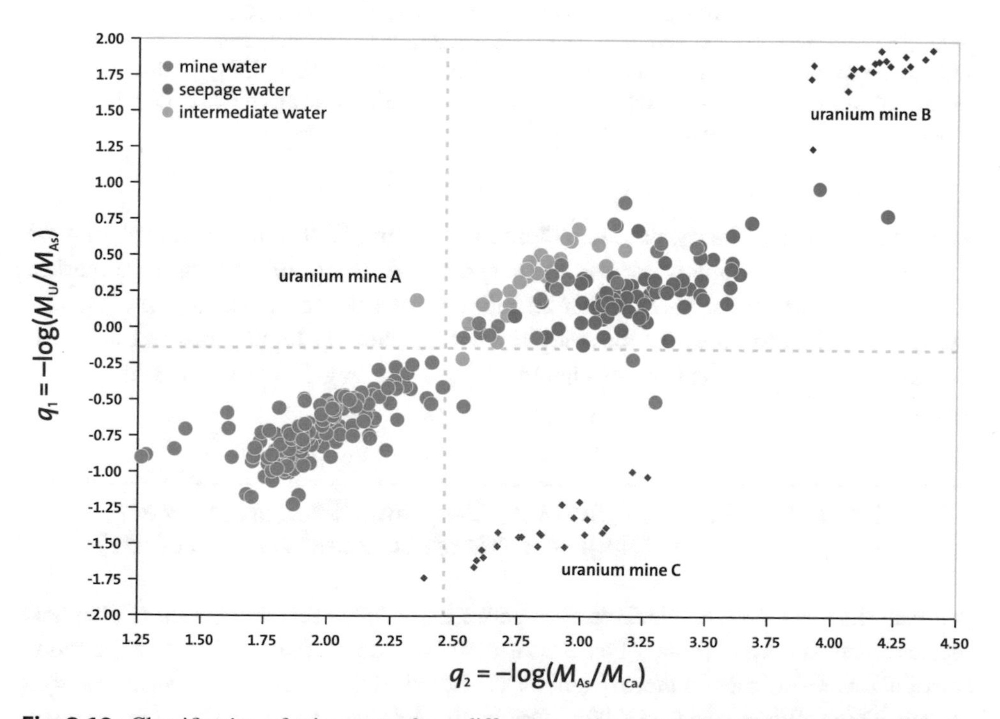


**Fig. 2.19** Classification of mine water from different uranium mines based on quotients of U, As, and Ca. Blue squares are seepage and infiltration waters, orange circles mine waters from uranium mine A. For comparison, the diamonds show mine waters from two other uranium mines (B and C)

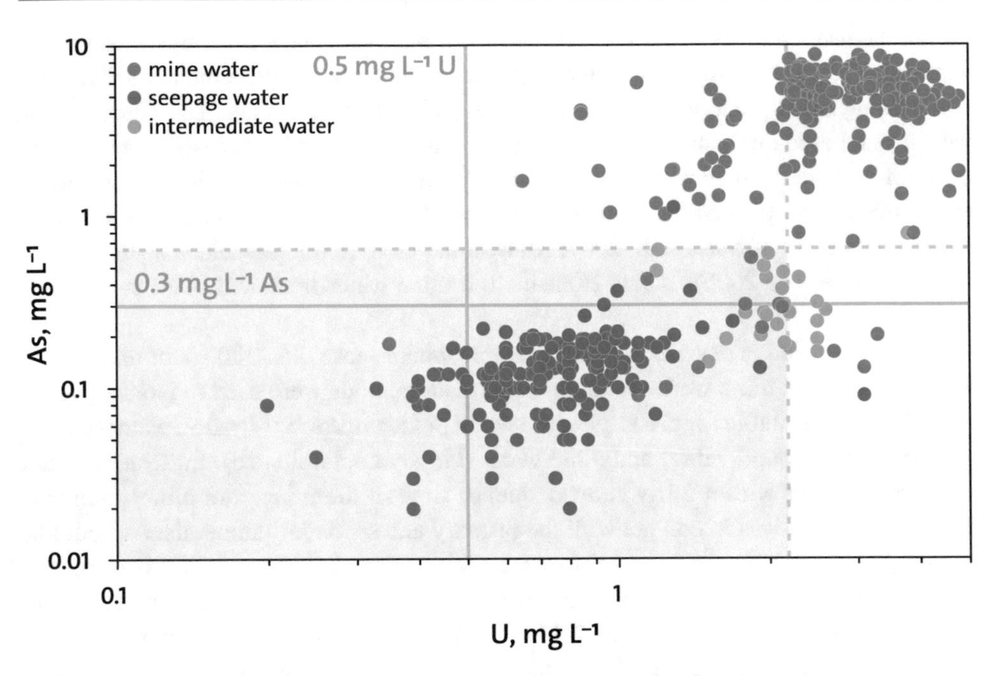


**Fig. 2.20** Scatter diagram of uranium and arsenic mass concentrations at uranium mine A. The discharge limits and monitoring concentrations of the mine water treatment plant are shown in green. The blue box corresponds to the uranium and arsenic mass concentrations of the seepage and infiltration waters based on the statistical evaluation given in Fig. 2.19. Initially, the discharge limit for arsenic was set at 0.2 mg $L^{-1}$. By increasing the discharge limit for arsenic to 0.3 mg $L^{-1}$, lower volumes of infiltration and seepage waters are treated

statistical analysis). Finally, based on these indices, Fig. 2.20 was constructed, and the boundaries between mine water and the seepage and infiltration waters were marked with yellow lines. To be able to plan an infinite mine water treatment plant, the discharge limits had to be set so that they lay within the field highlighted in blue. This resulted in technically feasible discharge limits of 0.5 mg $L^{-1}$ U and 0.3 mg $L^{-1}$ As being attained.

## 2.8 Pourbaix Diagrams (Stability Diagrams, Predominance Diagrams, $E_h$-pH-Diagrams, "Confusogram" *sensu* P. Wade)

Pourbaix diagrams are graphical representations that display the distribution of chemical species in aquatic systems as a function of the pH value on the abscissa and the redox potential on the ordinate. In these diagrams, neither kinetics nor microbiological reactions are considered (Whitehead et al. 2005). Originally, Pourbaix developed these diagrams in the 1940s to predict the corrosion of materials (Fig. 2.21), and he called them "potential–pH equilibrium diagrams" or simply "potential–pH diagrams" (Pourbaix 1966, 1973).

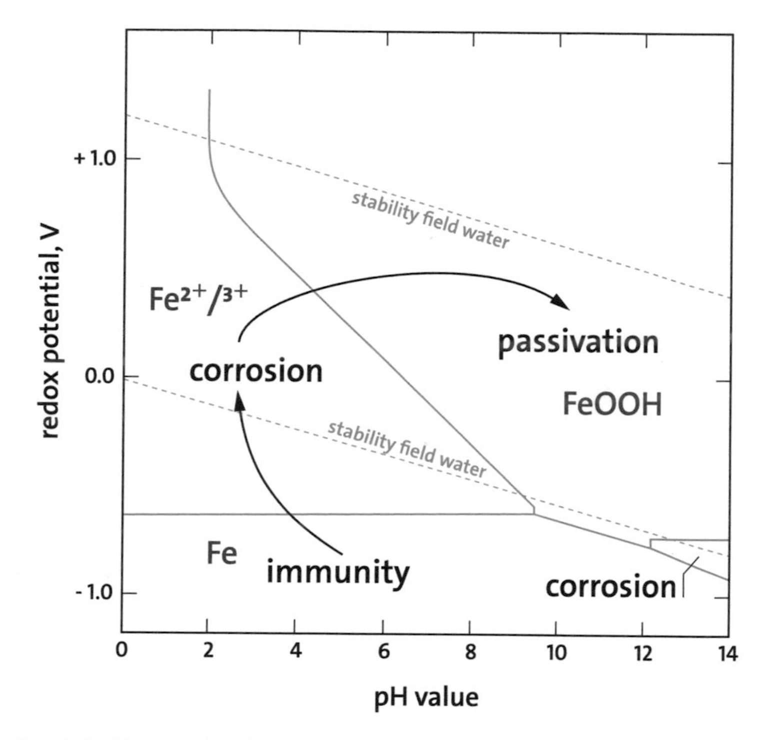


**Fig. 2.21** Pourbaix diagram showing the development of mining influenced water from the solid, immune phase to the dissolved, corroded species and the insoluble, passivated phase after mine water treatment. (Supplemented and modified after Pourbaix 1966, p. 314 ff.; Pourbaix 1973, p. 19)

Today, however, these diagrams are used in a wide variety of applications. Undoubtedly, the work of Garrels contributed to this further development (Garrels and Christ 1965; Hem 1985, pp. 78–82). In the mine water context and its treatment, these diagrams help to predict and understand the processes that will occur during mine water genesis and treatment. In the spirit of Pourbaix (1973, p. 19), it could be said that the diagrams help to understand the evolution of unreacted pyrite in the ore deposit, its corrosion in the mine workings or atmosphere, and finally its passivation in the mine water treatment plant.

Pourbaix diagrams are indispensable tools when it comes to determining chemical reactions or the stability of phases or species in aquatic systems. Furthermore, they are essential to understand redox reactions in aquatic systems (Ibanez et al. 2007). As Baas Becking et al. (1960) have shown, not any odd $E_h$-pH combination is possible in aquatic systems, especially in mine water. Rather, there is a relatively well-defined range of potential $E_h$-pH values (Baas Becking et al. 1960, p. 255). Pourbaix diagrams are most easily constructed using chemical thermodynamic codes such as The Geochemist's Workbench® software package or Hydra/Medusa chemical equilibrium software (Table 2.2). Simplified versions have appeared in the excellent work by Brookins (1988) and a comparison of computational results from different programs in Takeno (2005). Garrels and Christ (1965) describe their meaning and the manual construction of the diagrams and Nordstrom and Munoz (1994) the necessary geochemical thermodynamics.

# 3 Active Treatment Methods for Mine Water

## 3.1 Introduction

In this chapter, the most relevant commercially used processes of "conventional" mine water treatment are presented, as well as what I would call "modern treatment processes". Conventional mine water treatment includes neutralisation processes, whereas modern processes include nanofiltration, ultrafiltration, distillation and reverse osmosis. Consequently, "modern" does not refer to the fact that the method itself is "modern", but that its use in mine water treatment has only occurred on a larger scale in the past one or two decades. To help decide which treatment method can be used for which mine water, the load and several parameters relevant to mine water should be taken into account (Fig. 3.1).

In the context of mine water treatment, the term active implies that energy, chemicals and continuous monitoring of the treatment process are required to treat the mine water (Fig. 3.2). This means, for example, that an anti-scalant must be added to prevent the deposition of metal hydroxides on surfaces, an aerator must always be operated with electric power, or a technician must be on site, for example to carry out minor repairs or to clean pipes.

► I would like to take this opportunity to acknowledge the generosity of the inventor Paul D. Kostenbader. He invented the high density sludge process (HDS) in the late 1960s, applied for a patent in April 1971, was granted the patent in June 1973, and released it for future use by the public in January 1977: "Paul D. Kostenbader hereby dedicates to the public the entire remaining term of the patent" (U.S. Patent No. 3,738,932). This is an example worth imitating,


C. Wolkersdorfer, *Mine Water Treatment – Active and Passive Methods*,
https://doi.org/10.1007/978-3-662-65770-6_3

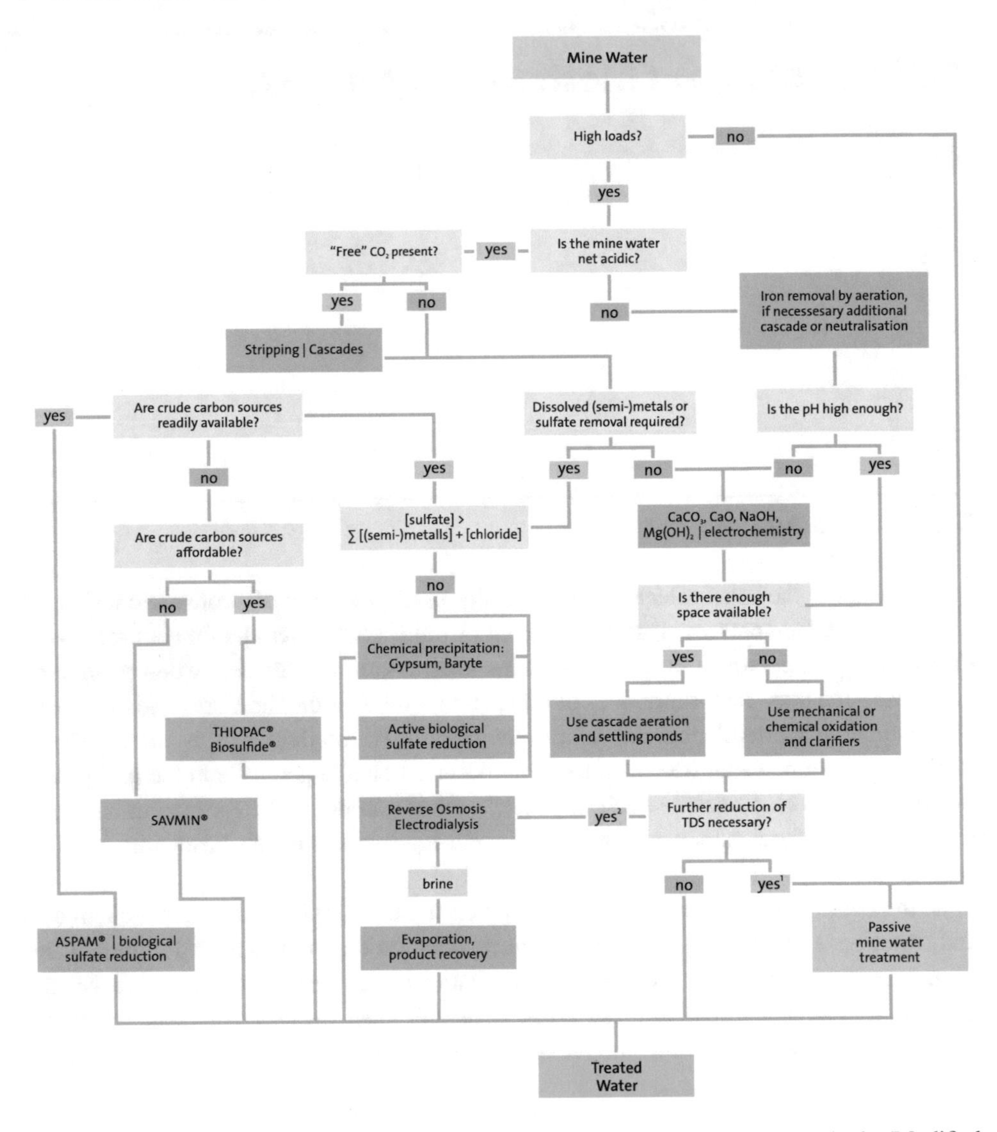


**Fig. 3.1** Flow diagram for the most relevant, mainly active mine water treatment methods. (Modified and supplemented after Jacobs and Pulles 2007; Younger et al. 2002) [1,2]different decision paths

especially since several processes presented below cannot be used because patent law prevents it – unless you would build your plant in China or North Korea. Consequently, patenting processes for mine water treatment may prevent the optimal method from being used in a particular case.

**Fig. 3.2** Pumps for chemical feed in an active mine water treatment plant (Straßberg/Harz, Germany)

## 3.2 Neutralisation Process

### 3.2.1 Principles and Historical Development

There are many neutralisation processes for mine water, most of which differ from each other only in details (Table 3.1). This is due to technological reasons, but also to patent law. Independent of minor details, the neutralisation processes can be divided into two types, namely low density sludge (LDS) and high density sludge (HDS; Table 3.2). These two processes are presented in Sects. 3.2.2 and 3.2.3 and the main characteristics of each are also discussed. Overall, neutralisation with the associated hydrolysis and precipitation of metal ions is still the most common method used to treat mine water. This is because the method is usually relatively easy to control, can handle a wide variation of flows and concentrations, does not necessarily require highly skilled personnel, and requires a manageable number of chemicals and maintenance. However, it does not mean that both processes could not be further optimised, as evidenced by the almost overwhelming amount of literature on the subject (e.g. Brown et al. 2002; Smith and Brady 1998; Zinck and Aubé 2000).

Neutralisation is a long-established process that can reliably treat a wide range of chemically variable mine waters. It is therefore still the most widely used process, partly because the mechanisms of neutralisation and aeration are well understood, and such

**Table 3.1** Compilation of treatment processes for the neutralisation of mine water

| Name of the procedure | Note | Literature |
|---|---|---|
| Conventional (low density sludge (LDS) process) | Addition of alkaline material to the water stream | – |
| High density sludge (HDS) process[HDS] | Low density sludge as crystallisation nucleus at the beginning of the process | Kostenbader and Haines (1970) |
| Aquafix | Dosing of calcium oxide in remote areas | Jenkins and Skousen (1993) |
| CESR (Cost Effective Sulfate Removal); Valhalla | Modification of the SAVMIN™ process; ettringite precipitation | Hydrometrics Inc. (2001), Jacobs and Pulles (2007, p. 49), Lorax Environmental (2003, pp. 4-11–4-14) |
| Cominco (CESL)[HDS] | Like HDS | Kuit (1980), Murdock et al. (1995) |
| Geco | Two reaction tanks; different configurations | Aubé and Payant (1997), Aubé and Zinck (2003) |
| Staggered neutralisation[HDS] | Neutralisation at different pH values | Aubé and Zinck (2003), Heinze et al. (2002), Märten (2006) |
| HARDTAC (High-Aspect Ratio, Draft Tube Agitated Crystalliser) | Apparatus for achieving improved crystal growth | Originally developed by DuPont, licensee today Veolia; e.g. Barbier et al. (2008) |
| Integrated limestone/ hydrated lime process[HDS] | Limestone is added in the first reaction tank | Geldenhuys et al. (2003) |
| Limestone neutralisation process | Limestone as neutralising agent | Deul and Mihok (1967), Mihok et al. (1968) |
| Hazleton® Iron Removal[a] | "Globular catalytic reaction" | U.S. Patent No. 7504030 of 17 March 2009, Brown et al. (2002, p. 52) |
| Keeco | Microencapsulation | Mitchell et al. (2000), Mitchell and Wheaton (1999) |
| NTC (Noranda Technology Centre) | Two-stage neutralisation process; similar to Geco | Kuyucak et al. (1995, 1999) |
| SAVMIN™ (see below) | Sulfate precipitation by addition of calcium; ettringite precipitation | Lorax Environmental (2003), Smit (1999) |
| Tetra (Doyon)[HDS] | Two-stage neutralisation process | Poirier and Roy (1997) |
| Unipure[HDS] | Addition of $Fe^{3+}$ at the beginning of the process; corresponds to the Geco or NTC processes | Coulton et al. (2003a, b, 2004a) |
| Virotec Bauxsol | Red mud from aluminium extraction; also used in reactive walls | McDonald et al. (2006), Munro et al. (2004) |

[a]Method does not appear to be used anywhere for mine water treatment; [HDS]in the broadest sense a modified high density sludge method; SAVMIN™ process developed by MINTEK

**Table 3.2** Characteristics of low- and high density sludge processes from seven water treatment plants using the neutralisation process

| Procedure | Particle density, % | Particle diameter ($D_{50}$), µm | pH, – | Redox, mV |
|---|---|---|---|---|
| Low density sludge (LDS) | 3.4…7.2 | 5.7…21.1 | 9.5…10.9 | 161…316 |
| High density sludge (HDS) | 16.1…60.0 | 2.9…5.9 | 8.9…10.0 | 166…301 |

Compiled from Aubé and Zinck (1999, p. 265)

plants are relatively straightforward to control. Depending on the sulfate concentration in the mine water, the sludge usually contains calcite ($CaCO_3$), gypsum ($CaSO_4{\cdot}2H_2O$), or bassanite ($CaSO_4{\cdot}0.5H_2O$) after neutralisation (Zinck et al. 1997). However, whenever the sulfate concentration exceeds the gypsum saturation concentration of 1500–2500 mg $L^{-1}$ $SO_4^{2-}$, gypsum will precipitate (Aubé and Zinck 2003; Barnes and Romberger 1968; Svanks and Shumate 1973; U.S. Environmental Protection Agency 1983):

$$\begin{aligned} &Ca(OH)_2 + H_2SO_4 \rightarrow CaSO_4.2H_2O \\ &Ca(OH)_2 + CO_2 \rightarrow CaCO_3 + H_2O \end{aligned} \tag{3.1}$$

The first known published use of calcium hydroxide (slaked lime) to neutralise mine water comes from the Königsgrube mine in Königshütte O.S. near Beuthen (Bytom) in Upper Silesia (at that time in Germany; Königshütte is now part of Poland and called Chorzów). There, in 1858, acid mine drainage from the Königsgrube mine was treated with calcium hydroxide, the resulting sludge settled in settling ponds, and the mine water treated in this way was used for steam boilers (Anonymous 1859; Schönaich-Carolath in Tarnowitz 1860). According to Teiwes (1916, p. 30 f.), the system was still in use at the beginning of the twentieth century on a cross-drift in the east field of the Königsgrube, with the sludge being taken back underground and the treated mine water being discharged into the receiving water course. Similar processes were used by the Meggen mine near Lennestadt, North Rhine-Westphalia and the Gottessegen colliery near Löttringhausen (König 1899, p. 423 f.). One of the earliest plants that added flocculants to the water to be treated was developed by René Auguste Henry in Belgium (U.S. Patent No. 1862265A) and was first installed in Germany by the Niersverband/North Rhine-Westphalia in the 1930s (Kegel 1950; van Iterson 1938). From 1934, such applications for mine water treatment can be found, and the first plant treating mine water on a large scale using the Birtley-Henry process started operation in 1935 at the Wallsend Rising Sun colliery in Newcastle-upon-Tyne, England (Anonymous 1934, 1935; Goette 1934). For this purpose, 265 mg of "lime" (on the problem of the imprecise use of "lime", see explanation in Table 3.3) and 6.5 mg of frozen potato starch were added as flocculants per each kilogram of water to be treated, in order to raise the pH to 11 and cause the suspended solids to precipitate. However, as the handling of the frozen potato starch proved to be cumbersome and sometimes did not produce satisfactory results, further development took place and synthetic flocculants or coagulants were also used (van Iterson 1938, p. 86 f.), of which a wide range is available today (Table 3.4).

**Table 3.3** Selected alkaline materials suitable for mine water neutralisation

| Alkali compound/material | Demand t t$^{-1}$ Acidity | Efficiency in % of material used | Relative costs € t$^{-1}$ Bulk material[g] |
|---|---|---|---|
| Calcium carbonate[a], $CaCO_3$ | 1.00 | 30…50 | 8…12 |
| Calcium hydroxide[b], $ca(OH)_2$ | 0.74 | 65…90 | 50…80 |
| Calcium oxide[c], CaO | 0.56 | 65…90 | 60…190 |
| Sodium carbonate[d], $Na_2CO_3$ | 1.06 | 60…95 | 150…270 |
| Magnesium hydroxide, $Mg(OH)_2$ | – | 80 | 300 |
| Magnesium oxide[e], MgO | 0.4 | 80…90 | Project-dependent |
| Sodium hydroxide[f], NaOH | 0.80 | 95…100 | 500…700 |
| Ammonium, $HN_3$ | 0.34 | 100 | 270…500 |
| Fly ash, $CaO·CaCO_3$ | Material dependent | – | Project-dependent |
| Cement kiln dust, $CaO·Ca(OH)_2$ | Material dependent | – | Project-dependent |
| Blast furnace slag | Material dependent | – | Project-dependent |

Modified and supplemented after Coulton et al. (2003b), MEND (2000), Skousen et al. (1988, 2000). t Acidity: $CaCO_3$ equivalents. aquaC (Kalka 2018) can numerically simulate 18 different alkaline-acting materials. This table is used by various authors without giving an exact source

[a]Limestone

[b]Hydrated lime, slaked lime, lime

[c]Unslaked lime, caustic lime, lime, quicklime

[d]Calcined soda, soda ash

[e]Magnesia

[f]Caustic soda

[g]The prices may be subject to considerable fluctuations depending on the market situation or the distance to the supplier: the prices listed here are only intended as a guide; the English term “lime” is unfortunately used inconsistently and refers to both calcium oxide and calcium hydroxide and sometimes calcium carbonate

In the USA, two plants can claim to be the first to have treated mine water industrially with neutralisation. The first is the mine water treatment plant at the Calumet Colliery (Westmoreland County, Pennsylvania, USA), which neutralised acid mine drainage from 1914 until the end of the war so that it could be used in the power station and steam engines there (Hebley 1953; Tracy 1921). It is relevant because some of the experience gained there went into the construction of the first plant built for environmental reasons in 1966. In 1965, Pennsylvania enacted Act 194, which classified mine water as an industrial waste. As a result, the Clean Streams Law applied to mine water and required compliance with discharge limits of pH 6.0–9.0 and a maximum iron concentration of 7 mg L$^{-1}$. At the Vesta No. 5 Colliery (Thompson well, Pennsylvania, USA; between Marianna and Beallsville on PA-2011), the Vesta Coal Company, a subsidiary of the Jones and Laughlin

**Table 3.4** Commonly used flocculants and coagulants

| Chemical | Comment |
|---|---|
| Alum, aluminium sulfate: $Al_2(SO_4)_3$[a] | Acidic material, forms $Al(OH)_3$ |
| Copperas, iron (II) sulfate: $FeSO_4$[a] | Acidic material, which usually reacts more slowly than alum |
| Iron (III) sulfate: $Fe_2(SO_4)_3$[a] | Iron (II) reacts faster than iron (III) |
| Sodium aluminate: $NaAlO_2$[a] | Alkaline coagulant |
| Polyaluminium chloride[b] | Removes various suspended solids |
| Polyferric sulfate (PFS)[b] | Decolourisation; also flocculates non-ferrous metals |
| Mercaptoacetyl chitosan (MAC)[b] | Flocculates non-ferrous metals |
| Anionic flocculants[a] (mostly polyacrylamide) | Negatively charged surfaces |
| Cationic flocculants[a] (mostly polyacrylamide) | Positively charged surfaces |
| Polyampholytes[a] | Positively or negatively charged surfaces depending on the pH value and the point of zero charge (PZC) |

After [a]Skousen et al. (1998) and [b]Fu and Wang (2011)

Steel Corporation, subsequently installed a "lime" neutralisation plant that was required to treat 390–950 L $min^{-1}$ of mine water with a pH of 3.1–6.5, a total iron concentration of 60–195 mg $L^{-1}$, and a base capacity of 4–19 mmol $L^{-1}$, achieving discharge values of pH 7.2–8.0 and total iron concentrations of 0.5–11 mg $L^{-1}$. The plant consisted of a stilling basin to allow uniform control of the influent (420 L $min^{-1}$), a mixing tank to mix the "lime" and mine water, a settling basin and a fishpond for quality control. To control the treatment, a pH meter was installed to regulate the inflow of the lime and to maintain the pH of the water at pH 8. All the sludge produced, which contained a solids concentration of 5–6%, was transported by truck to a nearby borehole at the Vesta No. 6 Colliery (McPhilliamy and Green 1973; Smith et al. 1970; Young and Steinman 1967). How long the plant was in operation could not be determined. No relics of the facilities can be found at either site today.

In principle, the neutralisation of mine water takes place in five steps, irrespective of the type of plant, whereby steps 2 and 3 can also be interchanged or run simultaneously:

1. Intermediate storage
   - Compensation of fluctuating water volumes in intermediate storage facilities.
2. Addition of alkaline material (Table 3.3)
   - The pH of the water is increased, and several metals hydrolyse to form poorly soluble compounds.
3. Aeration of the mine water
   - Oxidation of reduced species, especially iron and manganese; hydrolysis (e.g. aluminium, iron); sorption (Fig. 3.5); stripping of $CO_2$.

4. Coagulation and flocculation
   – Iron as a flocculant ensures coagulation and consequent flocculation of the hydrolysed compounds; flocculation is accelerated by the addition of flocculants (Table 3.4).
5. Separation of the sludge from the water
   – Solids and liquids are separated from each other by gravitational processes or by filtration.

Exchanging steps 2 and 3 is recommended if the mine water contains a large amount of "free" carbon dioxide. Stripping allows the gas to escape, and less neutralising agent is required compared to the mine water with carbon dioxide. Overall, these steps are consistent with what is considered "Best Available Technology Economically Achievable – BATEA or BAT" for contaminated mine water in Canada (Dinardo et al. 1991; Hatch 2014; Pouw et al. 2015).

In each of these steps, the plant manufacturers use different processes or chemicals, depending on the properties and volumes of the mine water as well as patent issues. In addition, the personal experience of the engineer or the company as well as preferences for one or another detail, play a role. In the end, the economic interests of the manufacturer will determine which of the technologies are recommended and finally selected. Consequently, a treatment system will not always represent the optimum of what would currently be technologically feasible – because the courage for something new is lacking or the knowledge of a suitable, possibly better technology is not available.

The aim of neutralisation is, on the one hand, to remove the proton acid from the mine water, but also to raise the pH to such an extent that the metal ions to be removed can hydrolyse and precipitate out of the mine water (Fig. 3.3). This requires different pH values depending on the target metal, spanning a relatively wide pH range, with different oxidation rates depending on the pH (Fig. 3.4). In particular, when, for example, aluminium, iron and manganese occur together in a mine water, a staged treatment process may be necessary (staged neutralisation process, U.S. Patent No. 5,672,280 of 30th September 1997: Aubé and Zinck 2003; Heinze et al. 2002; Märten 2006). This prevents the metal hydroxides or sulfates that have already precipitated (e.g. in the case of lead, pers. comm. Charles Cravotta 2013) from re-dissolving in the mine water. For example, the target pH to remove aluminium from the mine water as gibbsite ($Al(OH)_3$) would be around 6.5, but once the pH increases further to precipitate manganese at a pH around 11, the gibbsite can release the aluminium again. Which neutralising agent is used in individual cases depends on numerous factors. These result primarily from the costs of procurement, transport and storage as well as questions of handling, safety, reactivity (e.g. side reactions) and, finally, the quantity of sludge produced.

In many cases, the target pH will be between 8 and 10 to remove iron from the mine water. This results in the iron hydrolysing and forming colloidal suspended solids due to the addition of alkaline material and subsequent aeration. Further, it leads to flocculation, coagulation and finally precipitation of the iron oxyhydrates. Since the surfaces of the iron colloids have a slightly negative surface charge in the neutral to alkaline range, further

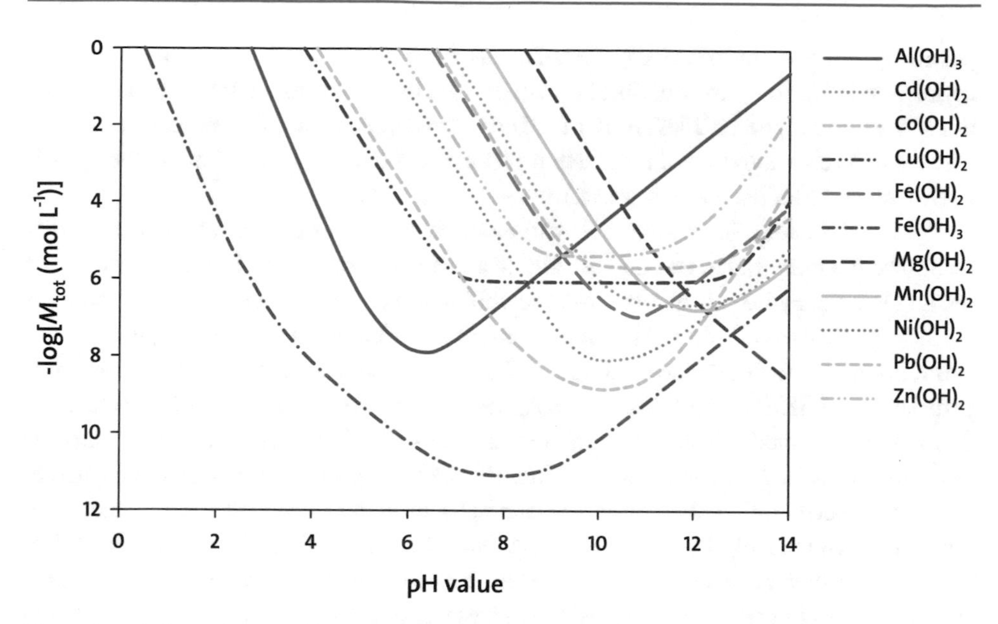


**Fig. 3.3** pH-dependent solubility of metal hydroxides. (Modified and supplemented after Cravotta 2008, original data obtained from Charles A. Cravotta III, pers. comm. comm. 2013)

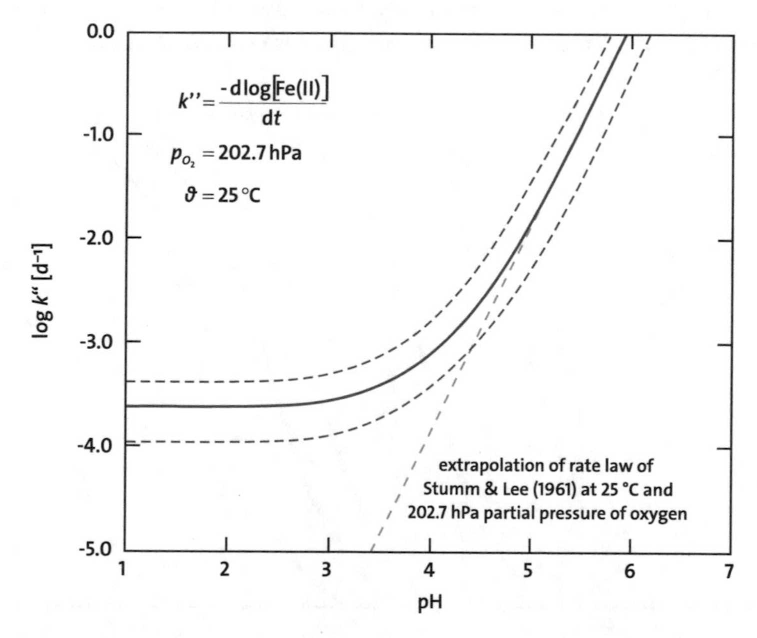


**Fig. 3.4** pH-dependent oxidation rate of Fe(II). (Modified from Singer and Stumm 1969, with data from Stumm and Lee 1961)

positively charged ions (e.g. $Cd^{2+}$, $Cr^{3+}$, $Cu^{2+}$, $Fe^{2+}$, $Ni^{2+}$, $Pb^{+}$, $Zn^{2+}$) sorb on the surface or coprecipitate *s.s.* (e.g. As, Mo, Sb, Ra) with the iron hydroxide in the settling basin or settler (Gusek and Figueroa 2009, p. 101 f.). There, the degree of sorption depends on the pH value and the concentration of the reaction partners involved (Fig. 3.5) (Smith 1999, p. 147 f.; Stumm and Morgan 1996, p. 542 f.).

When calculating the amount of alkali required, first consider the amount needed to neutralise the mine water, and additionally that needed to raise the pH to the point where the target metal hydroxide has its lowest solubility and can precipitate. Since the hydrolysis of iron releases two moles of protons per mole of iron, the alkaline material must also neutralise these. To determine the amount of alkaline material, a chemical thermodynamic code such as PHREEQC (Parkhurst and Appelo 2013), aquaC (Kalka 2018, the chemical thermodynamic model behind the program is also PHREEQC) or AMDTreat (Cravotta et al. 2015) can be used. When using PHREEQC, the respective mine water is titrated or mixed with the alkaline material to be potentially used, and PHREEQC calculates (not: "models" – you model) the processes taking place. Based on the results of the calculation, the plant can then be designed (Coulton et al. 2004b). This approach also considers possible complexation or other poorly soluble phases and is therefore more suitable for prediction than matching against tables or against Fig. 3.5. Croxford et al. (2004) used this approach to calculate the mine water treatment plants at the Frances (Fife, Scotland) and Horden (County Durham, England) collieries. In the case of Horden, where a high density sludge treatment plant was constructed in 2004 (Davies et al. 2012, p. 203), they modelled

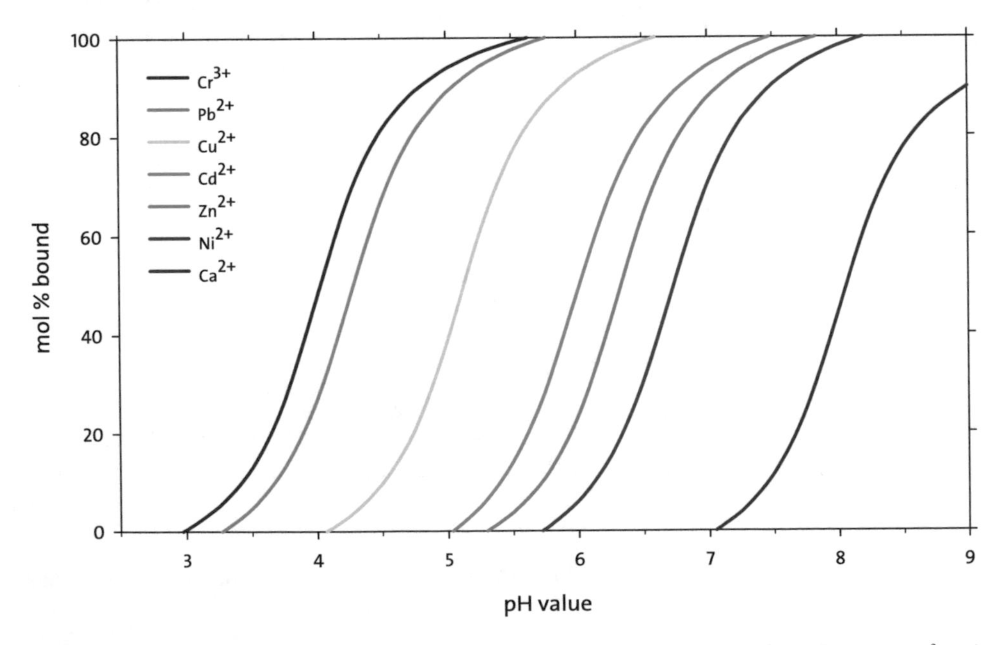


**Fig. 3.5** Examples of pH-dependent sorption of metal cations to iron hydroxides ([$Fe_{tot}$] = $10^{-3}$ mol, [Me] = $5 \times 10^{-7}$ mol; $I = 0.1$ mol $NaNO_3$). (Adapted from Dzombak and Morel 1990; Stumm and Morgan 1996, p. 543)

neutralisation using sodium hydroxide, calcium hydroxide and sodium carbonate. Based on the results of their modelling, $Ca(OH)_2$ was proposed as the neutralising agent, but for practical reasons sodium hydroxide was used (Davies et al. 2012, p. 204) (Table 3.5). Meanwhile, a passive mine water treatment plant is operating there.

On the question of whether limestone ($CaCO_3$) or "lime" (slaked $Ca(OH)_2$/unslaked CaO) should be used in neutralisation, there have already been numerous studies (e.g. Wilmoth 1977), which can be summarised in simplified terms as follows: Limestone produces a higher density sludge but takes more time to settle, whereas "lime" produces a lower density sludge (Mihok et al. 1968) but settles much faster. This behaviour is exploited, for example, by the integrated limestone/lime hydrate process (Geldenhuys et al. 2003), in which limestone is added first and then slaked lime to combine the two characteristics (e.g. Deul and Mihok 1967).

In finalising this section, a note on sludge stability: In many cases, the sludge generated in active or passive mine water treatment is not chemically stable, and re-dissolution of previously precipitated contaminants may occur (McDonald et al. 2006). Therefore, the sludge must be stabilised by immobilising the contaminants using one of the following methods: Fixation, encapsulation, or chemical immobilisation methods where specific chemicals are used (Jacobs and Pulles 2007). Zinck et al. (1997) recommend conducting laboratory tests for chemical stability even when it is understood that the results are subject to error, as the results may provide indications of relative sludge stability. Since each type of sludge has a different chemical and physical composition, it is a good idea to follow the recommendations in Zinck (2006) or Zhuang (2009) to stabilise the sludge. It could also be economically interesting to extract potentially valuable materials from the sludge (Demers et al. 2015; Zinck 2006).

### 3.2.2 Low Density Sludge (LDS) Process

Neutralisation of contaminated mine water using the low density sludge process is the oldest technical treatment technology for mine water (Fig. 3.6). It was intensively studied in the USA in the 1960s and 1970s to optimise the process (Skelly and Loy and Penn

**Table 3.5** Amount of alkali required to raise the pH of mine water at Horden Colliery (County Durham, England) to between 8 and 8.5 (from Croxford et al. 2004, Table 4). Meanwhile, the plant has been replaced by a passive water treatment system (Davies et al. 2012)

| | NaOH | $Ca(OH)_2$ | $Na_2CO_3$ |
|---|---|---|---|
| Necessary alkali concentration, g $L^{-1}$ | 0.4 | 0.37 | 0.53 |
| Predicted pH value | 8.27 | 8.23 | 8.24 |
| **Sludge volumes** | | | |
| $CaSO_4$, t per year | 2790 | 2760 | 2790 |
| $CaCO_3$, t per year | 40 | 40 | 40 |
| $Ca(HCO_3)_2$, t per year | 150 | 150 | 150 |
| Total, t per year | 2980 | 2950 | 2980 |

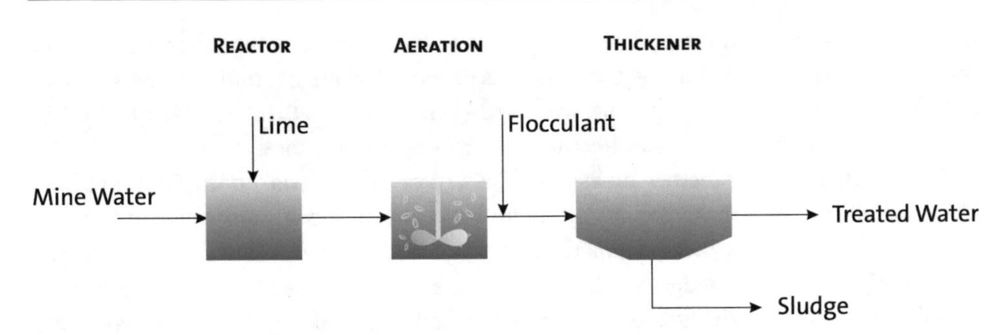


**Fig. 3.6** Principle of conventional mine water treatment (low density sludge process). "Hydrated lime" is used here to represent any usable alkaline material

**Table 3.6** Compilation of minimum pH values required to precipitate metals from mine water and to achieve concentrations below 1 mg $L^{-1}$ (without considering sorption or coprecipitation)

| To be precipitated metal | Minimum pH value required for precipitation |
|---|---|
| $Sn^{2+}$ | 4.2 |
| $Fe^{3+}$ | 4.3 |
| $Al^{3+}$ | 5.2 |
| $Pb^{2+}$ | 6.3 |
| $Cu^{2+}$ | 7.2 |
| $Zn^{2+}$ | 8.4 |
| $Ni^{2+}$ | 9.3 |
| $Fe^{2+}$ | 9.5 |
| $Cd^{2+}$ | 9.7 |
| $Mn^{2+}$ | 10.6 |

This table is repeated in a wide variety of publications without citing the original 1973 source (e.g., Brown et al. 2002; Jacobs and Pulles 2007; Skelly and Loy and Penn Environmental Consultants 1973, p. 269). Incidentally, the original authors point out that the results are based on only one study

Environmental Consultants 1973). The first step is to add alkaline material to the mine water (Table 3.3). This raises the pH of the mine water to a value at which the target constituent has its lowest solubility within the given $E_h$, pH, temperature, and pressure conditions (Fig. 3.5; Table 3.6). In the simplest case, the pH is adjusted by measuring the pH value and controlling the addition of alkali using this pH measurement.

In the next step, the mine water is aerated, whereby different methods can be used (details given in Sect. 4.9). In the simplest case, and if there is ample difference in elevation, a cascade is sufficient; in most cases, however, electric aerators are used (Fig. 3.7). If neither method provides sufficient oxygen to oxidise the iron in the mine water, strong oxidising agents (e.g. hydrogen peroxide) can be used (see Sect. 3.11). During oxidation, the iron hydrolyses and forms flocs, which take about 48 h to settle (PIRAMID Consortium 2003). To accelerate the process, mineral, organic or synthetic flocculants are added (Table 3.4) so that the iron flocs settle in the settler or lamella separator within a few hours. At the same time, further aeration improves the settling behaviour of the low density sludge in the settler by producing a more compact sludge (Coal Research Bureau 1971, p. 20 f.).

**Fig. 3.7** Aeration of mine water in a low density sludge plant (mine water treatment plant Schwarze Pumpe, Lusatia, Germany)

As the kinetics of the reactions taking place is often slow, the resulting sludge is usually still rich in alkaline material. This can lead to precipitates within the treatment plant, which must therefore be removed regularly (Younger et al. 2002). As a rule, the sludge from a low density sludge treatment plant has a solids concentration of a few percent (Aubé and Zinck 1999, p. 265; Coal Research Bureau 1971) (Table 3.2) and – as the comprehensive studies of the 1970s showed – can no longer be increased by process optimisation alone.

At the end of the process, the sludge must be dewatered, preferably by using filter presses (Fig. 3.8) or centrifuges to separate the water from the sludge and then the dried sludge can be disposed of. Normally, the sludge is not dewatered if it can be deposited in an underground mine or residual lake (see Sect. 6.4).

### 3.2.3 High Density Sludge (HDS) Process

There are many treatment processes that can be described as high density sludge processes in the broadest sense (Table 3.1). They all differ from one another in terms of detailed steps but are essentially based on the fact that part of the sludge produced at the end is reused at the start of the process. The classic high density sludge (HDS) process, which can have a sludge solids content of 20–40% (Aubé and Zinck 1999; Kostenbader and

**Fig. 3.8** Sludge dewatered by filter presses. (Acid rock drainage treatment plant Halifax Airport, Nova Scotia, Canada; image width approx. 1 m)

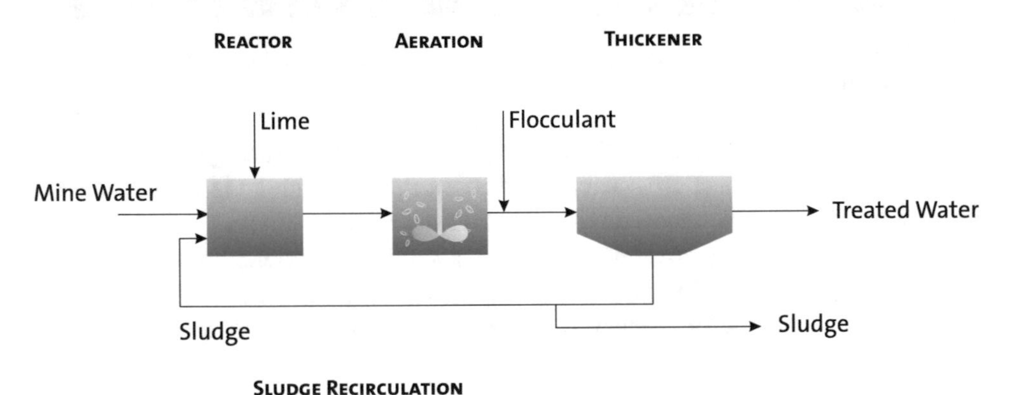


**Fig. 3.9** Principle of the classic high density sludge process with partial recirculation of the sludge. "Hydrated lime" is used here to represent any usable alkaline material

Haines 1970), is described here as an example. In general, the process offers advantages when the iron concentration of the mine water is relatively high or when higher concentrations of zinc, aluminium or copper occur in the mine water (Coulton et al. 2003b).

The hydrated lime and the recycled sludge are added to the mine water in the first reaction step (Fig. 3.9). This can either be done from separate silos (Fig. 3.10), or the recycled sludge is first mixed with the hydrated lime in a mixing silo and then added to the mine water in a first reactor (Coulton et al. 2003b). In this reactor, the mine water is mixed with the hydrated lime/sludge mixture, and there are different reactor designs that have slightly different mixing characteristics. The next step is to oxidise the mine water/hydrated lime/sludge mixture with atmospheric oxygen. This can be done in oxidation ponds, by cascades or aerators of various designs. Aeration prior to neutralisation is useful only if $CO_2$ is to be

**Fig. 3.10** Sludge collection tank of the former active Horden mine water treatment plant (County Durham, England, United Kingdom)

removed from the water; it does not accelerate the oxidation of $Fe^{2+}$ to $Fe^{3+}$ because the environment is too acidic to cause precipitation of iron oxyhydrates, as already recognised by Young and Steinman (1967).

Thereafter, a flocculant is added to the mixture, which is selected depending on the nature of the mine water. As a rule, anionic or cationic polymers are used today; these are usually supplied as a powder and mixed with water on site. However, there are also liquid polymers that can be added to the mine water without pre-treatment.

This approach increases the solids concentration of the sludge (typically to 20–40%), increases the particle size of the iron hydroxide particles, increases the settling velocity of the particles, and improves sludge dewatering. Due to this improvement in sludge properties, less space is required for liquid/solid separation ("dewatering"), which reduces the investment cost of the plant. In the case of the Wheal Jane active water treatment plant, for example, this reduced the footprint by 13% compared to an LDS plant (Coulton et al. 2003b).

Numerous studies in the past have shown that the high density sludge process can be further optimised. Above all, a gradual neutralisation has proven to be advantageous compared to a rapid increase in the pH value (Gan et al. 2005), but optimisation of the flocculant can also increase the effectiveness (Clark 2010).

► A low density sludge plant cannot simply be converted into a high density sludge plant in one afternoon. This was tried at a temporary low density sludge plant in northern Germany ("Northern Germany" from the perspective of a Franconian). There, a chief assessor had suggested once that the plant operator should take the sludge from the sludge settling basins with buckets, put the sludge into the aeration and mixing basin and see if this would increase the sludge density of the plant. In fact, the miner in charge of the plant spent an afternoon carrying sludge from the settling basins into the mixing basin, and after that afternoon determined that the plant could not be converted into a high density sludge plant and, consequently, that the mine water should continue to be treated by the low density sludge method. Even a larger high density sludge plant that is purpose-built for this type of treatment usually requires a certain run-in period before it operates optimally and meets all the design criteria. In the case of the New Waterford high density sludge plant on Cape Breton Island in Nova Scotia, Canada, the operators needed about 6 weeks to get the plant running optimally. This example from northern Germany shows that complex issues usually can only be solved with sufficient expert knowledge.

### 3.2.4 It Is Easier to Live in a Box

There are some process components in all the aforementioned neutralisation processes that would allow optimisation of mine water neutralisation. However, apart from the actual high density sludge process, these components are subject to patenting or licensing and can probably only be used by paying licensing costs. Why is it easier to live in a box? Because it's warm there, you're surrounded by the familiar, and you're largely protected from the atmospherics. Lift the lid, look over the edge of the box, and look around:

Let's consider five of these optimisation options: Hazleton Iron Removal System, staged neutralisation, the HARDTAC process, the OxTube of SansOx Oy, and glass beads.

In most cases, the neutralising agent is added to the mine water via a pipe. This can be the neutralising agent alone or a neutralising agent/sludge mixture. Hazelton Environmental have developed an apparatus, the MAXI-STRIP, which allows the mine water and neutralising agent to be intimately mixed (Brown et al. 2002). This passive apparatus breaks the neutralising agent into tiny droplets and swirls them with the mine water. As a result, slightly lower pH values than usual are required to precipitate the respective metals. There is not a single plant described in the literature that takes advantage of this effect to optimise the mixing of the alkaline material and the water. This may be because the process is called a "globular catalytic reaction", which almost sounds like something out of a magician's bag of tricks. Another component of the system is the passively operating Acid Mine Drainage Demineralizer (U.S. Patent 7,504,030 B2 dated March 17, 2009), which saturates the mine water with oxygen within a very short time. Apart from a few exceptions (e.g. Li et al. 2003b, p. 310), to date, this device has not been sufficiently mentioned in the technical literature, but it could considerably optimise the oxygen saturation of the mine water.

Staged neutralisation is used in a wide variety of treatment plants. In these, the pH value is increased in stages and different water constituents precipitate in each stage. This method proves to be optimal if different mine water constituents which have their solution minimum close together are to be removed simultaneously (Fig. 3.5). Overall, however, the process is rarely used because it is subject to a more intensive process control. Therefore, the goal must be to obtain a better process through optimised expert software and sensors. Many treatment plants could be optimised if it were possible to produce sensors that are reliable even in the harsh environmental conditions of a mine water treatment plant. One of the reasons why, for example, the technically complex F-LLX process (Sect. 3.12) has not as yet advanced beyond the pilot phase is probably that we are not yet able to use sensors for mine water that always operate consistently.

When neutralising mine water, it is important that the sludge settles well and dries easily at the end. In most cases, the particles in the sludge have a bimodal distribution, which you can interpret as an indication that it does not settle and dewater optimally. The HARDTAC process can be used to improve the behaviour of the sludge and to optimise dewatering (Barbier et al. 2008). However, the process is not only patented, but moreover details are kept secret. Figure 3.3a, b of the patent, as well as the explanations, do not really help to understand or reproduce the process (U.S. Patent No. 6355221 B1, March 12, 2012). In any case, the HARDTAC process increases the particle size (about 100 μm) and these are thus more easily filtered. See, you have another potential research focus now.

In water treatment, the supply of oxygen (aeration) is often inadequate as microorganisms use it and rapidly deplete the available supply. For these cases, the Finnish company SansOx Oy has invented and patented an inline device (patent EP 2796188 B1 of 6th April 2016) for saturating the water with oxygen. In the patent specification, it is called "Device for mixing an additive with liquid" and marketed under the name OxTube (Fig. 3.11). To date, the OxTube has been used in a Finnish vertical flow reactor, in fish farming (Hung 2017) at a German colliery for $H_2S$-removal and formerly in cleaning tin cans. The extent to which the OxTube and the MAXI-STRIP are based on the same principle cannot be said with any certainty – but the construction drawings of both apparatuses, which are based on the Venturi principle, suggest that they are.

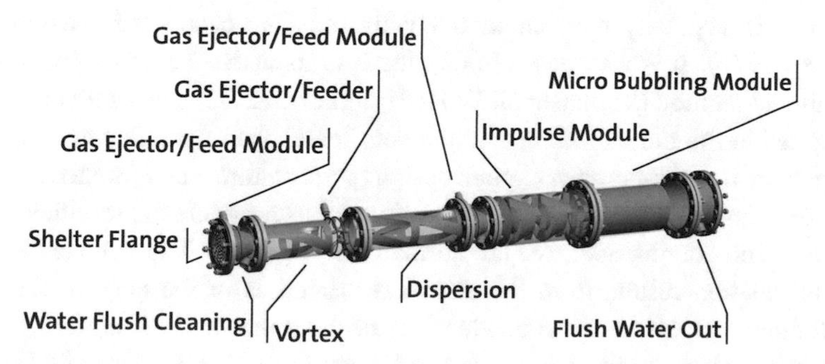


**Fig. 3.11** Schematic drawing of the OxTube of SansOx Oy. (Modified from the company brochure; see also Fig. 4.25)

Why glass beads? There are many treatment plants that use gravel filters at the inlet to exclude coarse particles, or as a final-stage polishing technique. However, this gravel is not always available in sufficient quantity or quality. In addition, the relatively rough surfaces of the gravel particles tend to accumulate microorganisms and lead to the formation of turbulent flows, which increase hydraulic resistance, thus preventing the optimum flow of the polluted mine water through the filters. Therefore, experiments have been carried out in well construction with glass beads as filter material, and consistently positive results have been shown to date (Treskatis et al. 2009). Perhaps you will succeed in devising a process that makes use of these properties of glass bead filters.

Now you've learned five possibilities to look for outside the box. Are you out yet? Very good, then we are a substantial step closer to a new procedure. Incidentally, I wonder how you ever got into the box in the first place. It's better to be outside the box, isn't it?

Let us now turn to processes that I personally consider to be forward-looking in mine water treatment. Not all the details of these processes are understood yet, but extensive research is currently being done and it is only a matter of time before industrial plants using these technologies for mine water are available. We are talking about electrochemical processes. And if your first thought should now be "too expensive", then please proceed reading, because these processes need not shy away from a comparison with the total costs of a neutralisation plant, including the final disposal of the sludge.

## 3.3 Electrochemical Processes

### 3.3.1 Electrocoagulation

Electrocoagulation or electroflocculation is understood here exclusively as an electrochemical method in which a current is passed through the mine water by anodes and cathodes (Fig. 3.12). The first descriptions of water treatment with electricity can be found as early as König (1899, p. 194). Although electrocoagulation is a common process in wastewater treatment to remove metals or turbidity from water (Koren and Syversen 1995; Mollah et al. 2001), it has rarely been used in the industrial treatment of mine water (see also Sect. 6.2.4). Typically, the comparatively high cost has been cited as a reason for this (Sisler et al. 1977). It is also possible that the less intensive scientific coverage of this water treatment method (Mollah et al. 2004; Siringi et al. 2012) means that it is less commonly applied in the mining sector – at the very least, plant manufacturers appear to be quite pragmatic in their approach to plant design (pers. comm. Kay Florence 2014; Hennie Roets 2016). Consumables in electrocoagulation are usually just the sacrificial electrode. Chemicals or anti-fouling additives are normally not required. The voltage necessary for optimal flocculation results from the electrical properties of the anodes, their distance from each other and the electrical conductivity of the mine water. Therefore, it is usually not necessary to carry out flocculation tests with variable voltages. A detailed comparison

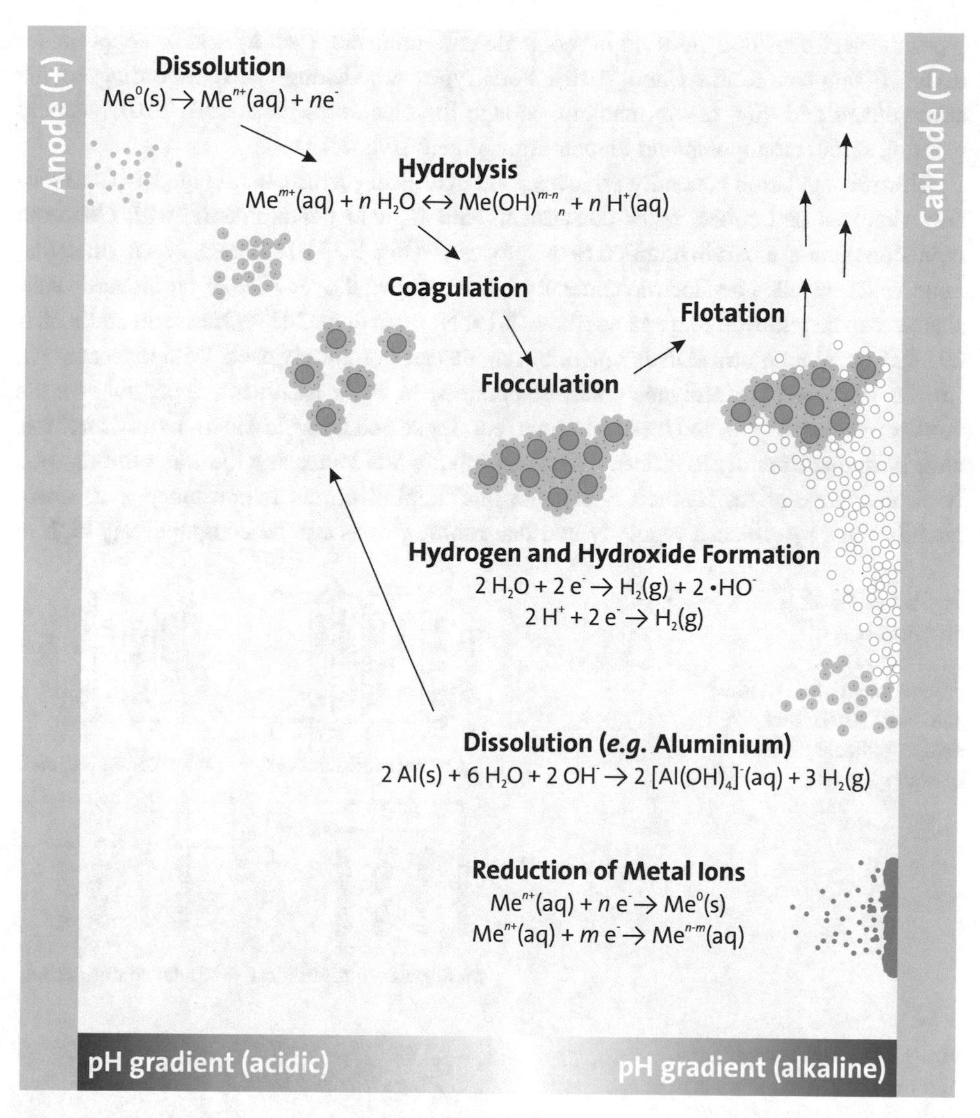


**Fig. 3.12** Principle of electrocoagulation. (Modified after Vepsäläinen 2012)

of the costs with other active mine water treatment processes is also still pending (pers. comm. Kay Florence 2013).

The electrodes used are usually plates, grids or tubes, more rarely rods or drums. These are classified according to their reactivity regarding organics and their arrangement in the reactor. In terms of reactivity, there are type 1 and type 2 electrodes, with type 1 corresponding to reactive electrodes and type 2 to non-reactive electrodes (Panizza 2010). Type 1 electrodes include carbon and graphite, platinum, iridium and ruthenium, and type 2 electrodes include doped antimony, lead and diamond. In addition, there are numerous types of electrodes that are used in inorganic environments, such as iron or copper electrodes (Comninellis and Chen 2010). Four types are distinguished according to the arrangement and flow of the medium through the electrodes (Liu et al. 2010), namely parallel, serial, monopolar, and bipolar arrangement (Fig. 3.13).

Electrocoagulation generally achieves good treatment performance (Fig. 3.14) and reliably removes particulate water constituents and organic contaminants, with Coulomb repulsion playing a part in addition to the processes (Fig. 3.12). Laboratory tests show that mine water can also be electrochemically treated and well over 80% of problematic substances can be removed (e.g., Mamelkina 2019; Nariyan et al. 2017; Orescanin and Kollar 2012). Iron, aluminium, zinc and platinum anodes are commonly used. With respect to Al, Cu, Fe, Zn, good performance could be obtained in water treatment, depending on the voltage, electrode type, and treatment time (e.g. Jenke and Diebold 1984). Lately, attempts have been made mainly to reduce the energy input while increasing the water throughput. A disadvantage of the method is that the sacrificial electrode is consumed and consequently must be replaced regularly and that running costs can be comparatively high in

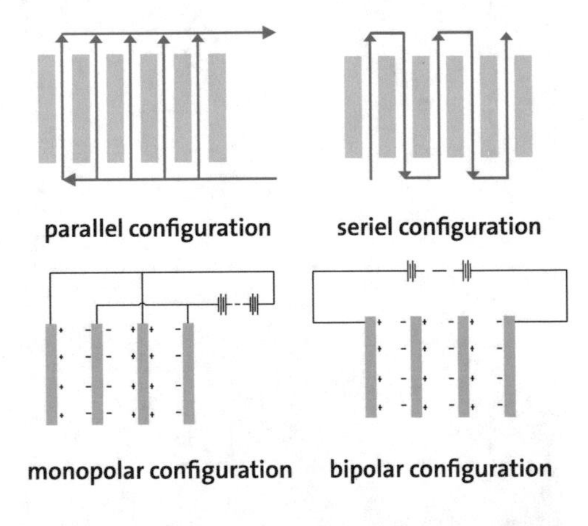


**Fig. 3.13** Possible arrangements of electrodes in electrocoagulation (After Liu et al. 2010). Blue arrows indicate the flow of water

**Fig. 3.14** Electrocoagulation of mine water on a laboratory scale. *Left*: Raw water; *right*: mine water after electrocoagulation. In reality, these plants are at least the size of garden sheds

areas with high energy costs. However, Rodriguez et al. (2007) showed that electrocoagulation can be operated cheaper by a factor of ten in direct comparison with sodium hydroxide neutralisation. Another advantage of electrocoagulation could be its use in the recovery of raw materials from mine water. Dinardo et al. (1991, pp. 19–22) summarise the results of successful experiments selectively recovering metals from mine water. However, detailed investigations are still necessary to apply the method on a larger scale.

Most commonly, acid mine drainage or synthetic acid mine drainage has been electrochemically treated to date (Chartrand and Bunce 2003; Jenke and Diebold 1984; Nariyan et al. 2017). In the Free Ukraine, a pilot plant was installed in 2006 at the "Glubokaya" colliery (Глубокая, Triplett et al. 2001) near Donetsk in the Donets Basin, which treats circumneutral mine water in a five-stage process (Kalayev et al. 2006). The process (Fig. 3.15) with an upstream ozonation of the mine water is patented (patent RU 2315007 C1 dated 20 January 2008) and the four-cell sedimentation tank used in the third step is an optimised Free Ukrainian invention (Horova et al. 2011). Information on how the plant is running today could not be found in the literature, e-mails to the authors remained unanswered, and – for understandable reasons – I did not want to visit the Russian occupied plant personally.

A pilot plant using the patented "Rigby Process" was running at Cement Creek in Colorado, USA, in September 2012. The mine water there is contaminated with zinc, copper, lead, cadmium, manganese and mercury and flows from the American Tunnel into Cement Creek and further into the Animas River (Mast et al. 2007; Sumi and Gestring 2013, p. 45). It has been shown that electrocoagulation can be used for treating this mine water, with sludge solids concentrations ranging from 30% to 70% (Rodebaugh 2012).

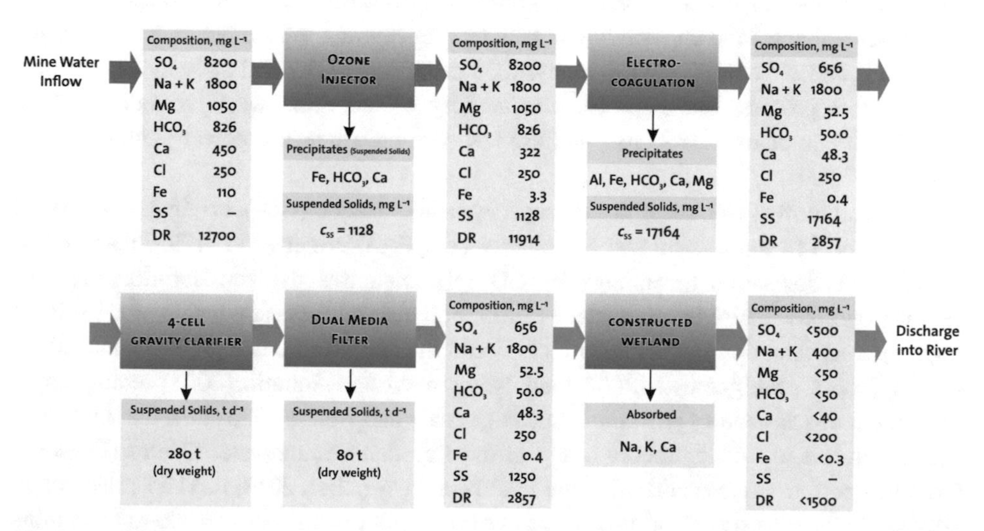


**Fig. 3.15** Block diagram of the mine water treatment plant with electrocoagulation at the *Glubokaya* (Глубокая) coal mine near Donetsk (Донецьк) in the Donets Basin in Free Ukraine. (Modified from Kalayev et al. 2006). *SS:* suspended solids, *DR:* dry residue without suspended solids

However, the plant only ran for about a week and then became so heavily clogged that it could no longer be operated. This was possibly due to the patent holder's insufficient experience with the process (pers. comm. William Simon, 11th August 2013; Fig. 5.2), in which iron oxyhydrates can easily precipitate between electrodes. A side note: On August 5th, 2015, 11,000 m$^3$ of acid mine drainage burst from the mine portal of the Gold King mine during remediation work, contaminating the receiving waters mentioned above (U.S. Department of the Interior – Bureau of Reclamation 2015, Clark and McCormick 2021).

In Serbia, it has been demonstrated at the "Robule" lake in the Bor ore district that mine water can be electrochemically treated (Orescanin and Kollar 2012; Stevanović et al. 2013).

### 3.3.2 Electrosorption (Condensation Deionisation)

Electrosorption is a relatively young technique in the field of mine water, which has been brought to industrial maturity within the past decade, mainly in China (Hwang and Sun 2012). Unlike reverse osmosis, which removes *water* from contaminated mine water, electrosorption removes *ions* from the mine water. To date, it has been applied primarily to desalination of brackish water or seawater (Mossad and Zou 2011; Seed et al. 2006). As with electrocoagulation, no chemical additives are required for the process, and since no membranes are used, there is for example no fouling. In electrosorption, water flows past anode and cathode plates, with cations being retained by the cathode and anions by the anode. To prevent electrolysis of the water, the system is operated at low voltages of between 1 V and 1.5 V. Periodically, the voltage is turned off and the electrodes are regenerated, knocking off the deposited ions with the concentrate. At the No. 3 colliery of the Chinese Yanzhou Coal Mining Company, 8000 m$^3$ of water is treated daily by electrosorption. In this process, the electrical conductivity of the mine water is reduced from 3.4 mS cm$^{-1}$ to below 2 mS cm$^{-1}$ and used as cooling water in a power plant (Sun and Hwang 2012).

A special form of electrosorption is capacitive deionisation (CDI), originally developed at Lawrence Livermore National Laboratory (LLNL) (Farmer et al. 1995, patented as Capacitive Deionisation Technology™ – CDT™). In contrast to electrosorption, this process uses porous electrode plates made of a carbon aerogel that are capable of selectively trapping ions at an optimal voltage of 1.2 V, and the concentrate is regularly knocked off in this process. Welgemoed (2005) and Welgemoed and Schutte (2005) attempted to implement this laboratory procedure on a large scale and were able to show that industrial use is possible. Another process tested at the Dresden Groundwater Centre (Dresdner Grundwasserforschungszentrum) is the DesEL process (Bilek 2013, p. 117 f.; Seed et al. 2006), which, however, could not yet be used on a pilot or industrial scale to treat mine water. According to Bilek (2012, p. 36), patent restrictions are said to hinder further development of the process. Consequently, there is as yet no full-scale experience regarding long-term use or the service life of the electrodes.

As experience in China shows, electrosorption can be used in industrial applications. Nevertheless, it is necessary to optimise the process in such a way that discharge limits are met even with high TDS concentrations.

### 3.3.3 Electrodialysis/Membrane-Based Electrolysis

Electrodialysis (ED) slowly seems to become established in mine water treatment. Overall, the number of plants in operation is still small (Clark and Muhlbauer 2010), but in special cases it is the economically and ecologically optimal solution. First experiences with electrodialysis of mine water were described from the USA, South Africa and others (Grebenyuk et al. 1979; Hill 1968; Volckman 1963). In the past two decades, the method has been intensively studied for its applicability to mine water (e.g. Biagini et al. 2012; Buzzi et al. 2011a, b; Schoeman and Steyn 2001; Turek 2003, 2004; Turek et al. 2005). In addition, an electrochemical process was used on the International Space Station (ISS), where the technology was copied from a mine water treatment system (Sparrow et al. 2012). In Lusatia, Germany, the Verein für Kernverfahrenstechnik und Analytik Rossendorf (VKTA) developed the trademarked and patented RODOSAN process based on the principle of membrane electrolysis (Bilek 2012; Stolp and Kiefer 2009). A special type of electrodialysis uses ion exchange membranes, separating cations and anions of the feed solution (Kinčl et al. 2017; Man et al. 2018). This technique eliminates the traditional problems encountered in reverse osmosis or electrodialysis systems. In electrodialysis, a current is passed into the mine water, causing the ions to move to either the cathode or anode, depending on their charge. Pairs of cation- or anion-selective membranes made of a polyelectrolyte are placed between the electrodes (in industrial plants there are usually several hundred pairs of membranes next to each other). Cation-selective membranes allow cations to pass, while anions are retained. The same applies, in reverse, to anion-selective membranes. When the current is switched on, a potential difference is created between the electrodes, causing the negatively charged anions to move to the anode and the positively charged cations to the cathode. Since the ions are retained at the selectively acting membranes, a concentrate with anions is formed at the anode and one with cations at the cathode. Due to the alignment of several alternating anion- and cation-selective membranes, a concentration of ions occurs on the way of the ions to the electrodes (concentrate). Between the electrodes, on the other hand, a mine water depleted of constituents is formed (diluent). Fouling can be prevented by changing the charge at the anodes, since the charge change produces a partial self-cleaning of the electrodes. However, sulfate or colloids can lead to deposits on the membranes and must therefore be separated at high concentrations in a preliminary stage. As with other membrane-based processes, pre-treatment of the mine water may be necessary, especially at high sulfate concentrations (compiled in part from Bowell 2004; Dill et al. 1998, p. 332; Strathmann 2012, p. 447 f.). In membrane-based electrolysis, it is not the migration of the ions that is relevant for pollutant removal, but the electrochemical reactions at the cathode or anode. The interposed membrane merely prevents undesirable chemical reactions at the electrodes (Hartinger 2007, p. 491 f.).

In a comparison with a mine water treatment plant at the Dębieńsko coal mine in Poland, it was shown that electrodialysis in addition to the existing processes requires considerably less energy than the respective process on its own for all alternatives investigated. Upstream electrodialysis prior to evaporation or crystallisation is ultimately the more cost-effective option (Turek 2004; Turek et al. 2005). For a synthetic mine water similar to that of the Bolesław Śmiały colliery, Poland, Mitko et al. (2021) showed that nanofiltration and reverse osmosis must precede the mine water electrodialysis. At the Beatrix gold mine in South Africa, it was possible to remove 80% of the total mineralisation of the mine water and to recover 84% as process water (Juby and Pulles 1990).

At Lake Ilse in Lusatia, Germany, the VKTA tested the RODOSAN process at the Rainitza mine water treatment plant, a membrane-based electrolysis process with coupled $CO_2$ injection that treats the acid mine drainage there. The development was financed by the Lausitzer und Mitteldeutsche Bergbauverwaltungsgesellschaft (LMBV), Vattenfall and the Saxon State Office for Environment, Agriculture and Geology (LfULG) from the EU project VODAMIN and was carried out in cooperation with Uhde GmbH in Dortmund. After basic research starting in 1995 and a pilot installation between 2006 and 2007 (Friedrich 2016; Friedrich et al. 2007), four different mine waters from Lusatia (Sedlitz, Tzschelln, Bockwitz, RL-107 [Plessa clay pit]) were treated in a second pilot phase at field scale from 2010 to 2012, revealing clear differences in the treatment performance. These were mainly due to the mineralisation, pH and base capacity of the different waters. At the same time, the process was intended to produce ammonium sulfate fertiliser. However, since ammonium passed several times into the treated mine water on the cathode side during the experiment, the LMBV and Vattenfall stopped further funding in February 2013 (process presentation on 26th February 2013 in Senftenberg, Germany).

## 3.4 Membrane-Based Processes

### 3.4.1 Introduction

When using membrane-based processes as a treatment method, the mine water is forced through filtration membranes under high pressure. In this process, the water (permeate) and the substances to be removed (retentate) are separated from each other in the feed, with separation efficiencies of 70–99.99% being achievable (Koros et al. 1996; Schäfer et al. 2006). Depending on the pore size of the membrane, a distinction is made between four basic membrane-based processes (Fig. 3.16), which are characterised by a number of properties and are not easily interchangeable: Microfiltration, ultrafiltration, nanofiltration and reverse osmosis (Melin and Rautenbach 2007; Peters 2010) (Table 3.7). While the membranes in microfiltration and ultrafiltration retain the constituents mainly due to size exclusion, the ions in nanofiltration and reverse osmosis migrate diffusively through the

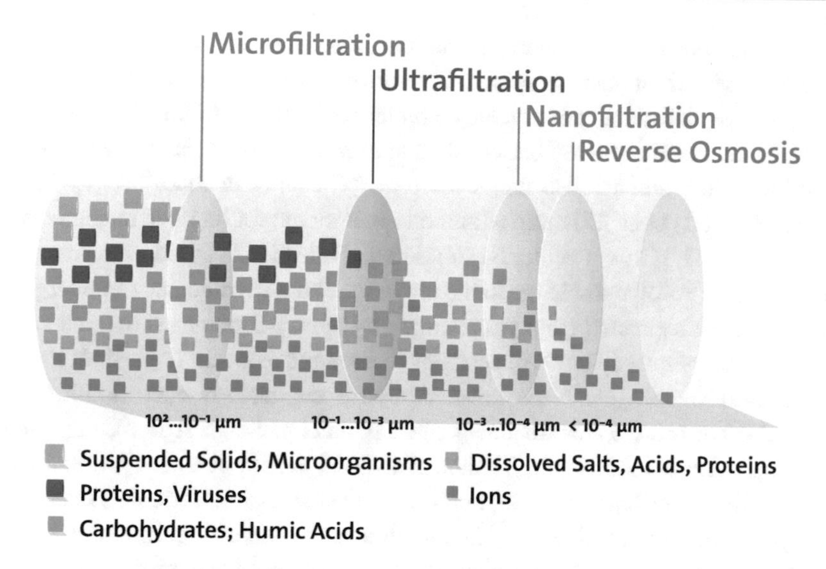


**Fig. 3.16** Separation limits of different membrane filters and selection of retained constituents. Spacing of the membranes in the graph corresponds to the logarithmic pore width. Retention of constituents based on Schäfer et al. (2006, p. 2)

**Table 3.7** Selected characteristic properties of membrane processes as well as electrodialysis (compiled from the sources mentioned in the text)

| Procedure | Pore width, µm | Working pressure, MPa | Retentate | Relative energy expenditure |
|---|---|---|---|---|
| Microfiltration | 0.1…10 | < 0.2…0.5 | Microorganisms, solids, colloids | Low |
| Ultrafiltration | 0.001…0.1 | 0.1…0.7 | Macromolecules, viruses, colloids | Low |
| Nanofiltration | 0.0001…0.001 | 0.3…2.1 | Organic compounds, polyvalent ions | Low to medium |
| Reverse osmosis | < 0.0001 | 1.6…6.9 | Monovalent ions | Medium to high |
| Electrodialysis | – | – | Monovalent or multivalent ions | Low to medium |

The working pressures and pore widths vary depending on the author. In the case of reverse osmosis, working pressures of up to 8 MPa are reported in the literature

solution-diffusion membranes (Strathmann 2012). To increase the lifetime of the membranes and to prevent fouling, chemical addition is often necessary (Melin and Rautenbach 2007, p. 336 f.). Antiscalants prevent carbonate, silicate or sulfate precipitation products from settling on the membranes (scaling), and biocides are often additionally necessary to prevent biofouling (deposition of biofilm on the membrane surface). Under optimal conditions, and if microorganisms are removed by microfiltration at the start of the process, the use of antiscalants or biocides can be omitted. However, the membranes must be regularly cleaned with treatment chemicals to remove other harmful contaminants such as colloids ("inorganic fouling") or organic components ("organic fouling"). At equally regular intervals, the membranes are backwashed to remove deposits on the membranes. Both processes are often summarised under the term "Cleaning in Place" (CIP) (Buhrmann et al. 1999; Chesters et al. 2016). In addition to membrane fouling, the retentate is a problem because it is a highly concentrated solution that must be disposed of in an appropriate manner. This may involve discharge to the sea, a receiving watercourse or underground, or the metals may be separated from the retentate by appropriate means (e.g. chemical precipitation) for further use in terms of sustainable resource recovery (e.g. Bilek 2012; Drioli and Macedonio 2012). At the New Vaal coal mine (Free State Province, South Africa) and at the eMalahleni mine water treatment plant (Gauteng, South Africa), the retentate is treated using freeze crystallisation (Nathoo et al. 2009; Reddy et al. 2010) (see Sect. 3.13). Membranes also contribute to the separation of substances in electrodialysis, so that this technology is often included among the membrane processes (e.g. in Koros et al. 1996, p. 1487 f.). However, since the electrical current is the decisive contribution in electrodialysis, a discussion can be found in the chapter "Electrodialysis/membrane-based electrolysis" in Sect. 3.3.3 (Hartinger 2007, p. 491 f.).

Membrane-based processes have gained considerable importance in mine water treatment in recent decades (Cartwright 2012; Drioli and Macedonio 2012). This can essentially be attributed to the following reasons:

- Optimised membranes
- Cheaper production of the membranes
- Reduction of energy consumption
- Higher water quality requirements
- Greater importance of environmental protection
- Purely physical separation process
- Mass-marketable
- Can be connected intelligently

The increase in installed capacity is likely to be similar to that in the case of seawater desalination plants. Whereas in the mid-1960s there were no commercial plants installed, today there are around 20,000 desalination plants with a total capacity of around $100 \cdot 10^6$ $m^3$ per day (www.Desalination.com, www.idadesal.org), with a much faster increase in installed membrane filtration plants evident from the 1990s onwards. From the number of publications related to mine water, it can be seen that this is also true for the mine water sector. Due to the immense number of publications about water treatment with membranes, it is not possible to provide a comprehensive overview in this book. In purely arithmetical terms, 18 articles on water treatment with membrane processes appeared weekly in the Clarivate Analytics Science Citation Index Expanded (SCIE) in 2020 (Fig. 3.17).

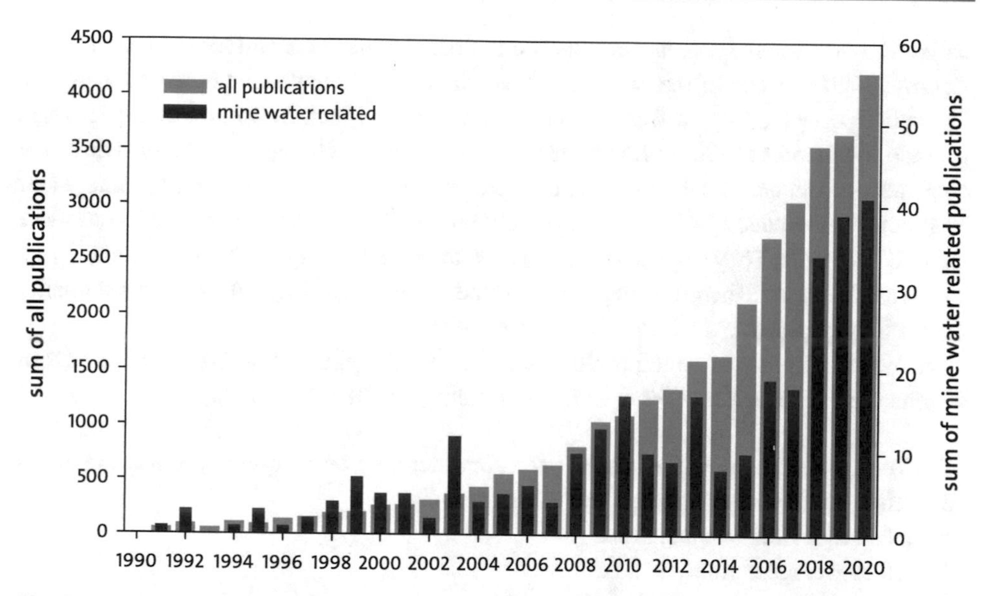


**Fig. 3.17** Number of publications on water treatment using membrane technology between 1990 and 2020. On the right-hand axis, the number of those that are relevant to mine water. (Source: Clarivate Analytics Science Citation Expanded [SCIE])

Until the late 1980s, there are few articles that discuss membrane technologies for mine water treatment. It is only since 1989 that the SCIE has listed 64 articles discussing this topic. This seems to reflect the development of the technology, as in the early 1990s the technology was making its way into mine water treatment, which until then was rarely used due to the high cost of energy (Schoeman and Steyn 2001, pp. 18–20, 29–30). Today, a combination of membrane-based processes is mainly used, with microfiltration and ultrafiltration usually being a precursor to nanofiltration, reverse osmosis or electrodialysis to protect the membranes from being damaged by larger particles. One of the main differences between the technologies, apart from the pressures required to separate substances, is the pore size of the membranes and hence their ability to remove substances from the water (Figs. 2.3 and 3.16). In the membrane-based processes commonly used today, the mine water is operated in tangential or cross-flow (cross-flow filtration). In this process, the water flow is pressure-driven *past* the filter and does not pass *through* the filter as in dead-end filtration ("coffee filter extraction"). In addition to mine water treatment, membrane-based processes are also used for seawater desalination for drinking water or process water as well as in ore processing in mine workings or open pit mining.

One of the main problems of membrane-based processes arises from the disposal of the concentrates and the chemicals used to flush or reactivate the membranes (Drioli and Macedonio 2012). For example, at eMalahleni or Kromdraai in Gauteng Province, South Africa, the concentrate is collected in large basins where natural evaporation is used to remove the water (see Sect. 7.2), a process which can be improved by granular activated carbon-based photothermal layers, attached to floats on the water surface (Monasmith and Zodrow 2020). At one of the two sites, research was underway some time ago using the HybridICE® technology, a freeze desalination process (Sect. 3.13) and an algae-based process (Nathoo et al. 2009; Reddy et al. 2010). These technologies aim to reduce the concentrate volume. In China, 77 coal mines use membrane processes for mine water treatment, thereunder 10 of the 13 largest coal mines with an annual production of more than 1.2 Mt. each. They regularly encounter water shortages, specifically in northern and western China, and therefore depend on mine water recycling with the water quality required to meet specific criteria (Wang et al. 2021).

A reverse osmosis or nanofiltration system is usually planned in stages, the details of which are as follows (Dow Water & Process Solutions 2013, pp. 83–86):

1. Consider feed source, feed quality, feed/product flow, and required product quality.
2. Select the flow configuration and number of passes.
3. Select membrane element type.
4. Select average membrane flux.
5. Calculate the number of elements needed.
6. Calculate number of pressure vessels needed.
7. Select number of stages.
8. Select the staging ratio.
9. Balance the permeate flow rate.
10. Analyse and optimise the membrane system.

Based on technological development and the increasing requirements for environmental protection, it can be assumed that, in addition to electrochemical processes and ion exchange, membrane-based processes will probably be the method of choice for treating mine water in the medium to long term. If it comes to total dissolved solids (TDS) being the discharge criterion for discharged mine water, neutralisation may no longer achieve the necessary discharge criteria.

Nearly all major membrane plant manufacturers have computer programs that can be used to design the plant (DuPont Water Solutions: Water Application Value Engine [WAVE]; SUEZ Water Technologies & Solutions: Winflows; French Creek Software: hyd-RO-dose). It is also a good idea to design integrative plants where pre-treatment and post-treatment of mine water are an integral part of the plant (Drioli and Macedonio 2012; Peters 2010). This results in the following advantages:

- Less fouling and thus longer service life of the membranes
- Lower energy consumption
- Less brine and therefore more environmentally friendly operation

By means of membrane-based processes in combination with conventional processes, the sulfate loads of mine water can be substantially reduced. The statement by Uhlmann et al.

(2001, p. 55) that "from a practical and economic point of view, no process is available for reducing the sulfate concentration for these dimensions" can therefore no longer be agreed with today. The examples of eMalahleni, Kromdraai or the Copper Basin Project show that mine water treatment is possible even for large flows (Faulkner et al. 2005). In addition, the SPARRO process is a pilot-scale method that can also treat acidic sulfate-rich mine water (details in Sect. 3.4.6).

Examples from copper, gold or uranium ore processing prove that membranes can optimise the ore processing process of polyvalent metals (Soldenhoff et al. 2006, p. 469 ff.). For example, nanofiltration membranes can selectively separate gold-cyanide complexes from other metal-cyanide complexes or to enrich copper in copper liquors.

In the past decade, there has been research with selective membranes capable of retaining specific metals (Arous et al. 2010; Fu and Wang 2011; Meschke et al. 2018; Werner et al. 2018). Industrial plants do not yet exist, but it is likely that in the future, these membranes could be used to recover some of our raw materials, while also providing a solution to the problem of high-salinity retentates. Should the development of these membranes proceed similarly to that of membrane separation as a whole, then around the year 2030 we should be extracting metal not only in the actual mines, but also in mine water treatment plants (Wolkersdorfer 2013; Wolkersdorfer et al. 2015, Fosso-Kankeu et al. 2020). Mine water would thus be transformed from a waste product into a raw material. In the next one to two decades, we therefore need research and applications on the following topics:

- How can metals be selectively extracted from the retentate?
- Development of ion-selective membranes
- Use of mine water as a raw material
- Reduction of membrane fouling
- Development of membrane technology using regenerative energy

Independently of the forecast given here for "Intelligent Mine Water Treatment", Bäckblom et al. (2010) have forecast a future mine for Boliden and LKAB in Sweden and KGHM in Poland, which they call "Intelligent Mine of the Future". Their target is "Vision 2030:> 30 by 2030", and the mine of the future should have over 30% less ore loss, use less energy, produce less $CO_2$, generate less waste and cause fewer accidents per tonne of ore extracted. For the time after 2030 they see mines with in situ production of non-ferrous metals, without waste and with fully automated mining without people at the mining face.

### 3.4.2 Microfiltration

Microfiltration removes particles between 0.1 and 10 μm in size from water. Working pressures of 0.2–0.5 MPa are generally used, and the membranes are relatively insensitive to external influences (Cartwright 2012; Peters 2010). Typically, the technology is used in the mine water sector to remove solids or bacteria before the mine water enters the reverse osmosis system (Buzzi et al. 2011a; Ericsson and Hallmans 1996; Harford et al. 2012; Lilley 2012).

### 3.4.3 Ultrafiltration

Ultrafiltration filters particles between 0.001 and 0.1 μm from the water to be treated and uses working pressures of 0.1–1 MPa (Cartwright 2012; Peters 2010). The process separates solids, bacteria and viruses and is often used in the mine water sector as a precursor to reverse osmosis (Karakatsanis and Cogho 2010; Knops et al. 2012; Fig. 3.18).

### 3.4.4 Nanofiltration

Nanofiltration refers to filtration processes that filter particles between 0.1 and 1 nm (1–10 Å) at working pressures of 0.3 to 2 MPa from the mine water. This makes the process suitable for removing multivalent ions such as sulfate (Schäfer et al. 2006; Visser et al. 2001), since the sulfate ion has a thermodynamic radius around 2.3 Å. If the membranes have additional positive or negative charges, charged ions will also be retained. The first attempts to treat mine water using nanofiltration date back to the 1950s in South Africa, whereas in Europe attempts have only been made since the mid-1970s (Motyka and Skibinski 1982; Turek and Gonet 1997). However, nanofiltration has been used on a

**Fig. 3.18** Ultrafiltration unit of the eMalahleni mine water treatment plant in Gauteng Province, South Africa

larger scale only since the early 1990s (e.g. Visser et al. 2001). Typically, membranes are made of polysulfone, polyethersulfone, cellulose acetate or a thin film composite. In addition, there are several ceramic membranes for various special applications (Cartwright 2012; Vankelecom et al. 2006). Before the term *"nanofiltration"* was introduced in 1984, the process was also known as "loose RO", "open RO", "tight UF" or "hybrid RO-UF" (R. J. Petersen in Schäfer et al. 2006, pp. xxf, 6).

Today, nanofiltration is often used in conjunction with reverse osmosis because the membranes used there must be free of major particulate contaminants. Biagini et al. (2012) describe a process in which brackish water for the drinking water supply of Alamogordo, New Mexico, USA, was treated in a multi-stage process that also included nanofiltration.

When the mines in the French Lorraine were closed and flooded at the end of the 1990s (Blachère et al. 2005), a mine water treatment plant was started in Jarny (Meurthe-et-Moselle, Lorraine), which works by means of nanofiltration (Bertrand 1997). This plant initially showed a rapid decrease in permeability, but this was resolved by cleaning the membranes every 6–8 weeks. In the years between 1995 and 1997, between 92 and 98% of the mineralisation was retained.

Nanofiltration can remove 90–95% of the most common uranium species from mine water, as Raff and Wilken (1999) demonstrated in a laboratory study and Benkovics et al. (1997) showed at an actually operating plant in Hungary. At the Bergakademie Freiberg and the BTU Cottbus, experiments with different mine waters were conducted in the years around 2008 to optimise the method with respect to mine water treatment (Härtel et al. 2007; Preuß et al. 2007). Preuß et al. (2012) showed that nanofiltration of acid mine drainage from Lusatia, Germany (membrane: Filmtec TM NF270; Dow Chemical Company) retains a large part of the TDS, including sulfate and calcium.

Rieger et al. (2010) used dead-end filtration to treat Chilean mine water at laboratory scale. They achieved retention rates of 87–99% for sulfate and 92–100% for iron. Visser et al. (2001) investigated whether acid mine drainage from South Africa could be treated using nanofiltration. Their results show that only some of the membranes studied can be used in acidic conditions. Among other things, they attributed this to a change in charge on the membrane surface. However, membranes with a lower throughput rate showed these limitations to a lesser extent at low pH values. In contrast, Preuß et al. (2007, 2010, 2012) investigated the retention of sulfate at circumneutral pH values (7.5–7.8) and obtained very good retention rates.

### 3.4.5 Reverse Osmosis (RO)

Reverse osmosis is a membrane filter technology that removes molecules and ions above 0.1 nm from mine water. Generally, the method operates at working pressures of 2–8 MPa (Cartwright 2012; Drioli and Macedonio 2012), with the required pressure depending on the osmotic pressure of the solution. For example, a mine water with TDS of about 30 g $L^{-1}$ requires a pressure of about 2.5 MPa for treatment (Bowell 2004). Reverse osmosis

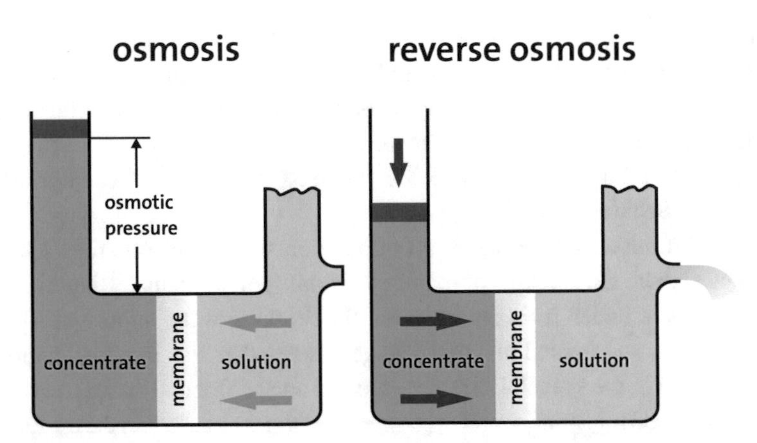


**Fig. 3.19** Simplified principle of osmosis (forward osmosis) and reverse osmosis. In osmosis, ions "migrate" through a semi-permeable membrane along a concentration gradient, thereby increasing the pressure on the concentrate side. In reverse osmosis, on the other hand, pressure is exerted on the side of the concentrate and the water ions diffuse ("migrate") through the membrane to the side of lower concentration

removes monovalent ions or smaller organic molecules from mine water, and retention rates of over 90% are usually achieved (Fig. 3.19). As early as the late 1960s, attempts were made in the USA to treat acid mine drainage by reverse osmosis (Wilmoth 1973). However, as with the other membrane processes, it took until the 1990s to establish the process. Today, it is the most widely used membrane technology for mine water treatment. One of the reasons for the success of the method is the approximately 60% lower energy costs of modern plants in relation to older plants (Melin and Rautenbach 2007).

In principle, every reverse osmosis system consists of the following four treatment stages:

1. Pre-treatment
2. Desalination (reverse osmosis)
3. Aftercare
4. Disposal or reuse of the retentate

Depending on local conditions and especially on the quality of the mine water, the above holistic concept must be adapted. Pre-treatment can thus consist of nano- or ultrafiltration, a multilayer filter, dissolved air flotation or neutralisation (Al-Zoubi and Al-Thyabat 2012; Drioli and Macedonio 2012; Karakatsanis and Cogho 2010; Knops et al. 2012; Schoeman and Steyn 2001).

As the plants installed thus far show, this approach has the advantage that it allows reverse osmosis to be combined with various other treatment technologies to form an integrated treatment system. In eMalahleni, South Africa (Fig. 3.20), for example, it is coupled with high density sludge precipitation and nanofiltration, whereas reverse osmosis

**Fig. 3.20** Reverse osmosis unit of the eMalahleni mine water treatment plant in Gauteng Province, South Africa. (See also Fig. 7.7)

in Kromdraai, South Africa (Optimum Coal Mine), manages with ultrafiltration only (Karakatsanis and Cogho 2010). However, combinations with other active treatment processes are also possible, as described by Drioli and Macedonio (2012). One of these combination processes is the SPARRO process described in Sect. 3.4.6.

Reverse osmosis will become very important where there is a shortage of freshwater for mines, and seawater can be desalinated and pumped to the mine. In Chile, for example, there has been a drought for some time, which requires building such seawater desalination plants for mining (Knops et al. 2012) and to pump the water several thousand kilometres inland. One of the first projects is the Minera Escondida (BHP Billiton, Rio Tinto, JECO Corporation), but a total of around 16 desalination plants are currently projected, and by 2026 half of all copper mines in Chile will probably rely on demineralised seawater. Another reason for their construction is the competition between mining, agriculture (e.g. vineyards, avocados) and tourism (World Water Assessment Programme 2012, p. 201 ff.).

Water shortages are also encountered in northern and western China, which requires coal mines to use RO for recycling their used mine water (Wang et al. 2021). At the Binchang coal mine, contaminated water is pre-treated with activated carbon and manganese sand filters before it enters the RO. On average, the two RO lines with a capacity of 150 $m^3$ $h^{-1}$ provide the mine with 5200 $m^3$ of freshwater per day. Desalination removes more than 99% of the contaminants, and the water recovery is above 75%, which saves the mine 0.7 million € of freshwater purchase costs annually. Another coal mine in northern Shaanxi uses UF pre-treatment and the annual savings of freshwater are 2.5 million €.

Reverse osmosis can also be used to remove explosive residues from mine water. For this purpose, Häyrynen et al. (2009) conducted tests with reverse osmosis and nanofiltration. While the reverse osmosis membranes recovered nitrate and ammonium at a rate of 92–98%, the nanofiltration membranes showed only low recovery rates.

Juby and Schutte (2000) observed hydrolysis of the membranes and suspected that this could have been caused by radionuclides. This may not be ruled out for the membrane they used (Membratek Tubular CA membrane), but the rather large number of successfully operating reverse osmosis plants that also remove radionuclides from the water argues against their hypothesis.

### 3.4.6 SPARRO Process (Slurry Precipitation and Recycle Reverse Osmosis)

A combination process developed in South Africa is the Slurry Precipitation and Recycle Reverse Osmosis (SPARRO) membrane process, patented in 1988, in which a wet slurry with gypsum seed crystals moves in the treatment circuit, acting as crystallisation nuclei for the supersaturated phases (the claim by Simate and Ndlovu 2014, and others, that the SPARRO process *s.s.* was developed in the USA must be strongly contradicted). This demineralisation process is based on a membrane process (reverse osmosis) where the wet sludge provides protection for the tubular cellulose acetate membranes as the solids crystallise on the gypsum nuclei. This has been successful in containing the undesirable deposits (scaling) on the membrane and maintaining a retention rate of 95% (Juby 1992; Juby and Schutte 2000; Juby et al. 1996; Lorax Environmental 2003; Pulles et al. 1992; Seewoo et al. 2004). The gypsum seed crystals can be recovered at the end of the process (Fig. 3.21). The reason for this procedure is that excessive concentrations of calcium and sulfate in the

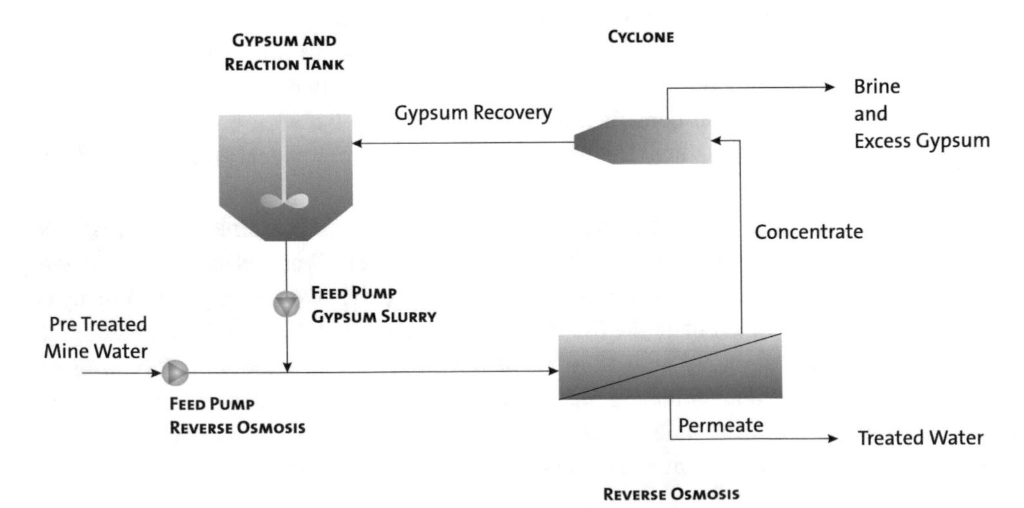


**Fig. 3.21** Flow chart for the SPARRO process. (Modified after Juby and Schutte 2000; Juby et al. 1996)

mine water can cause damage to the membranes. A precursor to this process is seeded reverse osmosis (SRO), which was actually developed in the USA and uses only the gypsum crystals as the crystallisation nucleus (Harries 1985; Juby and Schutte 2000).

As with other membrane processes, however, fouling of the membranes cannot be completely prevented. This leads to the fact that the minimum flow rate of 550 L $min^{-1}$ was not permanently achievable in the pilot plant and dropped to 300 L $min^{-1}$. One reason for the decrease in flow rate may have been suspended $SiO_2$. In total, this plant ran for about 5000–6000 h, testing four phases with different plant configurations.

In the SPARRO process, metals can be separated in an upstream separation by increasing the pH value to 10 with NaOH, oxidation with $KMnO_4$ and final filtration of the water after settling. For cost reasons, the pre-treatment had to be discontinued during the pilot test. Finally, to preserve the membranes in reverse osmosis, the pH must be lowered to 5–6 with sulfuric acid. Overall, it was possible to reduce the initial sulfate concentration from 5560 ± 1492 mg $L^{-1}$ to 279 ± 127 mg $L^{-1}$ (error margins are the standard deviations from 19 measurements).

Bowell (2004) concludes that the SPARRO process without pre-treatment of the mine water is currently only economically viable to a limited extent. The results of the pilot plant also confirm this finding; membrane failure was frequently shown in the pilot plant, which was due to the sometimes high suspension load of the mine water (Juby et al. 1996). However, this problem can be solved by upstream ultrafiltration or microfiltration, as shown by the eMalahleni example, where the concept of reverse osmosis with upstream treatment stage and current technology is used on an industrial scale. What a contrast to 1976 when Train et al. (1976, p. 251) wrote: "Meaningful reuse or contamination-free discharge cannot be achieved for mine water from coal mines."

Overall, the process has considerable potential for application, and through further research and process intensification, the problems encountered to date can be solved.

### 3.4.7 Forward Osmosis (FO)

Although forward osmosis is not a new process, it has not been widely used in the treatment of mine water. Recent examples exist in Finland and Australia (Thiruvenkatachari et al. 2016), as well as in pilot treatment of municipal wastewater (Wang et al. 2016) and a variety of other processes listed by Thiruvenkatachari et al. (2016). While reverse osmosis applies a pressure $\Delta p$ to move the contaminated water through a semipermeable membrane, forward osmosis uses the principle of osmosis itself (Fig. 3.19). Consequently, the driving force is the osmotic pressure, which results from the difference in the electrochemical potential (Melin and Rautenbach 2007, Ali et al. 2021) $\Delta\pi$ rather than the applied pressure difference $\Delta p$ (Cath et al. 2006). To generate this potential difference, a draw solution is used on the permeate side, which due to its properties does not permeate the membrane and produces a high osmotic potential (Fig. 3.22). A variety of draw solutions have been used to date, including NaCl, seawater, glucose, fructose or ammonium salts

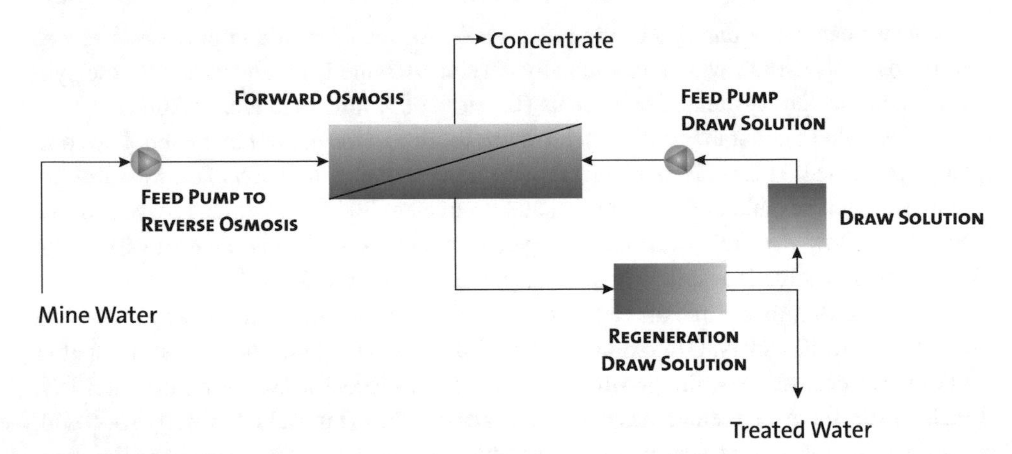


**Fig. 3.22** Schematic diagram of forward osmosis. (Modified from Cath et al. 2006; adapted from the idea of Weichgrebe et al. 2014)

(Cath et al. 2006). What they all have in common is that they must be recoverable to be able to carry out the process economically.

One of the methodological advantages is that many contaminants can be removed, lower or no pressures are used, and usually less biofouling of the membranes occurs (Cath et al. 2006). Whether forward osmosis can be used as a standard treatment method for mine water in the future cannot be predicted at this time, as too little experience is available to enable a reliable assessment. Weichgrebe et al. (2014) conclude that the methods for flowback fluid disposal in fracking "have substantial potential to be used on a large scale in the future," which they presumably infer from the increasing number of published articles. However, to move from laboratory to field scale, there is an urgent need to develop novel membranes and draw solutions (Zhao et al. 2012a).

## 3.5 Precipitation Methods for Uncommon Contaminants

Some contaminants require special precipitation methods, particularly for mine water that is less mineralised or contains little dissolved iron. These contaminants include As, Sb, Mo, Ra or Se, which do not usually precipitate as metal hydroxides or carbonates (Senes Consultants Limited 1994). However, provided sufficient iron concentrations are present in the mine water, coprecipitation of some of these contaminants is possible during neutralisation.

Arsenic, antimony or selenium can be precipitated with divalent or trivalent cations such as $FeCl_3$ or $FeCl[SO_4]$, as they coprecipitate with the ferrihydrite formed. Optimal conditions for arsenic occur when the Fe-to-As ratio is 2 to 1 or greater. To obtain a stable sludge, this ratio should tend to be greater (Golder Associates Inc. 2009; Senes Consultants Limited 1994).

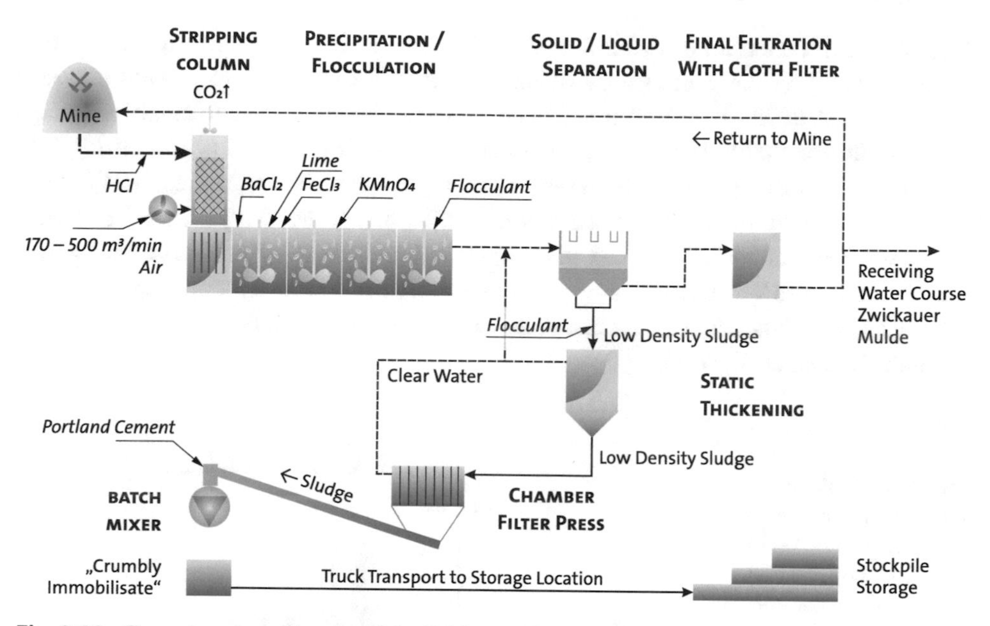


**Fig. 3.23** Current process diagram of the Schlema-Alberoda, Germany, mine water treatment plant including residue disposal. (After Fig. 3.4.2.1–23 in Badstübner et al. 2010)

Radium can be precipitated from mine water using $BaCl_2$ or $BaSO_4$, as at the Wismut GmbH plant in Bad Schlema, Germany (Badstübner et al. 2010, pp. 3.4.2–32), forming $Ba(Ra)SO_4$ (Fig. 3.23). However, the resulting sludge is unstable under reducing conditions, and the Ra can be released again (Jacobs and Pulles 2007). Chałupnik and Wysocka (2000) first used this method underground in Silesia to precipitate radium from saline mine water.

Zinchenko et al. (2013) removed uranium and molybdenum from mine water at Cameco Corporation's McArthur River uranium mine in Saskatchewan, Canada. They added lime milk and then sulfuric acid to the mine water in a two-stage pilot plant, adding $BaCl_2$ and $FeSO_4$ at each stage. They obtained treatment rates of 96% for uranium and 99% for molybdenum.

## 3.6 Ettringite Precipitation

### 3.6.1 SAVMIN™ Process

The formation of the tricalcium aluminate trisulfate hydrate ($Ca_6Al_2[(OH)_{12}|(SO_4)_3]\cdot 26\ H_2O$), commonly called ettringite, during mine water treatment was first reported by Bogner and Doehler (1984). They had observed that if pH values in a mine water treatment plant were accidentally set too high, precipitation of ettringite could occur, which they thus considered to be an indicator of consistently high pH values in the plant. The type locality of ettringite is the Ettringer Bellerberg in the Eifel region of Germany; this mineral forms naturally in the contact zone of metamorphosed carbonates with volcanic rocks (Anthony et al. 2003). It is also formed in concrete and can contribute to material damage there (Bollmann 2000). However, in the treatment process presented below, ettringite is deliberately formed by raising the pH to 11–14 to precipitate ettringite (Fig. 3.24) – consequently, the information in Uhlmann et al. (2001, p. 48) that the pH must be at least above 9.5 does not apply to a mine water treatment plant. One advantage of ettringite is that it incorporates "undesirable" ions into its crystal lattice by coprecipitation *s.s.* (Table 3.8) and thus removes, for example, Cr, As and Cd in addition to sulfate and aluminium (Gougar et al. 1996).

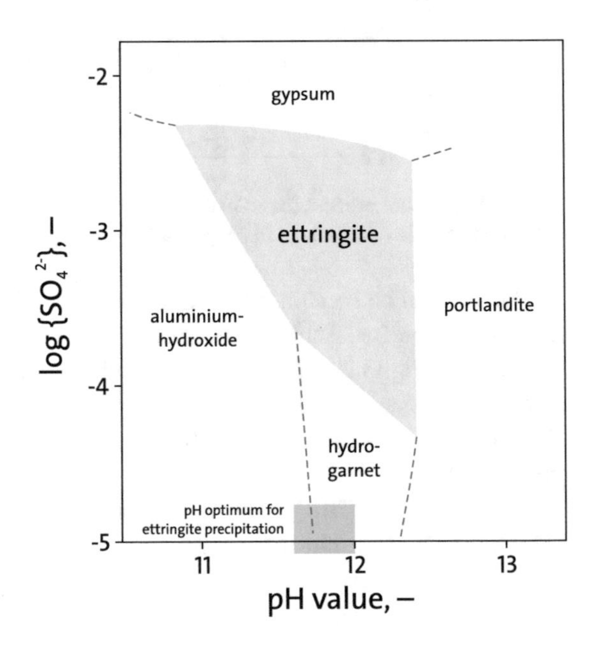


**Fig. 3.24** Stability diagram of ettringite in the system $CaO/Al_2O_3/SO_3/H_2O$; projected into the $Al(OH)_4$ plane not visible in the graph, $\{Al(OH)_4^-\} = 0$. (Modified after Hampson and Bailey 1982, Fig. 1; optimal pH range of 11.6–12.0 added after Smit 1999)

**Table 3.8** List of ions that can substitute lattice sites (diadochic substitution) in ettringite ($Ca_6Al_2[(OH)_{12}|(SO_4)_3]\cdot 26\ H_2O$)

| $(Ca^{2+})^{VIII}$ lattice site | $(Al^{3+})^{VI}$ lattice site | $OH^-$ lattice site | $SO_4^{2-}$ lattice site | |
|---|---|---|---|---|
| $Sr^{2+}$ | $Cr^{3+}$ | $O^{2-}$ | $B(OH)_4^-$ | $VO_4^{3-}$ |
| $Ba^{2+}$ | $Si^{4+}$ | | $CO_3^{2-}$ | $BrO_3^-$ |
| $Pb^{2+}$ | $Fe^{3+}$ | | $Cl^-$ | $NO_3^-$ |
| $Cd^{2+}$ | $Mn^{3+}$ | | $OH^-$ | $MoO_4^{2-}$ |
| $Co^{2+}$ | $Ni^{3+}$ | | $CrO_4^{2-}$ | $ClO_3^-$ |
| $Ni^{2+}$ | $Co^{3+}$ | | $AsO_4^{3-}$ | $SO_3^{2-}$ |
| $Zn^{2+}$ | $Ti^{3+}$ | | $SeO_4^{2-}$ | $IO_3^-$ |

From Gougar et al. (1996, Table 3)

The SAVMIN™ process is a patented, active mine water treatment system for sulfate-rich mine water that involves a five-step, sequential treatment process (Lorax Environmental 2003; Smit 1999). It is similar to the ettringite precipitation process patented by Walhalla-Kalk Entwicklungs- und Vertriebsgesellschaft mbH (patent EP 0250626, June 27, 1990) and the "improved" ettringite process patented by David J. Hassett and Jeffrey S. Thompson (U.S. Patent No. 5547588, August 20, 1996). Both Smit (1999) and Damons and Petersen (2002) have optimised the process flow in the SAVMIN™ process using the Aspen simulation program. In the SAVMIN™ process, mine water is treated to produce potable water, gypsum and calcite. In the first step, the pH is raised to 12 to precipitate metal hydroxides from the mine water and to enter the stability range of ettringite (Fig. 3.24), and a pH-range of 11.6–12 has been found to be optimal. In the second process step, gypsum crystals are added to the mine water as crystallisation nuclei to precipitate gypsum. Aluminium hydroxide ($Al(OH)_3$) is now added to the still gypsum-saturated mine water in the third step, resulting in the precipitation of ettringite. In process step four, sulfuric acid is added to recover the aluminium hydroxide. In the fifth and final step, $CO_2$ is added to the mine water to lower the pH, which remains high until then. In the process, pure calcite ($CaCO_3$) precipitates.

To date, the process has only been used at a few locations in South Africa (e.g. pilot plants at the Stilfontein and Mogale gold mines), as the plant and maintenance costs are currently still relatively high (Fig. 3.25). Nevertheless, it is possible to produce water that is safe for drinking (potable water) at the end of the process. Through cooperation with a leading plant manufacturer, the patent holder intends to optimise the process in the future and bring it to market maturity (further information can be found on the MINTEK and VEOLIA websites). Optimisation trials for an ettringite process have also been underway at a Finnish university since at least 2017, together with a globally active water treatment

**Fig. 3.25** SAVMIN™ pilot plant at Western Utilities Corporation's Mogale gold mine in South Africa. (From Roger Paul: "Presentation to the Parliamentary Committee on Water and Environmental Affairs" – 7/8 September 2011)

company. Details are – how could it be otherwise – subject to secrecy. If the problems arising from the relatively narrow pH range in which the process operates can be solved, then what Fernando et al. (2018) state could come true, namely that the SAVMIN process will become established for future sulfate removal.

### 3.6.2 Other Procedures

In an article whose title is at first misleading, Dvorácek et al. (2012) describe an ettringite process developed and patented at the VŠB Technical University of Ostrava, Czechia. At a pH of 12.4, the authors add slaked lime and sodium aluminate to mine water and, after flocculation, dewater the resulting sludge. They obtain industrially usable water with sulfate concentrations below 300 mg $L^{-1}$ as the product. An ettringite process is also used at the former Straž uranium mine in Hungary to remove sulfate from the mine water (pers. comm. Michael Paul).

Heviánková and Bestová (2007) use ettringite sludge to remove manganese from the mine water. However, this results in an increase in the sulfate concentration in the mine water. They conclude that this process may therefore be unsuitable for removing manganese. However, it is an indication that not only Mn(III) but also Mn(II) can be incorporated into the ettringite at the $(Ca^{2+})^{VIII}$ lattice site.

## 3.7 Schwertmannite Process

Ever since the mineral schwertmannite ($Fe^{3+}{}_{16}[O_{16}|(OH)_{10}|(SO_4)_3]\cdot 10\ H_2O$) was first scientifically described in 1990, its use and occurrence in connection with mine water have been studied extensively (Bigham et al. 1990; Paikaray 2021; Schwertmann et al. 1995). One would almost like to get the impression that it is the silver bullet of mine water investigation – or the "acid rain" of the 1980s, which no serious study should disregard. Be that as it may; schwertmannite has some excellent properties, or as Schwertmann casually put it: "poison catcher in mine tailings" (Schwertmann 1999). Therefore, the current investigations into whether the mineral is suitable as a "poison catcher" are of fundamental importance.

In addition to the name "schwertmannite process", the term "microbiological iron separation process" or "iron sulfate precipitation" can be found in the literature (Bilek 2012). The active treatment plant consists of an aeration tank followed by an oxidation chamber. The oxidation chamber contains biofilm carriers on which the microbially formed schwertmannite is deposited at pH values between 2.5 and 3.3 (Fig. 3.26). By means of a chain scraper at the bottom of the oxidation chamber, the sludge can be continuously discharged and is pumped to the sludge collection tank by means of a sludge pump. Of importance for the formation of the schwertmannite are not only the bacteria on the biofilm carriers, but also those in suspension in the oxidation chamber. The formation of the mineral takes

**Fig. 3.26** Schwertmannite on growth elements (Left: image width approx. 2 m) and microbially produced Schwertmannite (Right: image width approx. 15 cm). (Photos: Eberhard Janneck)

place wherever the bacteria succeed in forming a biofilm: consequently on the biofilm carriers and the reactor walls. On these surfaces it was possible to eliminate up to 35 g $Fe^{2+}$ per $m^3$ and hour. After a run-in period, a pH of 3.1–3.3 was established in the plant, with the pH of the mine water ranging between 4.2 and 5.1 (Janneck et al. 2007a). Overall, the two pilot plants for the treatment of mine water from coal mining in Lusatia, Germany, showed good success in treating the waters (Janneck et al. 2010) but are currently not being pursued further as a treatment method by LEAG (formerly Vattenfall).

Heinzel et al. (2009) and Hedrich et al. (2011) investigated the microbiological processes leading to the formation and precipitation of schwertmannite in one of these plants, the latter focusing their investigations primarily on the bacterium *Ferrovum myxofaciens*, which could potentially play an important role in mine water treatment in the future (pers. comm. Barry Johnson 2012) if the plants can be reliably inoculated with suitable strains. Large-scale application of the technology has yet to be established. In the case of the uranium- and arsenic-containing mine water in the Erzgebirge, Germany, the schwertmannite from the Tzschelln pilot plant can be used to remove arsenic and uranium from the mine water (Janneck et al. 2011).

At the moment, the number of publications on schwertmannite in mine water, which were first listed in the SCI in 1995, indicate a clear flattening of the curve. It seems as though the interest in the mineral and process is slowing down, which contradicts the leading idea in an article of the VDI-Nachrichten about the schwertmannite process, which gave the impression that the process has great potential (Ahrens 2013).

## 3.8 Bioreactors (Fermenters)

Microbiological treatment processes for mine water can be classified partly under active and partly under passive treatment methods (e.g. Gandy and Jarvis 2012; Gusek 2002). The transitions can be fluid, as also Weston Solutions (2004, pp. 4–4) note: "Microbiology reactors can be classified as active or passive, usually depending on whether they are tanks or ponds." This section will only present bioreactors that definitely fall under this chapter of active treatment processes. They have been studied since the 1990s to treat mine water using biological processes. For the time being, for practical reasons, I do not subscribe to the nice idea of calling constructed wetlands "bioreactors with a green toupee" (Wildeman et al. 1993a, pp. 13–1) – it would add to the confusion surrounding the term bioreactor, especially since there are specific terms for each of the other treatment types.

Bioreactors (fermenters, technical reactors) are containers filled with untreated mine water and organic material (e.g. straw, compost) as well as sulfate-reducing bacteria, whereby the organic material serves as a carbon source for the microorganisms (U.S. Environmental Protection Agency 2000, p. A-4). Molasses, syngas (also known as synthesis gas), or sewage sludge have been found to be suitable energy sources for the bacteria as well (Dill et al. 1994; Du Preez et al. 1991; Maree et al. 1991; Rose et al. 2004b). Generally, bioreactors constructed for mine water treatment are fixed-bed reactors in which reducing conditions prevail to chemically reduce dissolved metals (especially iron) and sulfate. This produces hydrogen sulfide and subsequently precipitation of low solubility metal sulfides in the digester, thus raising the pH of the mine water (Drury 1999). Sulfate removal efficiencies are highly dependent on the composition of the mine water, the substrate used, pH, and temperature, reaching daily rates of 2–30 g $L^{-1}$ sulfate (Dill et al. 1994; Lorax Environmental 2003, pp. 3–30) (Fig. 3.27). Drury (1999) added whey to the bioreactors (to stimulate the activity of sulfate-reducing bacteria), substantially increasing the removal efficiency of his reactors. It appears that bioreactors could only lower sulfate concentrations in mine water to less than 300–100 mg $L^{-1}$ under optimal conditions, as most investigators could only achieve decreases to about 200 mg $L^{-1}$ (Bilek et al. 2007; Drury 1999; Lorax Environmental 2003, pp. 4–29; Poinapen 2012, p. 241). Lorax Environmental (2003, pp. 5–11) conducted a comprehensive evaluation of

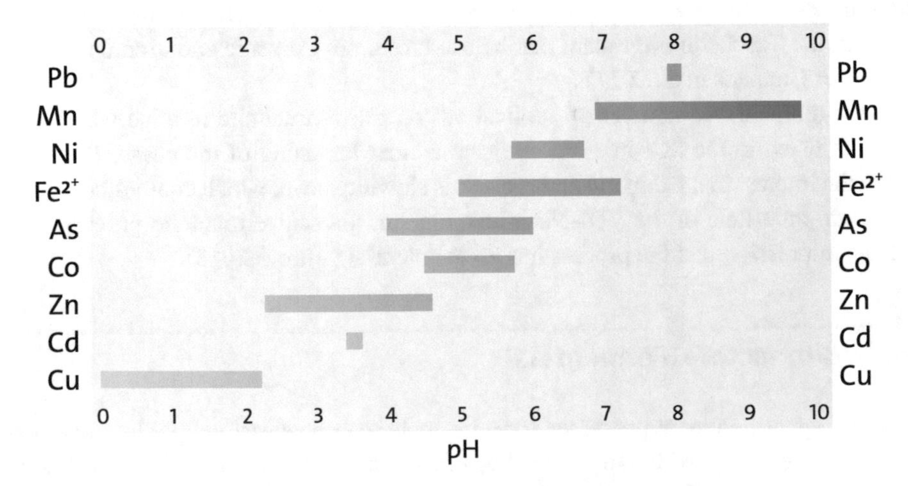


**Fig. 3.27** Used and recommended pH values for selective precipitation of metals and arsenic as sulfides. (Adapted from Kaksonen and Şahinkaya (2012) with data from Govind et al. (1997); Hammack et al. (1994a); Kaksonen and Puhakka (2007); Tabak et al. (2003))

methods that can remove sulfate from mine water. They conclude that bioreactors are among the most effective methods in this regard (presumably including "passive bioreactors" in their evaluation).

An industrial-scale plant went into operation at the Grootvlei gold mine in South Africa between 2004 and 2006. It treats 10,000 m$^3$ of sulfate-containing mine water daily (Holtzhausen 2006b) using the BioSURE® process. Sewage sludge from municipal wastewater is used as a carbon source, which thus represents a synergy effect. In the meantime, the original method (Rose et al. 2004b) has been improved so that higher treatment efficiencies are now achievable (Poinapen 2012).

Another bioreactor process is the THIOPAQ® process for biological sulfate removal patented by Paques Bio-Systems BV (Netherlands). With the help of sulfur obtained in the bioreactor, the metals are precipitated as sulfides and can be used commercially. In principle, the process consists of two steps: First, sulfate is anaerobically reduced to sulfide by sulfate reducers, and in another reactor, with the formation of alkalinity, is converted to sulfur by sulfate oxidisers. Usually, hydrogen, ethanol, butanol or organic waste are used as organic nutrients and electron donors. Since the THIOPAQ® process requires lower lime dosing than usual, produces smaller sludge volumes, and generates saleable by-products, this process is of interest to the mining industry, and even more so considering that contaminant concentrations after treatment are generally low (de la Vergne 2003; Jacobs and Pulles 2007; Lorax Environmental 2003). More than a dozen systems are in operation worldwide, for example at the Kennecott copper mine in Utah (USA) or since 1992 at the Budelco zinc refinery (Netherlands). An Anglo Coal Paques plant in South Africa had to shut down due to increased costs for ethanol and butanol (Expert Team of the Inter-Ministerial Committee 2010), and Simate and Ndlovu (2014) also report that the process had lost its attractiveness because of these cost increases.

In addition to the well-known bacteria such as *Acidithiobacillus thiooxidans* (Kelly and Wood 2000), the bacterium *Ferrovum myxofaciens* plays a decisive role in sulfate reduction at low pH values ($<$ pH 4) in the bioreactors (Rowe and Johnson 2009). The extent to which this bacterium is suitable for increasing the reaction rates in these reactors must be shown by future research (pers. comm. Barry Johnson 2012).

Membrane bioreactors (MBRs) are a special form of bioreactor in which drinking water can be obtained with the aid of ultrafiltration. Examples of such plants can be found in hotels or on ships and achieve daily treatment capacities of 48,000 m$^3$ of black or grey water (Peters 2010, p. 1234 ff.). Häyrynen et al. (2009) reported a membrane bioreactor treating ammonium and nitrate and observed treatment rates of 82–90%. Bijmans et al. (2009) and Şahinkaya et al. (2018) treated synthetic mine water in a membrane bioreactor and showed a decrease in the sulfate concentrations in the water. The problem with all applications thus far has been that the membranes exhibit rapid biofouling at high flow rates, which could only be removed by extensive cleaning. None of the publications on membrane bioreactors that I could find describe an actual existing system on a pilot scale or even on a field scale, so I will not go into detail about this process. Based on the results to date, it is not yet possible to assess whether the process will have potential for future mine water treatment.

## 3.9 Ion Exchange

Ion exchangers are mainly porous synthetic resins whose surface is provided with functional groups to which positively or negatively charged, water-soluble ions adhere (Fig. 3.28). An alternative to synthetic resins are membranes to which the functional groups are attached. In ion exchange, ions from the mine water are exchanged for these ions from the ion exchange material. As a rule, during the exchange reaction, higher-charged ions in the mine water (e.g. $Ca^{2+}$, $Mg^{2+}$, $Cu^{2+}$, $Fe^{2+/3+}$, $Cd^{2+}$, $SO_4^{2-}$) displace the lower-charged ions on the solid material of the exchanger (e.g. $Na^+$, $H^+$, $K^+$, $Ca^{2+}$, $Cl^-$, $SO_4^{2-}$, $OH^-$), which is thus "consumed" over time. Depending on the medium of the exchanger, cations or anions can thus be removed from the mine water. The exchanger materials are mostly synthetic resin beads of 0.3–1.2 mm diameter, more rarely natural zeolites, whereby the synthetic resins can be regenerated by acids or bases. The natural zeolites, on the other hand, must be disposed of. Today, the floating fixed-bed process is usually used, which allows the resins to be regenerated in counter current operation. Ion exchangers can be used to treat a wide range of industrial waters, and the process can be optimised in each case to remove cations or anions (Fig. 3.29). Suitable functional groups can be used to selectively remove metals from mine water for further recycling. Theoretically, it would even be possible to produce pure water through a suitable combination of anionic and cationic ion exchangers (Gusek and Figueroa 2009, pp. 97–100; Sincero and Sincero 2002; Skelly and Loy and Penn Environmental Consultants 1973, pp. 335–337; manufacturers' product information).

Regular, large-scale ion exchange rarely occurs in the field of mine water treatment, although the method was first applied to mine water in the late 1960s (Skelly and Loy and

**Fig. 3.28** Various synthetic resins for ion exchangers (LANXESS Deutschland GmbH)

**Fig. 3.29** Industrial ion exchange system for wastewater treatment from copper electrolysis at Montanwerke Brixlegg/Tyrol, Austria

Penn Environmental Consultants 1973; Sterner and Conahan 1968; Wilmoth et al. 1977) and many specialist applications are reported in the literature. This may be due in part to the fact that mine water is often highly mineralised and thus the ion exchanger must be regenerated at relatively short intervals. However, the large volumes of water may also have previously stood in the way of extensive use of the technology, as the cost of treatment is then comparatively high (Senes Consultants Limited 1994). However, new applications of methods developed in the past decade and those currently under development will establish ion exchange in the mine water sector in the medium to long term (e.g. Feng et al. 2000; Muraviev et al. 1995; Strathmann 2012). This would provide, in addition to membrane technology, another mine water treatment method whose product is a marketable raw material instead of waste.

One of the first applications of the method to mine water was the removal of uranium (George and Ross 1970; Hartley 1972; Himsley and Bennett 1985), where it is still used today (Botha et al. 2009; Braun et al. 2008; Jeuken et al. 2008). Other metals that can be selectively removed from mine water include copper, nickel, cobalt, zinc, cadmium, lead, thorium, radium, mercury, and the semimetal arsenic (Dinardo et al. 1991). For each group of elements, specific ion exchangers or processes are required to achieve optimal recovery of the respective element (Simate and Ndlovu 2014).

Ion exchange is particularly often used to treat mine water and leachate at former uranium mines, especially when the necessary infrastructure for regeneration, precipitation and further processing of concentrates is still available. The intention behind this, as also mentioned in Sect. 7.2, is to generate revenue to cover the costs of the treatment process (pers. comm. Michael Paul 2019) at least partially. At the Königstein, Germany, site of Wismut GmbH, an anionic exchange resin selective for uranium was used, which made it possible to produce a marketable uranium oxide. This was achieved by a multi-stage approach in which first iron and then uranium were removed from the exchanger. In a third step, the exchanger was regenerated (Braun et al. 2008). Other sites with ion exchangers for the removal of uranium exist, for example in the Czechia, Poços de Caldas in Brazil or Pécs in Hungary (Csővári et al. 2004; Rychkov et al. 2016; Smetana et al. 2002).

Currently, one of the largest ion exchange plants treats mine water from the Berkeley, Montana, USA, open pit mine, using the GYP-CIX process (Bowell 2004). The GYP-CIX process removes cations from mine water in a first step and anions in a second step (Everett et al. 1993; Lorax Environmental 2003). As with other processes, the concentrated residual solution is a problem in ion exchange and must be disposed of by appropriate means. According to Bowell (2004), the GYP-CIX process is said to be the only process that can be used for mine water without further pre-treatment.

A well-documented process is the Sul-biSul process (Schmidt et al. 1969; Wilmoth et al. 1977), in which first the metal ions are exchanged for protons and then the sulfate is exchanged for the bisulfate. For this purpose, the water is first passed through a cation exchanger and then through an anion exchanger. In subsequent reaction steps, the acidic water is neutralised again, and the ion exchangers are regenerated. It could be shown that water with a TDS concentration of up to 3 g $L^{-1}$ can be treated, but the plant operates optimally with neutral to alkaline mine water. However, to treat acidic water, the process is uneconomical (U.S. Environmental Protection Agency 1983, pp. 165–167). In Smith Township (Slovan, Pennsylvania, USA), a Sul-biSul plant was installed in 1969 and operated until 1971, treating 1900 $m^3$ of brackish mine water per day. After mandatory chlorination, the water was used as drinking water (Wilmoth et al. 1977, p. 820 f.).

Ion exchange by zeolites is described by some authors, but industrial use for mine water has yet to be established. Hamai and Okumura (2010) demonstrated that zinc can be selectively removed from mine water, although the treatment rate varied depending on the zeolite. Prasad et al. (2011) de-"hardened" mine water by treating it with zeolites obtained from fly ash. By dosing 40 g $L^{-1}$ of fly ash zeolite, they were able to lower the "hardness" by 72%. However, to treat the mine water into drinking water, further treatment steps would be necessary. In Chile, Fundación Chile developed and patented the ZEOTREAT™ process (Vidal et al. 2009), which removed molybdenum, arsenic, and sulfate from mine water with a low mineralisation. For the Reiche Zeche shaft in Freiberg, Saxony, Germany, Zoumis et al. (2000) investigated whether zeolites could be used for mine water treatment. They concluded that the reaction is relatively slow, but zinc can be removed relatively well. It was also investigated whether ion exchange could be used to treat mine water at the largest mining-related water pollution site in Germany, the Burgfey adit in the Eifel region (Heitfeld et al. 2012, p. 118).

This section has turned out to be much longer than planned. Why? Although the method has so far eked out a shadowy existence in mine water treatment, the literature on ion exchange is surprisingly diverse. Of 5469 literature citations that served as the basis for the GARD literature database, 128 are on ion exchange, but only 69 on membrane technologies. It would appear that the method has been waiting in the wings since the 1960s to be used as a selective resource recovery process from mine water, as evidenced by, for example, ion exchange membrane integrated solutions (Chartrand and Bunce 2003) or electrochemical methods (Bunce et al. 2001).

## 3.10 Sorption

Sorption involves the use of an organic or inorganic material capable of removing monovalent or polyvalent cations or anions from contaminated water by adsorption or absorption (see Sect. 1.2.20). Sorption may involve nano- or macro-particles. Ideally, the chosen sorbent is selective for a comparatively small number of ions and can therefore be targeted for decontamination. The list of sorbents that have been successfully used at laboratory scale to date is notably long and includes, for example, activated carbon, tea, banana peel, peanut shell, vermiculite, ash, titanosilicates, ettringite, zeolites, "green sorbent", mussel shell, schwertmannite or peat (e.g. Chang et al. 2016; Iakovleva and Sillanpää 2013; Lata et al. 2015; Oleksiienko et al. 2017). An extensive list of nanosorbents applicable to water treatment is discussed by Mishra (2014). Undoubtedly, sorption processes play a vital role in natural attenuation or passive mine water treatment, with oxides and biosorbents playing a crucial role (MEND 2000; Smith 1999). The sorption of mostly organic pollutants onto activated carbon in wastewater treatment is also a process that should not be underestimated (Cheremisinoff 2002). The same applies to the removal of arsenic from drinking water – a method that has proved extremely successful (Schlegel et al. 2005). However, coprecipitation of various pollutants, as described in U.S. Environmental Protection Agency (2014, p. 56 ff.), is not the content of this section. Here, the focus is exclusively on active mine water treatment using sorbents, such as those listed by Walton-Day (2003, p. 350).

For some time now there has been an avalanche of publications dealing with the sorptive removal of cations and anions from wastewater and mine water (Fig. 3.30). Almost every week, articles appear that use a new or known sorbent to remove metal ions or sulfate. In almost all cases, synthetic mine water is used, and a wide berth is given to real mine water. The experiments are almost exclusively batch or column experiments, and the overall transferability to real mine water is questionable. I will refrain from listing or selecting them because there are too many. Despite extensive research, I have not been able to find an industrial-scale plant that treats mine water using artificial sorbents. Also, one of my students was working on sorbents for mine water, and as her second supervisor,

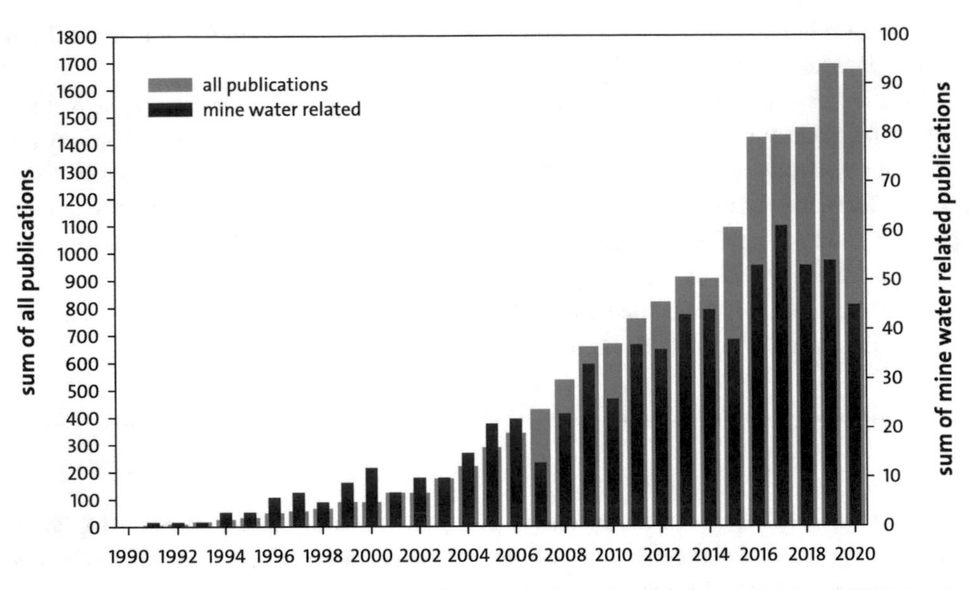


**Fig. 3.30** Number of publications on water treatment with sorbents between 1990 and 2020.On the right axis, the number of those relevant to mine water. (Source: Clarivate Analytics Science Citation Index Expanded [SCIE])

I had recommended that she repeat all experiments also with real mine water (Atiba-Oyewoa et al. 2016; Oyewo et al. 2019).

At the Rocky Flats Environmental Technology site in Colorado, USA, a pilot plant operated with the Colloid Polishing Filter Method (CPFM) and removed radionuclides from the mine water during the test period (U.S. Environmental Protection Agency 2014, p. 79). Details of the adsorbent or the method itself are not known. A successfully operating plant is located at the Wismut GmbH tailings pile in Helmsdorf/Thuringia, Germany (IAA Helmsdorf). There, arsenic is removed from the seepage and pore water by means of iron hydroxide (Kassahun et al. 2016). To the best of my knowledge, it is currently the only continuously operating plant in the world that uses adsorption *s.l.* to treat mine water, apart from plants using ferrihydrite, where the process is coprecipitation rather than adsorption (U.S. Environmental Protection Agency 2014).

Aquaminerals Finland Oy maintained a 5 $m^3$ $min^{-1}$ pilot plant under Arctic conditions for the sorptive removal of iron and manganese from mine and process water using the sorbent AQM PalPower M10. Commercial plants are found in West Africa where arsenic and manganese are removed from mine water. The same principle is used to remove arsenic for drinking water treatment. Further installations are planned in South America where copper, arsenic and manganese will be removed from mine water. The AQM PalPower M10 sorbent is based on thermally modified minerals. Furthermore, Aquaminerals Finland Oy is commercialising the sorption of ammonium, e.g. from explosives residues, on geopolymers (amorphous aluminium silicates) and its recovery as a synthetic fertiliser (Luukkonen et al. 2016).

**A Side Glance**

A few years ago, a colleague from the USA approached me and said that he thought it was great that I was working on mine water. Then I could publish three to four papers with him annually by taking different sorbents and synthetic mine water and then publishing kinetic ratios and sorption characteristics as well as XRD and SEM and FTIR analyses. When I told him that I would work exclusively with real mine water and that I wanted to focus entirely on methods that could be used industrially, he very quickly lost interest in collaborating. Of course, I am aware that science is always interested in gaining knowledge, and this of course includes exploring new methods and new mechanisms. But – hand on heart – don't we now have enough publications on "nanobanana" peels and "nanovermiculite" and whatever else is "in" at the moment? We should concentrate on the essentials again.

I am of the opinion that the increasing pressure on academics to publish leads to the creation of a plethora of publications that only help to increase their numbers, not knowledge. If you are doing your own work on sorbents for mine water, please try to develop a process to incorporate the sorbents into a fixed bed, limit the competition of the different ions in the water, and treat a water stream larger than 1 L $min^{-1}$. If not – just don't do it. After all, with the right choice of sorbent and ion to be sorbed in synthetic mine water, you will *always* get a result anyway.

In some publications, authors attempt to attribute the decrease in potential contaminant concentrations in mine water to sorption. However, a closer analysis of the experimental processes shows that it is in fact precipitation due to pH increase and that the sorbent did not play a substantial role. Alternatively, a sorbent is treated with NaCl, and the publication discusses how the sorbent removes the potential contaminant – although ion exchange probably played a more substantial role than sorption. However, conclusions are drawn from isotherms and kinetic results that are not always supported by further studies. Miller et al. (2013) have already found, through batch experiments and numerical modelling, that "when acid mine drainage is treated with limestone and the usual ratios of iron to secondary metals are used, the main treatment processes appear to be dominated by processes beyond adsorption".

Where can research with sorbents lead us to? I am convinced that with suitable sorbents and processes it will be possible to selectively remove metals from mine water – after all, nature is showing us how. The goal must be to find suitable biofilm carriers that allow the mine water to come into contact with the sorbents even in larger quantities. Currently, many sorbents fail because of poor selectivity due to the competition of the many ions in the water.

## 3.11 Advanced Oxidation

The advanced oxidation process (AOP) typically uses one of four technologies to accelerate oxidation in mine water treatment: Photolysis, radiolysis, chemical reactions (Fenton process, ozone, hydrogen peroxide), or photocatalysis (Attri et al. 2014; Rajeshwar and Ibanez 1997). Advanced oxidation has been particularly successful in the treatment of difficult-to-treat wastewater. To date, advanced oxidation processes have been used only to a limited extent in mine water treatment, since the "problematic" substances from wastewater treatment are usually present only in small quantities in mine water. Plasma technology could be used to remove organic contaminants from mine water (Attri et al. 2014), which were previously largely considered unproblematic (e.g. PCBs). However, plasma technology is still in its infancy and requires considerable research to be used as a standard method in mine water treatment. A combination of electrocoagulation with the Fenton process could also enable previously unknown technology combinations in mine water treatment, as described for example by Zhao et al. (2012b) for industrial wastewater from the electrical industry.

A common feature of almost all advanced oxidation processes is that they generate highly reactive, short-lived free radicals, for example HO• (Crittenden et al. 2012). The reactants are primarily, but not exclusively, $O_3$, dihydrogen trioxide (peroxone process: $H_2O_2 + O_3 \rightarrow H_2O_3 + O_2 \rightarrow HO\bullet + HO\bullet_2 + O_2$), $O_3$ and UV treatment, and $H_2O_2$ and UV treatment. The free HO•-radicals produced by these reactions react in oxidation reactions with the potential pollutants or microorganisms in the water, rendering them largely harmless. In essence, the substances that can be removed from mine water by advanced oxidation processes are organic substances such as BTEX, phenols, dyes, or dioxins, or inorganic substances such as cyanides, mercury, cadmium, manganese, or sulfites. Inorganic pollutants are mainly reduced to the elemental form or, as in the case of chromium, to a less harmful oxidation state. A particular advantage of the method is that both organic and inorganic and microbial pollutants can be removed in one step (Crittenden et al. 2012; Rajeshwar and Ibanez 1997).

Electrocoagulation is one of the processes that produces free radicals (Chen 2004), which can lead to direct oxidation at the anode. This is considered more effective than the production of oxygen alone (see Sect. 3.3.1). Florence (2014) conducted experiments with mine water from Ynysarwed in Wales, United Kingdom, and was able to show that initial $Fe^{2+}$ concentrations of 74 mg $L^{-1}$ could be completely removed from the mine water at higher currents (5 A). At the same time, the pH decreased from 6.7 to 4.6 and the redox potential increased by 90 mV.

Ozonation involves the addition of ozone ($O_3$), a strong and fast-reacting oxidant, to mine water, which, among other things, leads to the destruction of organic contaminants (Ødegaard 2004) and can decolourise the mine water (Jacobs and Pulles 2007). It is considered one of the most potent oxidants, and this technology has been used for mine water treatment since as early as the 1960s and in municipal wastewater treatment since the late nineteenth century (Charmbury et al. 1967; Cheremisinoff 2002). Furthermore, the addition of ozone to the sludge of the mine water treatment plant can help to reduce the amount of sludge. It has been shown that adding 0.05 g of ozone per gram of solids in the mine water can reduce the sludge volume by 25–35% (Böhler and Siegrist 2004). Approximately 1 g of $O_3$ is required to oxidise 2 g of $Fe^{2+}$ if an efficiency of 86% is assumed, resulting in greater sludge density compared to an HDS system (U.S. Environmental Protection Agency 1983). However, Davies et al. (2016) highlight that the ozone should be produced on site to minimise costs and Cheremisinoff (2002) also states that ozone is unstable and therefore should be produced and used on site. Moreover, the ozone generator must be physically separated for safety reasons, regular monitoring must be in place, and higher pH values of mine water must be avoided. In wastewater treatment, a 1–1.5% ozone solution is usually used and added to the water at a mass flow rate of 100 g $h^{-1}$ (Cheremisinoff 2002).

## 3.12 Flotation Liquid-Liquid Extraction (F-LLX; VEP; HydroFlex™ Technology)

Flotation liquid-liquid extraction is a patented process that removes cations and anions from mine water in a four-stage process. At the same time, the pH value is increased, and the mineralisation of the mine water is reduced (Fig. 3.31). To achieve this, a non-water soluble organic extractant is added to the mine water in a first reaction step and is used

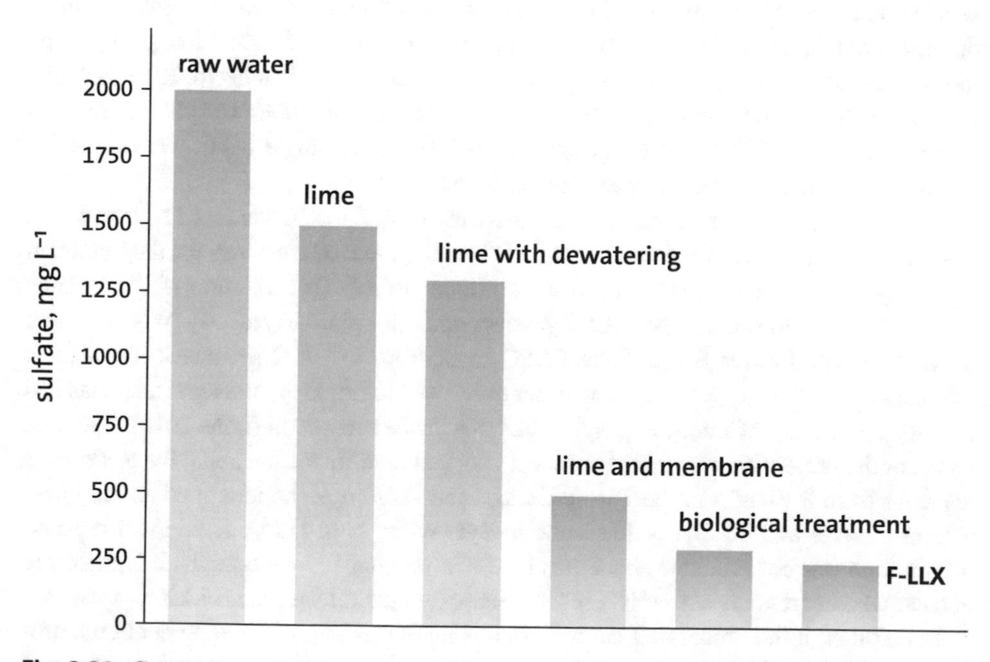


**Fig. 3.31** Comparison of sulfate removal between F-LLX and other treatment methods. (After Monzyk et al. 2010)

throughout the entire process and can be recovered and reused at the end. In the first two extraction stages, the metals and sulfate are removed, while the metals and sulfate are recovered during the last two stages in separate runs, with sulfuric acid and sodium carbonate being added in the final extraction stage. Finally, the expensive organic extractant can be recovered after sulfate elimination and reused in the first process stage. After treatment, the end-products are marketable iron, aluminium and sodium or ammonium sulfate. Treated water as well as carbon dioxide is produced, which can be used for process optimisation in the sulfate recovery. Although the exact extraction agents applied in the process are not specified, either quaternary ammonium or phosphate compounds ($R_4N^+$, $R_4P^+$) as well as alkyl-guanidine compounds ($R_5CN_3H^+$) play a decisive role as cation; alternatively a mixture of these extraction agents is used. No specific information is given about the anion. In the St. Michael (Pennsylvania, USA) pilot test, the chemicals Aliquat® 336, Exxal® 10 and Calumet® 400–500 were used (compiled from patent specifications: Fiscor 2008; Monzyk et al. 2010; Winner Global Energy and Environmental Services L. L. C. 2009). No commercial treatment plant exists to date.

The St. Michael pilot plant was in operation for more than 3 months (Fiscor 2008). It treated 38–114 L $min^{-1}$ of acid water depending on the experiment. This water had a pH of 4.6–4.9 and $SO_4^{2-}$ concentrations of around 1040–1220 mg $L^{-1}$, which could be reduced to below 38–987 mg $L^{-1}$ and the pH raised to 6–8 (Winner Global Energy and Environmental Services LLC 2009).

In 2008 and 2009, numerous newspaper articles appeared around the process, and it was claimed that this process would be the first cost-effective method to treat acid mine drainage containing metals and sulfate without creating a waste (Fiscor 2008, p. 96). Even the patent reads like an advertising brochure for the process (U.S. Patent No. 2010/0176061 A1 dated July 15, 2010 and EP 2118020 A2 dated February 14, 2008). In 2008, the process also received the R&D 100 Award (Anonymous 2008). No single publication about the process can be found in the Science Citation Index.

At the 2012 annual meeting of the American Institute of Chemical Engineers, the Battelle Institute gave a presentation entitled "Development of a process for the treatment of highly mineralised mine water from a copper mine" (presentation 417 f.). They announced plans to operate the F-LLX process at a pilot plant in Norway. Winner Water Services reported at the Shale Water EXPO in Stafford in late 2014 that it intended to establish the process to pre-treat mine water to be used in fracking, and reportedly has had a pilot plant in Butler County, Pennsylvania, USA, since early 2014 (Lane 2016). The final report of the project is anonymised in such a way that no information about the test sites is apparent from it (thanks to the Turing Galaxy, the sites could be identified as the Fawn Mine in Sarver and the Sykesville Mine in Sykesville, both USA). Additional detailed information was not available about the process, even after several e-mails. It is therefore unclear whether there is currently a reliably operating plant using the F-LLX process.

In no other of the processes I describe in this book was the information about the process as vaguely formulated, anonymous, presented in various disseminations, and sometimes contradictory as in the F-LLX process (also called the VEP process or

HydroFlex™ Technology). For example, the chemical analyses in the patent differ substantially from those during the pilot test, and the mean values in the final report also do not reflect the actual range of variation in effluent values (on average < 500 mg $L^{-1}$ sulfate, then again <100 mg $L^{-1}$, but extreme values of almost 1000 mg $L^{-1}$ also occur)· The name has also changed in the meantime and was announced as "Hydro Flex Technology", in 2014 at the R3 Conference in Philadelphia. Before commercial use, it is necessary to optimise the complicated treatment process and to enable better process control so that the process parameters can always be maintained.

## 3.13 Eutectic Freeze Crystallisation

In eutectic freeze crystallisation (EFC), also known as freezing-out, an aqueous solution is cooled to such an extent that freezing of the water occurs at the eutectic point (van der Ham et al. 1998). Since the ice incorporates only minute amounts of "foreign" matter into the crystal lattice (Shone 1987), water constituents are separated from the ice and can be disposed of or reused. The ice can be thawed and used as drinking or process water. Common applications of the method are in the food, pharmaceutical or chemical industries. Its use in water treatment, however, especially in mine water treatment, is relatively new, although first investigations were carried out in the late 1960s (Hill 1968). The method is interesting in that cooling of water consumes about one-sixth of the energy compared to evaporation (Fernández-Torres et al. 2012; Gnielinski et al. 1993, pp. 150–157; Nathoo et al. 2009; Skelly and Loy and Penn Environmental Consultants 1973, p. 349 f.): While evaporation of water consumes 41 kJ $mol^{-1}$, freezing consumes only 6 kJ $mol^{-1}$ (Shone 1987).

A HybridICE® pilot plant with a daily throughput of 36 $m^3$ was installed at the Tshwane University of Technology, Pretoria, South Africa in cooperation with the company Aqua-Simon from Flensburg, Germany (Adeniyi et al. 2013, 2016) to treat a wide variety of industrially contaminated waters (patent DE 60028806 T2, dated 18th January 2007). The system was able to further concentrate brine from reverse osmosis and substantially reduce the brine volume. The special process made it possible to clean the ice without fresh water (Fig. 3.32). These authors were also able to demonstrate that the yield is dependent on the working temperature. In a further intensification of the process, a combination of ion exchange and subsequent freeze crystallisation was carried out, which made it possible to remove up to 1 g $L^{-1}$ of iron from synthetic mine drainage (Malisa et al. 2013). Regarding the cost factor energy, Mtombeni et al. (2013) point out that the HybridICE® process can operate economically, especially in times of cheap electricity.

Investigations in more recent times to treat brines from reverse osmosis of mine water using freeze crystallization took place at the University of Cape Town (South Africa) (Nathoo et al. 2009; Randall et al. 2009; Reddy et al. 2009; Reddy and Lewis 2010). Among other things, they built on the findings of Shone (1987), who investigated whether ice or cooling water could be produced from mine water to cool mine ventilation. Typical

**Fig. 3.32** Former water treatment plant for eutectic freeze crystallisation at the Soshanguve campus of the Tshwane University of Technology, Pretoria, South Africa. (HybridICE® process)

concentrations of the brines studied ranged from 2.3 to 80.8 g $L^{-1}$ Cl, 7.4–72.9 g $L^{-1}$ $SO_4^{2-}$, and 5.0–75.8 g $L^{-1}$ Na (Nathoo et al. 2009; Reddy and Lewis 2010), up to 92% of which could be removed. A substantial influence on the separation of ice and the dissolved salts is the concentration of the resulting solids (i.e., ice and salts), which is complicated by the fact that salt crystals are often embedded in the ice. Subsequent washing processes can substantially increase the purity of the ice, and thus the yield of salt crystals (Reddy et al. 2009). The authors were also able to show that yield is a function of temperature and that it may be possible to recover dissolved salts sequentially by staged cooling (Reddy et al. 2010), as is the case at the Proxa Water Eutectic Freeze Crystallization plant at the New Vaal coal mine (Free State Province, South Africa) (Fig. 3.33).

Fernández-Torres et al. (2012) compared the energy cost of the method and its environmental impact with evaporative crystallisation. Their results show that eutectic freeze crystallisation is preferable to evaporative crystallisation from both energy and environmental points of view since it consumes six to seven times less energy than the latter. Further energy savings are possible by converting the heat energy generated during freeze crystallisation into electrical energy.

At South Africa's deepest gold mine, Mponeng (formerly Western Deep Levels – South Shaft), ice is produced to cool the mine air, as at many other South African mines. Shone (1987) has proposed treating water simultaneously with ice production. The freeze crystal-

**Fig. 3.33** PROXA Water Eutectic Freeze Crystallization plant at the New Vaal Colliery, an open pit mine of Anglo American Coal in South Africa. (Photo: Jochen Wolkersdorfer)

lisation methods would provide an opportunity to satisfy the demand for ice for cooling, as process water or drinking water on the one hand, and to treat mine water on the other. A first industrial plant has existed since 2015 at the New Vaal Colliery open pit mine of Anglo American Coal in South Africa and was built by the company PROXA (Fig. 3.33).

In my opinion, freeze crystallisation has considerable potential to separate the water and the dissolved salts from reverse osmosis brines. Since the disposal of these brines is a largely unsolved problem in mine water treatment, a combination of reverse osmosis and freeze crystallisation could contribute to solving this problem. Furthermore, valuable substances could be recovered from the mine water or brine if the process were further optimised. However, it is absolutely necessary to switch to real mine water and real brine for process intensification in order to be able to model real conditions. To date, nothing has been published on using a combination of reverse osmosis and freeze crystallisation to treat brines prior to discharge.

Let me move from this highly complex technology, as you can also see from the photos, to methods that are far less technical. In the following chapters I will describe the most common passive mine water treatment methods.

# 4 Passive Treatment Methods for Mine Water

## 4.1 Note

As I said on the previous page, in this chapter, I will present the currently known "passive" as well as "semi-passive" treatment methods for mine water. Passive mine water treatment is not identical to natural attenuation (Table 4.1). Rather, passive mine water treatment is like raising a child: If you guide it properly, hold your hand over it a little, and always keep a watchful eye so that it can develop its own character traits, it will grow into a confident adult. Natural attenuation, on the other hand, is to put the child into the world, stop worrying about it, and see what becomes of it; whereas monitored natural attenuation is the process from which I release the child into adulthood, but watch that everything goes smoothly during the transition from being a child to being an adult. So let us raise the child "Passive Mine Water Treatment" and start first with a foray into the required procedures.

The focus of this chapter is on methods whose success has already been proven internationally. A definition of the different passive treatment methods will introduce the chapter. Since there are a wide variety of detailed variants in individual cases, these will not be discussed in detail. A method will be classified as "passive" if it largely follows the definition given in the introduction. Although it may seem so; this chapter is in no way intended to call for every mine water to be passively treated – because it cannot be. It is, however, intended to raise awareness of passive treatment systems and to provide guidance on how to consider whether a passive treatment system is feasible in a particular case. Even those buying a car are looking at the diverse options: electric motor, fuel cell, hybrid system, gasoline engine, diesel engine, Wankel engine, car sharing, or riding a bicycle after all? Many active mining operations or mine water discharges with highly elevated pollutant loads will still need to be actively treated in the future – that's not in question here. But what about the 10,000 or so mine water discharges around the world that have thus far



C. Wolkersdorfer, *Mine Water Treatment – Active and Passive Methods*,
https://doi.org/10.1007/978-3-662-65770-6_4

**Table 4.1** Classification of passive and natural processes for mine water treatment

| Procedure | Meaning |
|---|---|
| Passive mine water treatment | New infrastructure (ex situ) is built and must be operated; usually no supply of chemicals or energy (Chap. 4) |
| (monitored) natural attenuation (NA) | Happens at the site of contamination within the existing system (in situ); and monitoring (MNA: Monitored natural attenuation) should always take place (Sect. 5.2) |
| Enhanced natural attenuation (ENA) | The processes of pollutant reduction must be understood to be able to use and stimulate them. Milieu conditions may need to be adjusted by adding chemicals to enhance, stabilise or support the target process. |

**Fig. 4.1** Part of the passive mine water treatment system for the tailings pile at Enos Colliery (Pike County, Indiana, USA). The operation of the wetland is described in Behum et al. (2008). (The photo composite shows the Canal Aerobic Wetland in the 6.5 ha constructed Enos East wetland)

been classified as "natural sources" and for which no one feels (or may feel) responsible? There would be possibilities in all countries to use passive treatment methods in addition to active treatment measures. A small improvement in water quality is still better than doing nothing at all. We need to have more courage to try different things (cf. Sect. 1.1). This means that scientists, engineering companies, operators and authorities should work hand in hand to find the best solution. And this requires mutual trust, optimism, a desire to experiment, the courage to face the unknown, "a little" time and money and a suitable good Samaritan legislation.

This chapter is thus also a call to authorities, responsible operators or scientists to show more self-confidence when it comes to passive mine water treatment. The first systems in the USA and the United Kingdom were built because there were initiative-taking colleagues and private persons who said, *mutatis mutandis*: "It is better to improve the water quality a little with a passive system than to do nothing at all." In the USA there are several thousand passive treatment systems (Fig. 4.1), in the United Kingdom several dozen, and worldwide there are probably several hundred passive treatment systems operating reliably. In many other countries it is sometimes said that there is still truly little experience with passive mine water treatment systems; this may be sometimes correct, but internationally there have been numerous positive experiences since the 1980s. Of course, there are also negative examples, which are essentially based on design errors. These could have been avoided in all cases with sufficient knowledge of the subject matter. To begin with, every mine water is different, there are hardly any identical mine waters, and as with our living organism "child", the treatment system must be adapted to the particular needs of the mine water. It is often said that passive mine water treatment is not possible because the space required is too large. In many cases this is correct, but before someone makes such a claim, it should be calculated how much space is needed and whether it might not be cheaper to build a passive treatment system instead of an active one.

Sometimes, during the discussion phase of my presentations, colleagues have told me that the example of Wheal Jane in Cornwall, United Kingdom, clearly shows that passive mine water treatment would not work. This is partly true, as Jeff Boyd of the Environment Agency said in a BBC interview in 2008:

> The passive treatment [at Wheal Jane] was an experiment to see if we might do the treatment in a different way with a minimum amount of energy and chemicals and, hence, have a more sustainable way of doing it. Unfortunately, it would have required a reed bed area which was so enormous that there just wasn't sufficient room to put enough in.

It is often claimed that passive mine water treatment is only suitable for small flow rates. This results in claims such as "The process is predominantly suitable for small water volumes due to the large area required." (allow me to refrain from providing a source) or in Rosner et al. (2012, p. 302): "Passive measures are not suitable for the treatment of heavily contaminated mine water and only achieve the desired results after the system has been standing and adjusted for a long time. This can take several years (see example of the Hachen pond, Meggen ore district)." Laypeople believe these statements to be true and therefore they are not checked before being published on the Internet. However, such claims are only partially correct, because less than the flow rate, the load is the criterion for the size of the passive treatment system. It is therefore primarily a question of water quality whether a passive treatment system will be large or small. Unless the space is available to build a reducing and alkalinity producing system (RAPS) or constructed wetland – then it is just not possible, and active treatment must be done. When reading some publications, I have a suspicion that there is often the thought in the back of the author's mind that passive mine water treatment does not work, or that it is a method in development – which, after more than four decades of international experience with these systems, is not true. As early as 1990, only less than a decade after the first passive treatment pilot plant was constructed, 400 constructed wetlands for mine water treatment existed in the USA (Kleinmann 1990a).

It is often stated in the literature that passive treatment systems would only work under warm climatic conditions, but not when it is cold in winter. As numerous studies in northern Canada or Scandinavia show (e.g. the results of the MiMi project), passive treatment systems can also be installed and operated successfully in cold climates. One of them is, for example, the vertical flow reactor (VFR) in Metsämonttu (Finland), the in situ experiment in Kotalahti (Finland), various systems in Sweden (Bomberg et al. 2015; Höglund et al. 2004; Wolkersdorfer and Qonya 2017) or the Neville Street Systems on Cape Breton Island, Canada. Pilot studies by Nielsen et al. (2018) for mine water from the Keno Hill mine (Yukon Territory, Canada) also demonstrated that passive mine water treatment is possible at average temperatures around 4.5 °C.

Unfortunately, in the context of mine water, it is often incorrectly stated that "biological water treatment" is a process to "accumulate certain pollutants in the roots or leaf mass." However, this is not correct for the treatment of mine water, as will be discussed below. Rather, the plants provide growing surfaces for microorganisms and reduce the flow velocity of water in the system, giving the hydrolysed metals sufficient time to precipitate in the passive treatment system.

In quite a few lectures about passive mine water treatment in the past years you could observe that someone stood up and said to the lecturer: "That all sounds nice, but as you know, it doesn't work." Exactly – it's just that in some countries these systems do not as yet operate as they should, and this chapter is intended to help ensure that future systems work everywhere and that these systems can be implemented in all countries. It is often said that it is not possible to build passive treatment systems because of the large space required. This may well be correct for the active operation of mines, but one could also think of building an active system and for 'final treatment' building a smaller passive treatment system or polishing pond to polish the mine water after the end of operation, as at the Ynysarwed plant in South Wales, United Kingdom (Brown et al. 2002; Walder et al. 2005). In addition, there are a variety of passive treatment system types, and if the available space is small, it is still possible to build a RAPS/SAPS – just not like in South Korea, where, because of translation errors, every passive treatment system was a RAPS/SAPS for a long time (pers. comm. Bob Kleinmann). With the knowledge that engineers and scientists have now, it should be easy to build and optimise passive treatment systems that take up less space than before. In the beginning, Mr. Otto's engines also blew up in his face – and today we reliably control high-performance engines with sophisticated computer systems that fly me to St. Helena or Walvis Bay, and the speeds would surprise even Mr. Otto.

As this is not yet the case, the question arises at the outset: what is passive mine water treatment, what systems are there, and how do they work? This is explained in Sect. 4.2. The focus is on methods that are already being used successfully. I will not discuss experimental processes that have never progressed beyond a pilot stage (e.g. SCOOFI reactor: PIRAMID Consortium 2003, p. 47), which does not mean that they are unsuitable for passive mine water treatment in individual cases.

As the cost of mine water treatment is often a crucial deciding factor, Eppink et al. (2020) compared the costs and design criteria of six passive treatment systems in New Zealand with each other and with active treatment. Under highly variable flow conditions, such as at the Blackball system with a relative standard deviation of 50% for the flow, passive treatment has a disadvantage in terms of costs in comparison to active treatment systems. If a planned system is on the rule of thumb boundary between active and passive treatment systems, the most important financial factors are the site-specific conditions.

Finally, two quotes from colleagues regarding passive mine water treatment:

> All stakeholders need to take a fresh look at an old problem. Nature's repair mechanisms may be slow but they're thorough. We must find ways to assist and expedite them (Kalin 2004a).

Earlier, Davison (1991) commented as follows on the same subject:

> We believe that there are limitations, but the courage to try new techniques and change out-dated methods is the most difficult hurdle we face in developing new demonstration sites. I think it will finally come down to not 'what does it cost?', but 'can I afford not to?', as construction and chemical costs continue to soar and compliance becomes tougher to maintain and afford. The industry is slow to change, but when they're ready, we'll be ready too. It can't be done by lunch yesterday and it can't be done cheaply, but it can be done, whenever you're ready to do it.

So let's get on the passive treatment road trip in the next sections.

## 4.2 What Is Passive Mine Water Treatment?

The publications by Wildemann et al. (1993a, b) and Hedin et al. (1994a) form the basis for all passive mine water treatment; the early developments of these systems are summarised in Kleinmann et al. (2021). They described and summarised the most important principles of passive mine water treatment for the first time. A very good overview of passive methods for sulfate removal is given by Kaksonen and Puhakka (2007; including the always wrong Fig. 4.1e for a RAPS). Gusek (2009) and Gusek and Waples (2009) have designed a periodic table of passive mine water treatment and described its application (Fig. 4.2, Table 4.2). From this, it can be seen which passive treatment technology is appli-

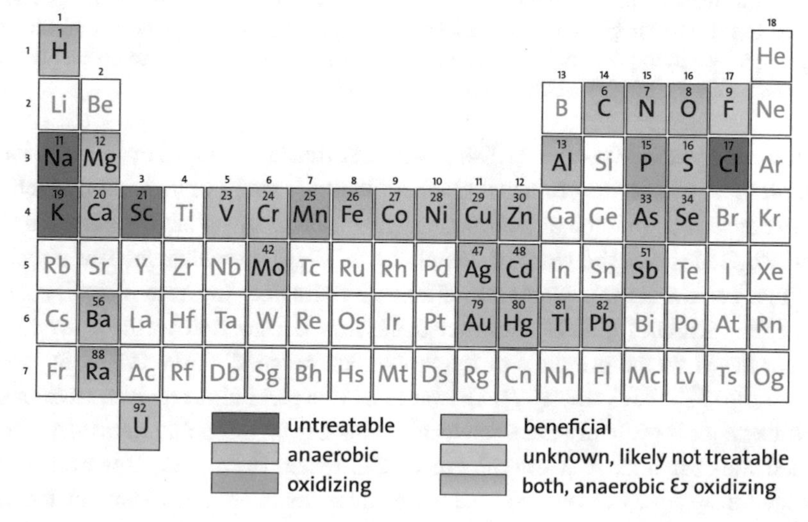


**Fig. 4.2** Periodic table of elements for the passive treatment of mine water. Explanations are given in Table 4.2

**Table 4.2** Oxidation and reduction conditions in "conventional" passive treatment systems

| Passive treatment technology | Aerobic ($E_h > 0$ mV) oxidising conditions | Anaerobic ($E_h < 0$ mV) reducing conditions |
|---|---|---|
| Biochemical bioreactors | • Upper 2–3 cm | • Bulk of the cell |
| Aerobic wetlands | • | |
| Oxidation & settling ponds | • | |
| Anoxic limestone drains | | (•) |
| Successive alkalinity producing systems | Upper 2–3 cm | • Bulk of the cell |
| Vertical flow reactor | | • In substrate |
| Open limestone channels and limestone beds | • | |

Modified from Gusek (2009), Gusek and Waples (2009). Biochemical bioreactors are also referred to as sulfate-reducing bioreactors. $E_h$ values refer to the standard hydrogen electrode; bullets mean that the conditions are met in the respective system

cable to which element, but their list does not include technologies that show promise but may still be under development. Although an anoxic limestone drain (ALD) may have slightly reducing conditions, its function is to add alkalinity, and not to remove metals. Therefore, it should not be inferred that ALDs are suitable for precipitation of metals other than aluminium and iron. The most important basis for the use of passive treatment technology in Europe is provided by the final report of the EU project PIRAMID (contract number EVK1-CT-1999-000021), which was published in 2003 (PIRAMID Consortium 2003). It explains in detail how to prepare, plan and construct a passive mine water treatment system. According to the definition given there, it is passive mine water treatment if the following conditions apply:

> Passive in situ remediation (PIR) signifies an engineering intervention which prevents, diminishes and/or treats polluted waters at source, using only naturally available energy sources (such as topographical gradient, microbial metabolic energy, photosynthesis and chemical energy), and which requires only infrequent (albeit regular) maintenance to operate successfully over its design life (PIRAMID Consortium 2003, p. vi).

However, it was not Hedin et al. (1994a) who first treated mine water in a passive treatment plant; there was an experimental plant in Germany many years earlier which treated mine water as well as municipal wastewater. The basis for this plant in Othfresen north of the Harz Mountains was the work of Seidel (1952, 1966), in which she was able to demonstrate that certain aquatic plants can eliminate pollutants from municipal wastewater. Based on this, Kickuth (1977) constructed the first constructed wetland for municipal wastewater based on the principle of the "root zone process" (Brix 1987a, b; Bucksteeg 1986; Ebeling 1986; Kickuth 1977). The fact that this plant also received smaller quantities of mine water is probably rather an irony of history (presumably mine water from the abandoned Ida-Bismarck iron ore mine near Othfresen, Germany). However, there is a fundamental difference between passive mine water treatment and a constructed wetland for municipal wastewater: municipal wastewater wetlands treat wastewater with elevated concentrations of organic water constituents and nutrients, whereas a passive mine water treatment system must treat largely inorganic constituents of mine water. This difference is specifically important in countries where both systems are given different names in local language, but the translated English names result in an identical translation (e.g. in Germany, where a municipal wastewater wetland is called “Pflanzenkläranlage”, plant treatment system, whereas a passive mine water treatment plant should be called “Konstruiertes Feuchtgebiet”, constructed wetland).

To date, no comprehensive experience with passive mine water treatment is available in German-speaking countries, although a German-language publication on the subject appeared as early as 2002 (Wolkersdorfer and Younger 2002). The reason for this may be that most installations have been constructed in principle like constructed wetlands, without sufficient consideration of the special features for mine water. Furthermore, aerobic systems were constructed for net acidic mine drainage and the treatment performance consequently did not meet the predicted expectations. In addition, the negative example of Lehesten/Thuringia is repeatedly used to prove that passive mine water treatment does not work. Also the plants in Pöhla/Saxony (pilot and full-scale plants), Paitzdorf/Saxony or Schlema/Saxony, which did not meet the expectations, do not help to inspire confidence in the method. Whether the experiences of Opitz et al. (2019) with the pilot plant in Wackersdorf, Germany can contribute to the development of a full system, must be shown in the future.

Often, basic principles, such as those outlined by Hedin et al. (1994a) or Wildeman et al. (1993a), have been insufficiently considered, which can either lead to the failure of the treatment plants or that their full performance is only attained to a limited extent. However, experience at the Urgeiriça uranium mine in Portugal (Fig. 4.3), where passive mine water treatment plants have been in place since 2012, shows that uranium-containing mine water can be passively treated (Pinto et al. 2016).

**Fig. 4.3** Sedimentation and macrophyte basin and – no longer necessary – activated carbon filter bed and aerobic wetland at the Urgeiriça uranium mine, Portugal. Mine water ($Q \approx 90$ L min$^{-1}$) flows from right to left, and pH is raised from about 6 to 8 on average. Iron, uranium and radium are retained in the system

The definition given in the second paragraph of this section therefore includes passive treatment as well as passive prevention, for example by covering or flooding tailings or waste dumps. Processes that do not meet this definition are therefore either active treatment systems (e.g. conventional mine water treatment) or semi-passive treatment systems (e.g. installations requiring the addition of alcohol or molasses).

Skousen and Ziemkiewicz (2005) made an initial attempt to compare the effectiveness of 116 passive treatment systems. Several passive treatment systems such as aerobic and anaerobic constructed wetlands, sulfate-reducing bioreactors, anoxic limestone drains, RAPS, open limestone drains, and open limestone drains were subjected to analysis, and their design specifications were compared with actual treatment performance. They discuss the performance and reliability of the various systems in numerous tables and examples. They conclude that "passive systems with low iron or that have had the iron removed before introduction into a passive system have a much greater chance of remaining viable and effective" (Skousen and Ziemkiewicz 2005, p. 1129).

Extensive hydrogeological investigations are necessary before selecting an appropriate passive treatment system. Based on the results, the system is then selected using the flow diagram of Hedin et al. (1994a) and Skousen et al. (1998, 2000) or the one in PIRAMID Consortium (2003) (Fig. 4.4).

So let's start by describing these systems, but I will limit myself to the basic features of each method. If you want to set up one of these systems described, you should read the specific literature cited in each case – but in any case, please read the three sources cited at the beginning of this Sect. 4.2: Wildemann et al. (1993a, b); Hedin et al. (1994a) or Kleinmann et al. (2021, p. 818), who state that they "wanted to help newcomers to the field avoid repeating the mistakes of the past or rediscovering what had already been observed and documented."

## 4.3 Limestone Drains and Channels

### 4.3.1 Classification of Limestone Drains and Channels

There are three types of limestone drains and channels: anoxic limestone drain, open limestone drain and open limestone channel. While the first two are closed carbonate dissolution systems that do not have extensive contact with the atmosphere, the limestone channel is a system that is open to the atmosphere. This also explains the different flow velocities and residence times of the mine water in the respective system: low flow velocities and long residence times in the drains and high velocities with short residence times in the channel.

### 4.3.2 Anoxic Limestone Drain (ALD)

Anoxic limestone drains are covered (buried) limestone-filled drains in which alkalinity is added to the acidic mine water by dissolving the limestone. Or in other words, acid is buffered in an anoxic limestone drain and the acid capacity ($k_A$ value) of the mine water is

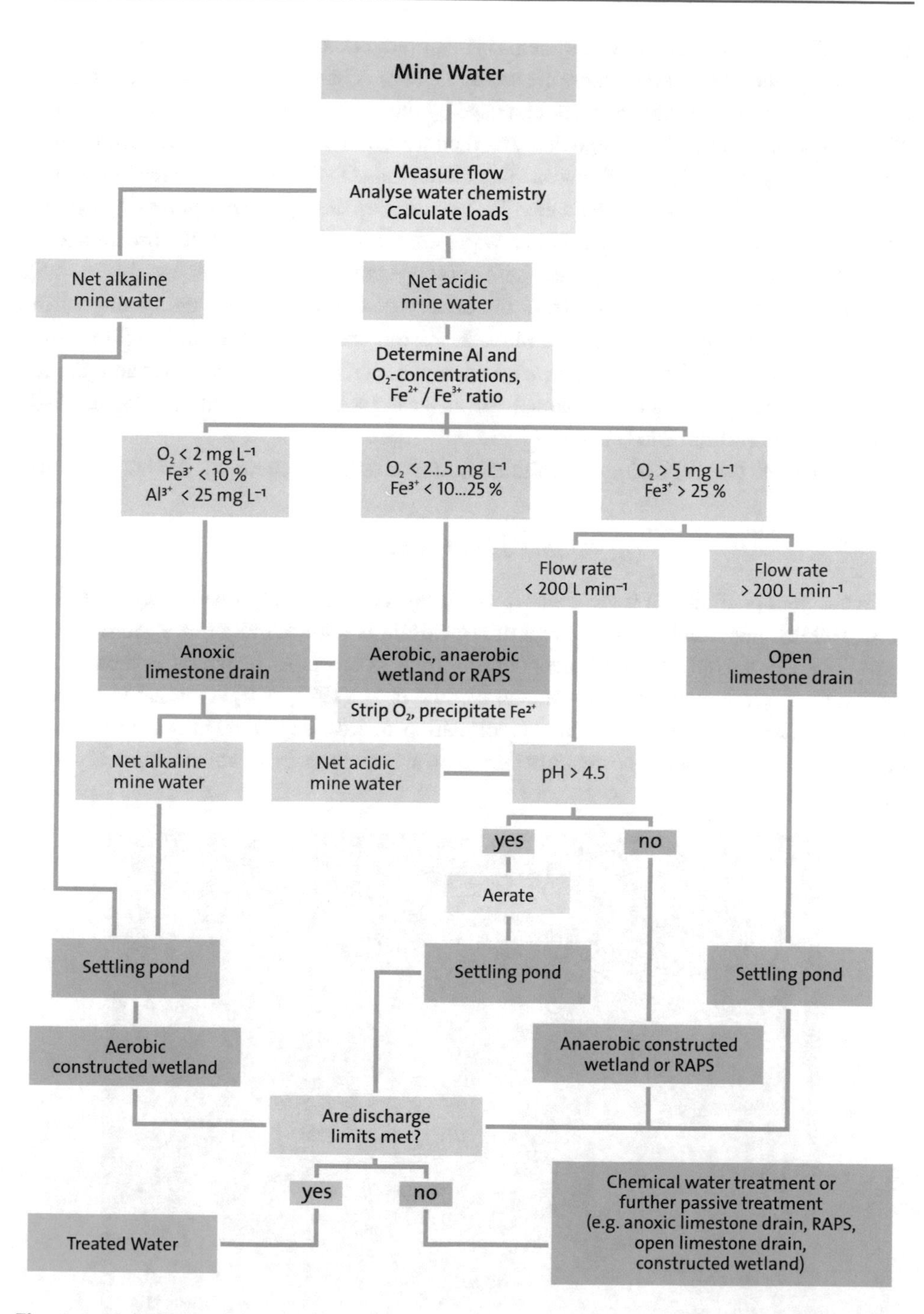


**Fig. 4.4** Flow diagram for passive mine water treatment processes. (modified after Hedin et al. 1994a; Skousen et al. 1998, 2000; PIRAMID Consortium 2003)

increased. The advantage of these systems is that precipitation of iron or aluminium oxides is generally prevented in the limestone drain, as there is hardly any oxygen present in the airtight conduit. These systems were first used by the Tennessee Valley Authority, USA, in 1988 to buffer acid mine drainage that was then treated in aerobic wetlands (Brodie et al. 1991). Already by 1991, over 50 anoxic limestone drains had been constructed in the USA to increase the alkalinity of mine water. Provided the system has been properly designed and the chemical parameters of the mine water are within the limits for the process, a life span of 20–80 years can be expected before the system needs to be renewed (Fig. 4.5). This is approximately calculated from the acidity of the mine water, the quality of the limestone and its hydraulic permeability. A design method that would facilitate this exchange is not yet known. To achieve an adequate increase in acid capacity and pH, the mine water must reside in the channel for at least 14 h (Hedin et al. 1994b, p. 1343; PIRAMID Consortium 2003, p. 69).

Chemically, the following, simplified buffer reaction occurs in an anoxic limestone drain:

$$CaCO_3 + H^+ \rightarrow Ca^{2+} + HCO_3^- \quad (4.1)$$

To avoid precipitation, the mine water must almost be completely free of oxygen (Hedin et al. 1994b). Importantly, an anoxic limestone drain is not a stand-alone system (Brodie et al. 1991, p. 9), but serves to precondition mine water, most of which is then further treated in an aerobic constructed wetland. Brodie et al. (1991, p. 4) state that in properly constructed anoxic limestone drains, precipitation of iron oxyhydrates can be predominantly prevented. Because the construction of an anoxic limestone drain is relatively inex-

**Fig. 4.5** Anoxic limestone drain during the construction phase. The liner sits in the open drain which is already filled with limestone at the rear end. (Photo: Jeff Skouson)

pensive compared to a conventional mine water treatment system, Brodie et al. (1991, p. 5) recommend that the drains be oversized to provide sufficient alkalinity to the mine water and have a large enough safety factor.

For an anoxic limestone drain to function reliably, the water quality must meet certain criteria. For this purpose, the pH value, oxygen saturation, redox potential, acid and base capacity, and iron and manganese concentrations must be known. The oxygen concentration of the mine water should not exceed 2–3 mg $L^{-1}$, the redox potential should not exceed 300 mV, and the pH should not exceed 6 (Hedin et al. 1994b). At Al concentrations above 2 mg $L^{-1}$, or if the mine water has $Fe^{3+}$ concentrations above 2 mg $L^{-1}$, the limestone drain may clog within a few months. This effect was previously reported by Hedin et al. (1994a, p. 28) and PIRAMID Consortium (2003, p. 48). To remove possible aluminium precipitates, Kepler and McCleary (1997) proposed a system known as the "aluminator", which allows the passive treatment system to be backwashed (see also Vinci and Schmidt 2001). As the example of Lehesten/Thuringia, Germany, with 14 mg $L^{-1}$ aluminium has shown, the drains can become clogged relatively quickly if flushing is not carried out regularly (Janneck and Krüger 1999, p. 52 ff.).

To construct the drain, PVC liners with a thickness of 10 or 20 mm should be used (Fig. 4.5); these liners are strong enough to prevent damage during and after the construction phase. However, not all systems have a liner (PIRAMID Consortium 2003, p. 69). When constructing the drain, the sides can be sloped, but it is better if they are nearly vertical. The limestone used should contain more than 90% $CaCO_3$ and have a grain size of 16/32 mm (in the original work: ¾ × 1 ½ inches: Brodie et al. 1991, p. 5), since at this grain size the ratio between surface area and hydraulic permeability is optimal. The thickness of the limestone layer is determined by the expected maximum flow, the lifetime of the drain, and a sufficiently large factor of safety for both variables. At least 60 cm of impermeable material (e.g. clay) should be placed above the limestone drain to largely prevent oxygen from entering.

The required volume of limestone $V_g$ (in the equation it is assumed that the limestone has a purity of 100% – if necessary, the amount must be increased) and the cross-sectional area $A$ of the anoxic limestone drain can be calculated using a simplified rule of thumb as follows (PIRAMID Consortium 2003, p. 69):

$$V_g = \frac{Q_e \cdot \varphi}{n_e} \tag{4.2}$$

$$A = \frac{Q_e}{k_f \cdot I} \tag{4.3}$$

where:

$V_g$ volume of limestone, $m^3$
A cross section of drain, $m^2$
$Q_e$ expected flow, $m^3$ $s^{-1}$

$\varphi$ factor (for 14 h mean residence time: 50400 s), s
$n_e$ average effective porosity of the packed bed, usually around 49% (0.49)
$k_f$ hydraulic permeability, m $s^{-1}$
$I$ hydraulic gradient

However, Cravotta and Watzlaf (2002) concluded that a larger number of parameters is required for a correct calculation of the necessary limestone amount. This has mainly to do with the fact that the dissolution rate of the limestone is not linear but changes exponentially over time. Overall, they propose a larger number of different equations based on various laboratory experiments and experience with existing limestone drains. To allow for different mean residence times, they also introduce time $t$ as a variable, not just the 14 h of the above equations:

$$M_0 = e^{k \cdot t} \frac{Q_e \cdot t_d \cdot \rho_s \cdot (1 - n_e)}{n_e} \tag{4.4}$$

where:

$M_0$ mass of limestone, kg
$k$ solution rate of limestone, $h^{-1}$ (on average $2.7 \cdot 10^{-6}$ $h^{-1}$)
$t$ lifetime of the limestone drain, h (usually 14–15 years)
$Q_e$ expected flow, $m^3$ $h^{-1}$
$t_d$ average residence time in the drain, h (usually 12–22 h)
$\rho_s$ density of limestone, kg $m^{-3}$ (between 2600 and 2900 kg $m^{-3}$)
$n_e$ average effective porosity of the packed bed, usually around 49% (0.49)

All the above constants are based on the experience gained from 13 limestone drains and extensive laboratory experiments as indicated in the publication of Cravotta and Watzlaf (2002) and the rock densities provided by Prinz and Strauß (2018). However, it is recommended that the constants be determined on a site-specific basis. Hence, from the mass and the bulk volume of the limestone, its volume can then be calculated. Based on a comparison of six passive treatment systems in New Zealand, Eppink et al. (2020) concluded that the rule-of-thumb methods used to plan passive treatment systems are a suitable design approach.

Downstream of the limestone drain and upstream of the aerobic wetland, a settling or polishing pond should first follow to allow the iron to hydrolyse and precipitate under the now elevated pH. If the iron concentration is below 50 mg $L^{-1}$, the treated mine water can be discharged directly into an aerobic constructed wetland.

### 4.3.3 Oxic Limestone Drain (OLD)

Oxic limestone drains fulfil a similar function to anoxic limestone drains: they increase the alkalinity (acid capacity) of the mine water. Unlike these, however, the mine water does not have to be anoxic, but it is necessary to keep the velocity of the water sufficiently high to prevent limestone clogging. Cravotta (1998) and Cravotta and Trahan (1999) recommend mean residence times of less than 3 h and mean effective velocities of 0.1–0.4 m $min^{-1}$. The mine water must not be too highly mineralised, and the base capacity must not be too high: $O_2$ > 1 mg $L^{-1}$; base capacity <1.6 mmol $L^{-1}$; total $Al^{3+}$, $Fe^{3+}$, $Fe^{2+}$, $Mn^{2+}$ and trace elements <5 mg $L^{-1}$. Cravotta (2010) indicates that oxic limestone drains in the Swatara Creek Basin (Pennsylvania, USA) were relatively expensive compared to the treatment capacity. In another case, clogging of the limestone drain occurred relatively quickly, requiring the installation of a backwash system (Cravotta 2007). The design criteria for the limestone quantity correspond to those used for the anoxic limestone drain.

However, based on previous experience with oxic limestone drains, at present it cannot be recommended to install such systems, as the risk of clogging is too high if the design parameters are not precisely adhered to.

### 4.3.4 Open Limestone Channel (OLC)

Open limestone channels are characterised by larger limestone aggregates over which the mine water flows and is buffered. They are suitable for mine water with higher oxygen and TDS concentrations. However, the sulfate concentration should be below gypsum saturation. The slope of the channel must be sufficiently large (10–20%) to prevent clogging of the limestone (PIRAMID Consortium 2003; Ziemkiewicz et al. 1996; Ziemkiewicz et al. 1997), otherwise the treatment efficiency is no longer guaranteed.

Even with open limestone channels that are clogged, buffering of the mine water can still be observed in ideal cases (Fig. 4.6), and the performance drops to 20% of that of unclogged limestone channels. This effect can be accounted for in the design by using five

**Fig. 4.6** Open limestone channel with iron hydroxide precipitates at the mine water discharge of the former Dominion No. 11 coal mine on Cape Breton Island, Nova Scotia, Canada

times more limestone than calculated. However, if the residence time of the mine water on the limestone is too short, no positive effect can be demonstrated. This is the case, for example, at the Steinbach rivulet in Gernrode, Germany (Menzingteich), where mine water flows over a limestone cascade a few metres high and the average residence time is a few seconds. An increase in the pH value could not be detected there. Rather, a longer, open limestone channel with a corresponding gradient could possibly have made a considerable contribution to improving the water quality there.

Over a period of 5 years, Alcolea et al. (2012) studied the functioning of an open limestone channel in Sierra de Cartagena-La Unión (south-eastern Spain), a 2500-year-old mining region with lead and zinc deposits. The length of the open limestone channel was approximately 2 km, the gradient was 4.6%, and the channel aimed to passively treat acid mine and tailings leachate from a 9 km$^2$ area. After 5 years of operation, the treatment performance of the channel had decreased substantially, which the authors attribute to the lack of maintenance, the relatively low gradient, and thus the flow velocity of the mine water.

However, as soon as an open limestone drain becomes too clogged, it is no longer able to increase the acid capacity of the mine water. Furthermore, if the mine water has iron concentrations above about 2 mg L$^{-1}$, a pH drop of the mine water through iron hydrolysis occurs. An example of this was the open limestone drain at Mina de Campanema/Minas Gerais in Brazil (Fig. 4.7). There, in 2001, slightly acid (pH 6.4), net acidic mine water was discharged from a drainage water catchment into the receiving watercourse. After passing the channel, the pH decreased to 6.2 and the acid capacity decreased from 0.10 mmol L$^{-1}$ to 0.04 mmol L$^{-1}$. At the same time, the total iron concentration decreased from 6.5 mg L$^{-1}$ to 0.7 mg L$^{-1}$, and the redox potential increased from 250 mV to 330 mV (Wolkersdorfer 2008, p. 249). Wherever acid mine drainage needs to be buffered before it reaches a receiving watercourse, open limestone channels are a suitable option as they are relatively inexpensive to construct and maintain.

## 4.4 Constructed Wetlands

### 4.4.1 Prologue (I Have Always Wanted to Write This)

Leaving Othfresen aside for the moment, we can clearly establish the date of the first passive mine water treatment system in Europe and its initiator: it was Paul Younger (Fig. 2.7) who set up the first European pilot plant for passive mine water treatment at Quaking Houses in Northumberland, England on 20th February 1995 (Younger et al. 1997, p. 205). Since then, much water has flowed down the Stanley Burn, a full system has been constructed (Fig. 4.8), and in the meantime passive mine water treatment is considered a standard solution in the United Kingdom (Crooks and Thorn 2016). This has been primarily based on the design criteria in PIRAMID Consortium (2003) and, in the meantime, on proprietary solutions based on these. As recently as 1997, Price (1997, p. 133) wrote: "The

**Fig. 4.7** Former open limestone channel at the Mina de Campanema/Minas Gerais, Brazil. The limestone clogging is clearly visible

**Fig. 4.8** First European anaerobic constructed wetland Quaking Houses (County Durham, England, United Kingdom). Plant cover is absent, as the sludge had been replaced after 10 years of operation

use of passive Wetland Systems to treat polluting discharges from abandoned mines is still a relatively new science in the UK, which has had limited practical application to date." In 1995, I first pointed out the use of constructed wetlands for mine water treatment in Germany: "In addition, as practised in other mines (Norton 1992, 1995), the effluent waters can be routed through a wetland to reduce the contaminants below the concentrations of today's seepage waters" (Wolkersdorfer 1995, p. 806). More than 20 years later, I feel that a similar development to that in the United Kingdom is currently taking place in Germany and that constructed wetlands and passive treatment technology in general, can be established as an alternative to traditional mine water treatment.

Importantly, and I point this out again, aerobic and anaerobic constructed wetlands are not interchangeable: Aerobic systems require net alkaline mine water, and anaerobic systems can also treat net acid mine drainage (see Fig. 4.4). However, in both systems, the sludge and substrate must be replaced after an operating period of 10–20 years. To date, the sludge can only be reused to a limited extent (cf. Sect. 7.2), and therefore a suitable final disposal method (either in a landfill or by incineration) must be identified. Shepherd et al. (2020) investigated the chemical and physical properties of the spent sludge that was used in the Jennings, Pennsylvania, USA, RAPS for 15 years. They found that the results from the leach tests did not show an exceedance of the RCRA criteria, and it is therefore not considered as hazardous material. It is often believed that the plants in constructed wetlands or in a RAPS would take up the metals. However, this is only marginally the case, as demonstrated, for example, by Batty and Younger (2002). Rather, the plants in the constructed wetland contribute to lowering the flow rate of water in the system. This results in longer mean residence times and correspondingly a support for the treatment process (Opitz et al. 2021). In addition, the plants provide a complex community for microorganisms and fungi in their rhizosphere ("mycorrhization"), since relatively high carbon concentrations are present there. The fact that some of the precipitated iron oxides can be dissolved by siderophores and are available in the plant as nutrients (Blume et al. 2010) is beyond any doubt.

### 4.4.2 Aerobic Constructed Wetland (Reed Bed)

An aerobic constructed wetland is a planted wetland with a water depth of 15–50 cm that treats net alkaline mine water (Sects. 1.2.3 and 1.2.13) with a pH above ≈ 6. Since protons are released during the hydrolysis of iron, aerobic systems are not suitable for net acid mine drainage because the pH would drop even further, and the iron would not precipitate to the desired extent. In general, the systems are simple to produce and blend well into the landscape (PIRAMID Consortium 2003). Emergent plants (rooted in the substrate but with firm stems and leaves extending above the water surface) are used and suitable plants are metal-tolerant plants such as cattails (*Typha* sp.), reed grass (*Phragmites* sp.), and rushes (*Juncus* sp.), which require about one to two growing seasons to acclimate to the particular mine water (Fig. 4.9). They can be considered state of the art for the treatment

**Fig. 4.9** Inlet (left) and outlet (right) of the aerobic constructed wetland Neville Street (Cape Breton Island, Nova Scotia, Canada). Water flows into the constructed wetland in an ochre and metal-rich state (left) and exits in a crystal-clear state with a substantially reduced metal load (right)

of net alkaline mine water, as plants of various sizes have been successfully used to treat iron-rich mine water since the 1980s in the USA and since the 1990s in the United Kingdom. In total, there are probably several thousand aerobic constructed wetlands worldwide. In addition to iron, aerobic wetlands can also remove manganese and zinc, although the effectiveness of the latter two elements is less than that for iron. Sorption or coprecipitation also removes a number of other metals and metalloids, which are described in more detail in Gusek (2013). To account for the amount of sludge produced in the wetland, a sufficient freeboard must be planned for, which depends on the metal load that is to be treated.

It is important that plant species appropriate for use in a constructed wetland are selected; they should be sourced locally to ensure that they are adapted to the local climatic conditions. There are cases where the “foreign” plants did not grow well enough under the local climatic conditions. The role of the plants is essentially to provide growing surfaces for microorganisms, to lower the flow rate of the mine water in the wetland, and to maintain hydraulic conductivity in the substrate so that biologically catalysed reactions can occur (Batty 2003). In addition, the plants help to outgas $CO_2$ dissolved in the mine water, thereby increasing the pH of the water (Wolkersdorfer and Younger 2002). Therefore, the emergent vegetation should be planted in a largely statistically distributed manner to prevent hydraulic short circuiting in the system. The substrate used for the plants depends on local conditions and can be any organic material that is available. At the Neville Street aerobic wetland in Nova Scotia, Canada, for example, residues from crab processing were used and mixed with humus.

Although wetland plants can accumulate metals, their primary role is not to actively remove metals from mine water. However, it has been shown that metal removal in wetlands increases with the maturity of the plant cover (Batty 2003). Yet, at lower iron concentrations of up to 1 mg $L^{-1}$, metal uptake into plants can account for a large proportion of metal elimination, and this accumulation can be relevant when aerobic wetlands are used as a final treatment step.

As with active mine water treatment plants, the question of sludge disposal in the aerobic constructed wetland arises. It must be solved in a similar way as for active mine water treatment: The sludge must be removed, the substrate rebuilt, and the plants replanted. However, because the sludge is about ten times denser than for active mine water treatment (Dempsey and Jeon 2001), its volume and thus its disposal costs are lower. Experience suggests that the sludge must be removed at least every 10–20 years; if the sludge is used to extract pigments (Hedin 2016; Kirby et al. 1999), the treatment frequency may be shorter (pers. comm. Bob Hedin). The sludge may need to be stabilised as described at the end of Sect. 3.2.1.

To determine the area required for an aerobic wetland, there are numerous approaches that can be assigned to zero- or first-order equations (Hedin et al. 1994a, p. 25 f.; Tarutis et al. 1999; Younger et al. 2002). However, it has been shown that zero-order equations can be used to calculate the area requirements of an aerobic constructed wetland sufficiently well (Younger et al. 2002, p. 327 f.), although Zipper and Skousen (2010) elaborated that the effectiveness of passive treatment systems depends on several parameters rather than just one. It is important, however, that two systems should be connected in parallel so that the parallel system can be used during maintenance or cleaning. According to PIRAMID Consortium (2003, p. 65 f.), the area required in an aerobic constructed wetland is calculated as follows:

$$A = \frac{Q_\mathrm{d}}{R_\mathrm{A}} \cdot \left(c_\mathrm{i} - c_\mathrm{t}\right) \tag{4.5}$$

where:

$A$ wetland area, $m^2$
$Q_d$ average flow, $m^3\ d^{-1}$
$R_A$ area-adjusted treatment rates of the water constituent, $g\ m^{-2}\ d^{-1}$
$c_\mathrm{i}$ average daily concentration of the water constituent in the influent, $mg\ L^{-1}$
$c_t$ average daily desired concentration of the water constituent in the effluent, $mg\ L^{-1}$

The magnitude of $R_A$ depends on whether the discharge criteria are to be met at all times (required treatment rates) or whether the mine water is to be treated only as a matter of principle (acceptable treatment rates) and is based on empirical values from numerous wetland systems (Table 4.3).

Before a aerobic constructed wetland is to be built, it is essential to conduct sound hydrogeological monitoring. For this purpose, the flow rate over a hydrological year must be known, and at least the iron, aluminium and manganese concentrations must be measured regularly. It is also important to know the iron speciation and the $k_\mathrm{B}$ and $k_\mathrm{A}$ values of the mine water to be treated. Without this information, a constructed wetland cannot be designed with sufficient accuracy. Once again, experience shows that systems usually do not work if sufficiently accurate mass loads were not known in advance.

### 4.4.3 Anaerobic Constructed Wetland (Anaerobic Wetland, Compost Wetland)

Anaerobic constructed wetlands treat net acid mine drainage or mine water with elevated sulfate concentrations with pH values below ≈ 6. They are also known as compost wetlands (PIRAMID Consortium 2003; Wildeman et al. 1993a, b). Unlike aerobic wetlands, the choice of substrate plays a crucial role in anaerobic constructed wetlands, because the mine water in the latter must flow through the substrate rather than the plant cover. Therefore, the substrate must have a good hydraulic conductivity, and the water level in the wetland must reach a maximum of one decimetre, preferably less. The reason for constructing the first anaerobic wetlands was to counteract the pH decrease of the mine water due to the hydrolysis of iron and aluminium and the formation of acid. As Hedin et al. (1988) were able to show, microbially catalysed reactions occur in the anaerobic wetlands that are responsible for fixing the metal ions as sulfides in the substrate of the wetland. In principle, these reactions correspond exactly to those in oxygen-deficient marine sediments, where sulfate reducers contribute to the formation of monosulfides and eventually pyrite (Berner 1972, 1984, 1985). Optimal conditions for the growth of sulfate reducers range between pH values of 4.2 and 10.4 and redox potentials of −0.5 to +0.35 V (Trudinger 1979, p. 269). Yet, Hedin et al. (1988) were not able to distinguish during their initial studies whether microbial sulfate reduction or hydrolysis was causally responsible for the decrease in metal and sulfate concentrations (Table 4.4,

**Table 4.3** Empirical treatment rates $R_A$ of aerobic ("alkaline") and anaerobic ("acidic") constructed wetlands to be used for the area calculation of constructed wetlands (Hedin et al. 1994a)

| | Acceptable treatment rates $g\ m^{-2}\ d^{-1}$ | | Required treatment rates $g\ m^{-2}\ d^{-1}$ | |
|---|---|---|---|---|
| Water classification | Net alkaline | Net acidic | Net alkaline | Net acidic |
| Fe removal | 20 | – | 10 | – |
| Mn removal | 1 | – | 0.5 | – |
| Acidity removal | – | 7 | – | 3.5 |

Acidity is expressed in units of g $CaCO_3$ equivalent ($c_{azi\text{-}eq}$[g $L^{-1}$] = 50.04 × $k_B$ [mol $L^{-1}$]). The water classification is the classification explained in Sect. 1.2.13
– not applicable

**Table 4.4** Change of mine water chemistry in the course of a anaerobic constructed wetland in Westmoreland County, Pennsylvania, USA (Hedin et al. 1988) when sampled on 7th October 1987 (sampling locations in Fig. 4.10); units in mg $L^{-1}$, pH without unit, $k_B$ in mmol $L^{-1}$

| Sampling point | pH | $k_B$ | $SO_4^{2-}$ | $Fe_{tot}$ | $Fe^{2+}$ | Al | Mn | Na | Ca | Mg |
|---|---|---|---|---|---|---|---|---|---|---|
| A: Inflow | 2.86 | 17.4 | 2050 | 181 | 73 | 50 | 36 | 4 | 194 | 133 |
| C: Black zone | 4.75 | 0.2 | 1725 | 30 | 28 | <0.4 | 58 | 5 | 404 | 159 |
| D: Black zone | 5.39 | 0.1 | 850 | 3 | 3 | <0.4 | 11 | 6 | 214 | 82 |
| B: Orange zone | 2.96 | 15.5 | 2000 | 147 | 39 | 47 | 36 | 5 | 223 | 139 |
| E: Black zone with white precipitate | 4.60 | 3.2 | 1000 | 87 | 84 | <0.4 | 20 | 4 | 171 | 75 |

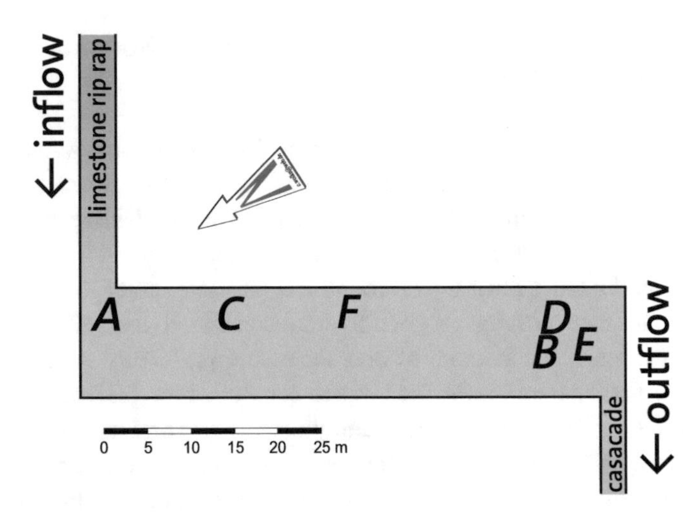


**Fig. 4.10** Sampling locations of Table 4.4 in the anaerobic constructed wetland Westmoreland County, Pennsylvania, USA. (After Hedin et al. 1988)

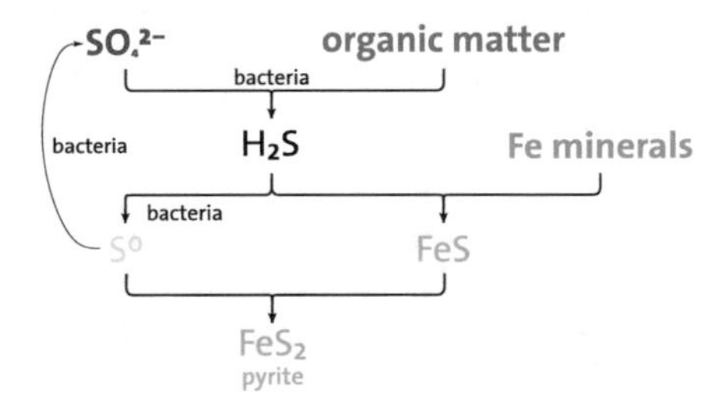


**Fig. 4.11** Illustration of the overall process for sulfate reduction and formation of monosulfides and pyrite in sediments. (Modified after Berner 1972)

Fig. 4.10). However, as Berner (1984, p. 606) points out, "the first step in the overall process of sedimentary pyrite formation is the bacterial reduction of sulfate" (Fig. 4.11). In addition to compost, anaerobic wetlands contain alkaline material to buffer acidity. Limestone is used in the vast majority of cases, and sometimes dolomite.

The substrate for anaerobic wetlands can be almost any organic substrate that serves as an electron donor and carbon source for the microorganisms and provides a suitable environment for the growth of sulfate-reducing microorganisms. It is often said that constructed wetlands would only work with mushroom compost. In fact, the first constructed wetlands for mine water were constructed using this waste as a substrate (Hammack and Hedin 1989; Hedin et al. 1988). But what is mushroom compost commonly made of? It is nothing more than an organic substrate "consisting of composted manures, straw, corncobs, and an occasional horseshoe or syringe" (Hammack and Hedin 1989). It is composed of about 25% organic matter with moisture of around 60%, has a circumneutral pH, and nitrogen concentrations around 1%, including 0.1% ammonium nitrogen. Its electrical conductivity is usually relatively high (0.5–13 mS $cm^{-1}$), often due to the concentrations of sodium, iron, manganese and aluminium in the compost (Van Houten 2018). Consequently, since mushroom compost is essentially organic material, there is no need to use mushroom compost, but any suitable organic substrate, with composted horse manure being particularly suitable (Younger et al. 2002) – a side note here: modern mushroom farms are not using manure any more due to hygenic reasons; you might need to source more suitable substrate nowadays. A good compilation of possible organic substrates and their effectiveness has been given by Skousen et al. (2017). The thickness of the substrate does not need to exceed 0.5 m, as diffusion of mine water to greater substrate depths is quite slow. In addition, construction costs increase with increasing excavation depth (Younger et al. 2002). In some anaerobic wetlands, limestone has been added to the compost to induce higher alkalinity (e.g., Hedin et al. 1988; PIRAMID Consortium 2003). In addition, like root penetration, this increases the hydraulic conductivity of the substrate.

The same plants as in aerobic wetlands are suitable as plant cover. In addition, however, the dead plants provide a carbon source for the sulfate-reducing bacteria (Batty 2003). Furthermore, enzymes released by the roots play a role in the mineralisation of the dead plants. Another advantage of plants is their function of stabilising the substrate. They thus help to counteract erosion of the substrate at higher flow rates.

PIRAMID Consortium (2003, p. 50) recommend that an anaerobic wetland should only be constructed if the hydraulic gradient is not high enough to construct a RAPS (see Sect. 4.5). The performance of a RAPS is generally better than that of an anaerobic wetland, and therefore it should be given preference over the latter. Care should be taken in the design to ensure that the flow through the wetland is as uniform as possible, so that areas where reactions do not proceed as planned are minimised (Hedin et al. 1988). Areas with oxidising conditions should be avoided to a large extent to obtain optimum treatment performance (Table 4.4, Fig. 4.10).

The first European anaerobic wetland was constructed in 1995 (Fig. 4.8). It treats sulfate-rich, aluminium- and iron-containing mine water from a spoil heap at Morrison Busty Colliery near Quaking Houses (County Durham, England, United Kingdom). Its pH is 4.4–7.0, iron concentration 2–40 mg $L^{-1}$, aluminium concentration 6–35 mg $L^{-1}$, and sulfate concentrations vary between of 170 mg $L^{-1}$ and 1800 mg $L^{-1}$. On average, at a mean flow rate of 127 L $min^{-1}$ in the wetland, 50% of the iron and 50% of the base capacity is removed (range of variation for both parameters 0–100%: Jarvis and Younger (1999); Younger (1998); Younger et al. (1997); the figures in the last publication do not agree with those in the other two, and the figures listed here are compiled from the three publications).

## 4.5 Reducing and Alkalinity Producing Systems (RAPS); Successive Alkalinity Producing Systems (SAPS); Sulfate Reducing Bioreactor, Vertical Flow Wetlands

Reducing and alkalinity producing systems (RAPS) are a combination of an anoxic limestone drain and an anaerobic constructed wetland (Rose and Dietz 2002) and are now also popularly referred to as vertical flow wetlands (Skousen et al. 2017). They combine the advantages of the two systems and circumvent the often observed precipitation of oxyhydrates at aerobic limestone drains. Provided a sufficiently high hydraulic gradient is available, they are preferable to anaerobic wetlands. The first systems of this type were constructed between 1991 and 1994 (Kepler and McCleary 1994, 1995). Today, their treatment (or operating) mechanisms are relatively well known (Matthies et al. 2009), and therefore the criticisms made in Bilek (2012, p. 30) do not apply. Worldwide, there are now probably well over 100 such systems treating net acid mine water; in Pennsylvania (USA) alone, 114 of these systems are installed, treating around 40 million $m^3$ of acid mine drainage annually (www.datashed.org).

A RAPS consists of two elements: a limestone layer and a compost layer, through which mine water flows vertically. First, the mine water flows through the compost bed, where microbial processes consume oxygen, reduce $Fe^{3+}$ to $Fe^{2+}$, and produce $CO_2$ or methane (Fig. 4.12). In the next process step, the mine water flows through the limestone bed where it is neutralised (PIRAMID Consortium 2003). Reducing alkalinity systems are therefore particularly suitable for the treatment of net acid mine water with elevated $Fe^{3+}$,

**Fig. 4.12** Channels caused by gas upwelling on the surface of the organic substrate covered with iron oxyhydrates in the RAPS Bowden Close, County Durham, England. Diameter of the channels in the centimetre range; image width in the centre about 3 m

aluminium and oxygen concentrations. To ensure a steady flow of water, and to be able to store the resulting sludge, a head difference of 2.5 m between inlet and outlet is necessary and a freeboard of at least 1 m is required. In the large RAPS at Pelenna II (Wales, United Kingdom), this was not always the case, so that the treatment performance of the system fell far short of the design criteria (Rees and Connelly 2003).

Because of the functional similarity of a RAPS and an anoxic limestone drain, the amount of limestone is calculated in exactly the same manner. Equation 4.4 can be applied for this purpose. The bulk volume of this limestone quantity can then be used to calculate the quantity of organic substrate, which is just as large in terms of volume.

At the Bowden Close RAPS, unintentional mixing of limestone and compost occurred during filling by heavy equipment. Therefore, the RAPS was eventually designed as a single layer system (Fabian et al. 2006, p. 17 f.). Subsequently, it was found that the effectiveness of the system had increased compared to the layered pilot system. Consequently, unlike described in the original work, there is no need to separate the two substrates.

An essential design criterion is the avoidance of short circuits between the inlet, the flow path and the outlet. This can be ensured by height-adjustable drainage pipes in the outlet chambers (Fabian et al. 2006, p. 18). It is also important that the chambers are designed in such a way that they have sufficient space for measuring instruments, data loggers or autosamplers.

A RAPS should not be understood as a single system for treatment but should always be followed by an aerobic wetland to aerate the water and to remove any remaining iron from the mine water by hydrolysis (Figs. 4.13 and 4.14). Furthermore, a RAPS is not primarily a sulfate-reducing system, but rather it is designed to add alkalinity to the mine water (Gusek 2002). To date, only one pilot plant has been operational in Germany at the Hohe Warte mine near Gernrode in the Harz Mountains. Once the flow rate of the plant had been optimally adjusted, the iron concentration was reduced from 20 mg $L^{-1}$ to 5 mg $L^{-1}$ and the base capacity from 1.1 L to 0.6 mmol $L^{-1}$. At the same time, the pH increased from 5.5 to 7 and the acid capacity from 0.27 mmol $L^{-1}$ to 0.7 mmol $L^{-1}$. The initial nitrate concentrations of 4 mg $L^{-1}$ in the RAPS effluent decreased to 0.3 mg $L^{-1}$ during the 7 months of the trial (Hasche-Berger and Wolkersdorfer 2005; Hasche-Berger et al. 2006; Hasche and Wolkersdorfer 2004).

## 4.6 Settling Pond (Settling Basin, Settlement Lagoon)

Settling ponds are used in both passive and active mine water treatment systems. They are appropriate whenever hydrolysis of the iron occurs rapidly enough after the mine water discharge or when the mine water is already alkaline enough to allow rapid iron hydrolysis (Geroni et al. 2009; Train et al. 1976, pp. 144–168; Younger et al. 2002). Also, if the mine water has an elevated suspended solids concentration, it is a good idea to place a settling pond upstream of the mine water treatment system (Fig. 4.15). In the ARUM system, for example, settling basins protect the macrophytes from clogging (Sect. 4.10).

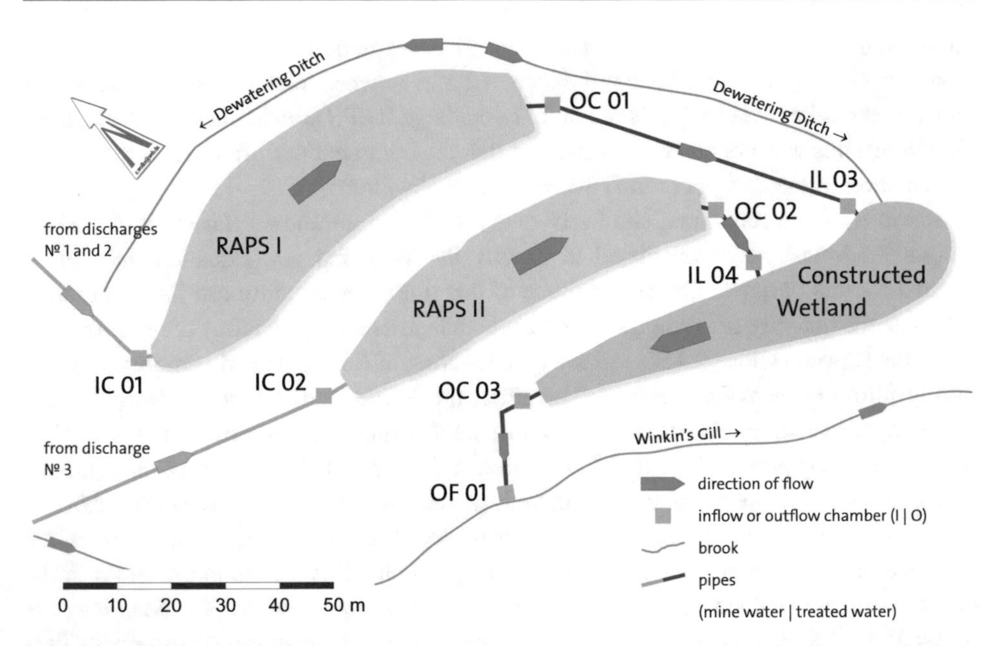


**Fig. 4.13** Site plan of the Bowden Close RAPS, County Durham, England. (Modified from Wolkersdorfer et al. 2016)

**Fig. 4.14** RAPS II at the Bowden Close site, County Durham, England, looking west. Typical wetland plants are growing by succession on the substrate of limestone and horse manure. Outside the image on the right is RAPS I and on the left is the constructed wetland

**Fig. 4.15** Settling pond at the former Straßberg/Harz, Germany, mine water treatment plant. The green colour is mainly caused by suspended limestone. Since the relinquishment of this treatment system, the pond is overgrown with plants

Already at iron concentrations above 5 mg $L^{-1}$ it is recommended to set up a settling pond, since 50–70% of the hydrolysed iron can accumulate therein. The mean residence time $\tau$ should be calculated so that it is at least 48 h. At lower mean residence times, the hydrolysed iron does not have sufficient time to flocculate. In such cases, the settling pond has no effect because the iron hydroxide would only settle in the subsequent wetland (Wolkersdorfer 2011). The construction of a settling pond upstream of the constructed wetland can also be omitted for net acid mine drainage, since a sufficient amount of iron hydroxide cannot usually be formed. Settling ponds for solids (e.g. sand, silt) in mine water can be calculated for average residence times of 3–4 h.

Settling ponds for contaminated mine water have been known in the literature since the sixteenth century. Lazarus Ercker v. Schreckenfels (1565) mentions them for mining in the Rammelsberg, Germany, where the ochre sumps still contribute to improving the water quality in the receiving streams Abzucht and Oker, and where the settled ochre was used for paint production (Brauer 2001; Roseneck 1993).

In Lusatia, Germany, the LMBV constructed several settling ponds to ensure that the elevated iron concentrations are removed from the rising groundwater (Benthaus et al. 2015a). Examples include the Wüstenhain or Vetschau mine water treatment plants, for which Bilek et al. (2013) discussed potential uses or final disposal of the iron sludge produced there. The installed earthen dams at the Vetschau plant achieve a mean calculated residence time of 21 days and reduce the iron concentration of the water from 6–14 mg $L^{-1}$ to below 2 mg $L^{-1}$ (Bilek et al. 2013, p. 16). Other settling ponds for oxidised iron used by the LMBV in addition to the Vetschau plant are the former fishponds of the Eichow plant. The LMBV is currently discussing whether additional fishponds there should be converted into settling ponds.

Particularly important are guiding foils (baffles) made of PVC or earth walls, which are intended to prevent hydraulic short circuiting from occurring in the settling pond. Short circuits, and thus an insufficient mean residence time, result in the inability to meet quality criteria at the discharge (Hilton 1993). As Wolkersdorfer (2011) has shown using the settling ponds of Neville Street, Nova Scotia, Canada, the installation of baffles allowed the mean residence time of mine water to be increased from 10–18 h to 35 h (Fig. 4.16). This in turn led to a much better sedimentation rate and consequently better mine water quality. Chamberlain and Moorhouse (2016) also demonstrated at the Clough Foot mine water treatment plant, Yorkshire, United Kingdom, that water quality in a settling basin can be improved by diverting the water with baffles, although the degree of improvement varies. Let me summarise loosely with Hilton (1993), "Well, since we are stuck with these ponds or ditches, what do you do? BAFFLE–BAFFLE–BAFFLE".

In addition, settling ponds should be constructed upstream of a constructed wetland or ARUM basin so that the rapidly hydrolysed iron does not settle in the constructed wetland itself – where it would be difficult to remove – but one step upstream. If the settling pond

**Fig. 4.16** Settling pond at the Neville Street passive mine water treatment system, Nova Scotia, Canada, before and after installation of baffle sheets. (Details in Wolkersdorfer 2011)

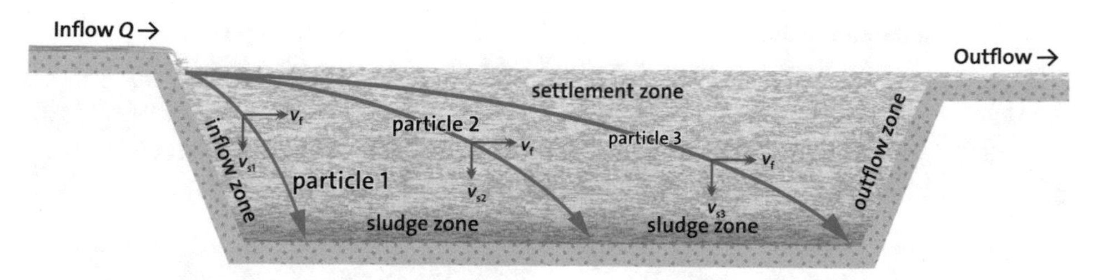


**Fig. 4.17** Schematic diagram of the settling process in a simple settling pond. (Modified after Crittenden et al. 2012)

is well constructed, the iron sludge and coprecipitated cations can be mechanically removed from the pond fairly easily. For larger plants, it is always advisable to construct two settling ponds running in parallel so that the treatment plant does not come to a standstill when the pond needs to be cleaned.

Each particle in a settling pond moves along a path which is a function of the forward velocity $v_f$ and the settling velocity $v_s$ (Fig. 4.17). While $v_f$ *is* usually considered constant, although this is hardly true in normal cases, $v_s$ depends on the density of the particles. Whether a linear movement path of the particles results or a curved one as in Fig. 4.17 depends on the geometry of the basin and the time-dependent coagulation of the particles, from which the velocity distribution in the basin is determined. In any case, the larger or heavier particles will settle in the inlet zone and the progressively lighter or smaller particles will settle along the sludge zone. If the velocity $v_f$ is too high compared to the necessary settling time, the particles will no longer settle within the settling basin, but will pass over the outlet zone into the outlet. To reliably prevent this, the velocity in the settling pond must not be below the critical velocity $v_c$ for each particle, which is calculated as follows:

$$v_c = \frac{h_0}{\tau} \tag{4.6}$$

Where $h_0$ *is* the height of the water level in the settling basin and $\tau$ is the mean residence time in the settling basin (Crittenden et al. 2012).

If it is necessary to prevent waterfowl or other animals from landing on the water surface, floats or photothermal material can be applied in the ponds, basins or lagoons. Should, at the same time, evaporation of the potentially contaminated water be required, fibreglass floats have been developed to serve both purposes (Monasmith et al. 2020).

## 4.7 Permeable Reactive Walls

Permeable reactive walls are vertical in situ systems placed in the subsurface across the groundwater flow direction; they contain reactive materials in sufficient quantity to remove potential contaminants (Fig. 4.18). Their original purpose was to remove organic contami-

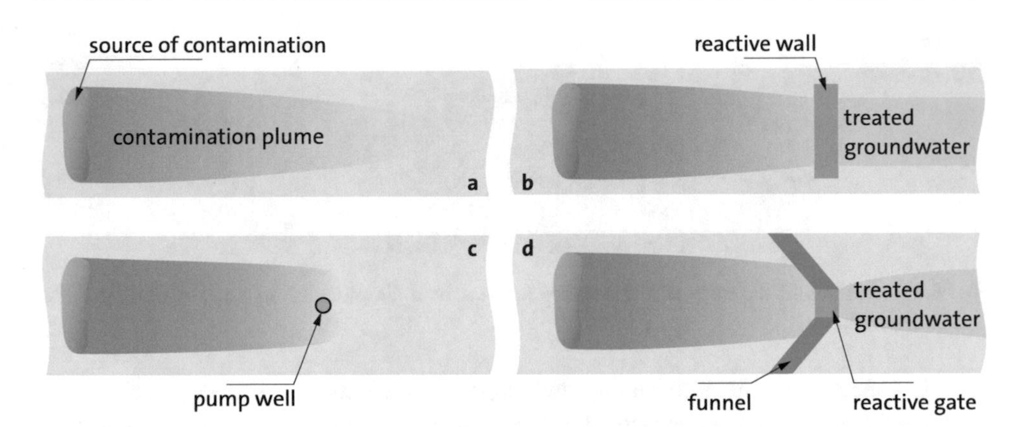


**Fig. 4.18** Remediation options for groundwater contamination: (**a**) untreated contamination; (**b**) reactive wall; (**c**) pump and treat; and (**d**) funnel-and-gate system. (After Starr and Cherry 1994)

nants; however, they have been further developed for the removal of inorganic contaminants (Blowes et al. 2000; Naftz et al. 2002). Waybrant et al. (1995) and Blowes et al. (1995) first used a permeable reactive wall at the Nickel Rim mine in Ontario, Canada, to treat mine water from tailings. By testing a large number of reactive materials, they succeeded in lowering the sulfate and iron concentrations of the water and increasing the pH and redox potential. The physico-chemical processes that take place in the substrate can be sorption, redox reactions or microbially catalysed sulfate reduction (Blowes et al. 1995; Noubactep 2010; Younger et al. 2003; Zoumis 2003).

Permeable reactive walls have several advantages over other treatment systems: Generally, no complex technology is required for installation, energy consumption is low to non-existent, groundwater treatment occurs in situ without having to operate equipment at the surface, and interference with the groundwater regime is generally minimal. A substantial disadvantage is that the reaction products within the reactor lower the system's permeability and groundwater can flow around the reactor (and I really do mean *substantial,* because reactive walls can only be used for mine water under optimal conditions). They also have a limited lifetime, depending on the nature of the reactive materials, the water to be treated and the reaction products generated (Blowes et al. 2000). Therefore, careful planning is necessary to be able to exclude these effects to a large extent. Considering all the advantages and disadvantages of the method, permeable reactive walls must be considered to have great potential for treating contaminated groundwater, although the number of published and SCI-listed publications has been declining since 2010. Long-term experience with the Nickel Rim (Ontario, Canada), Monticello (Utah, USA) and Fry Canyon (Utah, USA) permeable reactive walls is reported in Blowes et al. (2003). They also briefly review the basics of the process. Because contaminated groundwater was often found to flow past the reactive wall, systems were developed that directed the groundwater toward a reactive gate by means of a funnelled impermeable wall (funnel-and-gate). They have the advantage that bypasses are largely avoided (Fig. 4.18d).

In Europe, field experiments have been conducted with mining contaminated groundwater in Hungary (Roehl 2004), Spain (Gibert et al. 2004), Germany (Schöpke et al. 2006), Portugal (Pinto et al. 2016, in which case I am not sure if one can speak of a reactive wall *s.s.*, even if the authors state that), and in the United Kingdom. Of these systems, the one that works most effectively is the reactive wall in Shilbottle in Northumberland, north-east England (Bowden et al. 2005; Samborska et al. 2013), whereas little detail is available about the other systems in terms of their current effectiveness. Reactive materials can be, for example, $Fe^0$, limestone or compost. In addition, there have been experiments with fly ash or with cement kiln dust and blast furnace slag (Sartz 2010). One of the main problems of reactive walls is that they become clogged by ochre relatively quickly, so the selection of material with sufficiently large permeability is of major importance (Munro et al. 2004). Numerous strategies exist to avoid or slow down the process of ochre build-up, and these are discussed in detail in the literature (e.g. Bolzicco et al. 2003; Gozzard et al. 2005; PIRAMID Consortium 2003). The substrate used can also reduce the permeability to such an extent that the flow is almost prevented, as was clearly shown by the test facility at Skado dam in Lusatia, Germany: "The tests further showed that the walls introduced have a hydraulic sealing effect" (Schöpke et al. 2006, p. 171). The same has already been reported by Blowes et al. (2000), who, in addition to the above-mentioned advantages of reactive walls, also list in detail their disadvantages, some of which have been mentioned above.

On October 1st, 2011, patent protection for permeable reactive walls expired in the European Union because the licence fee was no longer paid (Patent DE 692 31983.2; Patentblatt 132 (5) 2012, p. 3194). Strictly speaking, all previously constructed systems are subject to the patent protection of Waterloo University, which has patented all conceivable processes of "reactive walls" or "funnel-and-gate systems" (Starr and Cherry 1994; pers. comm. David Blowes 2011).

## 4.8 Vertical Flow Reactor (VFR)

A vertical flow reactor (VFR) is a passive treatment system for treating ferruginous mine water in which the contaminated mine water flows vertically through a 10–20 cm thick gravel bed (Fig. 4.19). In this process, the gravel bed acts both as a filter for larger particles and as a growth medium for microorganisms that contribute to the autocatalytic reactions occurring in the VFR (Barnes 2008; Sapsford and Williams 2009). Previous sizes of vertical flow reactors range from less than 1 $m^3$ to 65 $m^3$, thus occupying relatively little space. Some authors also refer to SAPS or RAPS systems as vertical flow reactors (e.g. Bhattacharya et al. 2008; Vinci and Schmidt 2001) – but these I describe in Sect. 4.5. Although this method for mine water treatment can be considered new, the physicochemical principles date back to earlier studies of autocatalytic iron removal (e.g. Best and Aikman 1983; Burke and Banwart 2002; Zhang et al. 1992). The latter have found their way into various treatment systems described by Younger et al. (2002, p. 344), called

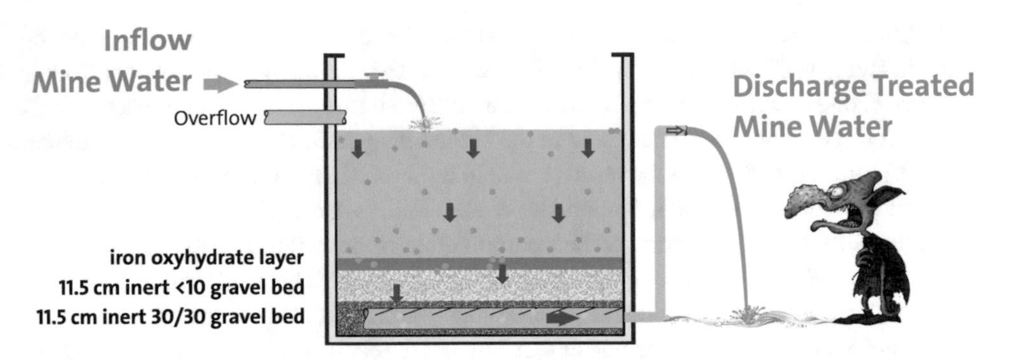


**Fig. 4.19** Schematic diagram of a vertical flow reactor (modified after Florence et al. 2016). Prolotroll (*Stollentroll*, Adit Troll) by Walter Moers (2000), in *The 13½ Lives of Captain Bluebear* © Verlagsgruppe Random House GmbH

surface-catalysed oxidation of ferrous iron (SCOOFI) reactors in this case (see also Burke and Banwart 2002). Best and Aikman (1983, p. 563 f.) state: "it is apparent that precipitated iron particles will readily 'grow' under aerobic conditions on suitable surfaces to form a robust sludge of relatively high density, which should have good dewatering properties."

Building on this knowledge, the first experimental vertical flow reactor was constructed. In a narrower sense, the experimental vertical flow reactor at Taff Merthyr, Wales, United Kingdom, is considered the first attempt to treat mine water using this method (Sapsford et al. 2006). This was preceded by experiments using a 1000 L intermediate bulk container (IBC), which showed that iron could be removed from mine water without the need for growth media, unlike the SCOOFI reactor (Dey et al. 2003; Sapsford et al. 2005). The Taff Merthyr mine water is circumneutral mine water (pH 6.7) with an acid capacity of 2.33 mmol $L^{-1}$ and an iron concentration of 7 mg $L^{-1}$ (Sapsford et al. 2007). After an initial iron removal rate of 53% (Sapsford et al. 2005), the rate increased to 99–100% by the end of the test and after optimisation of the gravel bed (Sapsford et al. 2007). In the final year of the experiment, it ranged between 42% and 100%, with the lower rates occurring when flows were too high. As shown by Barnes (2008) and Geroni (2011), the main mechanisms contributing to the removal of iron are autocatalysis and filtration by the iron hydroxide and gravel bed. Dey et al. (2003) conclude that about 79 g of iron per square metre per day can be removed with a vertical flow reactor.

Using an IBC, Florence et al. (2016) investigated whether mine water from the former Cwm Rheidol lead mine, Wales, United Kingdom, could be treated using a vertical flow reactor. Unlike the Taff Merthyr circumneutral mine water, this mine water is acidic and has a pH of 2.9 with iron concentrations ranging from 24 mg $L^{-1}$ to 126 mg $L^{-1}$ (average 95 mg $L^{-1}$). Over the course of 1 year, iron removal rates of 26–84% could be achieved (average 67%), with no dependence on mean residence times in the system. It was shown that autocatalysis in acid mine drainage plays no role in iron removal in a vertical flow reactor. Rather, the critical mechanisms are filtration, heterogeneous precipitation, and microbially catalysed oxidation followed by precipitation. Another key aspect of the Cwm

Rheidol experimental system was the long transport route from the mine adit to the vertical flow reactor (Florence 2014, pp. 189–191): first through a 600 m long abandoned adit and then through an 800–900 m long pipe. Already on this path, oxidation of the mine water, hydrolysis and precipitation of the iron occur. Therefore, average iron removal rates of 67% cannot be considered representative for acid mine drainage.

A mine water similar to that of Cwm Rheidol is emerging from an abandoned coal mine near Carolina in Mpumalanga, South Africa. It is also strongly acidic and contains elevated iron concentrations. So I installed an IBC there as well and upgraded it to a vertical flow reactor. However, unlike in Cwm Rheidol, removal rates of only about 20–40% could be achieved, which can be seen as evidence of the special position of Cwm Rheidol. While an ochre layer formed relatively quickly in Cwm Rheidol, hardly any ochre layer was observed in Carolina even after several months. I installed another vertical flow reactor at the former polymetallic Metsämonttu mine in Finland where multiple metals were mined (Zn, Cu, Pb, Au, and Ag). There, as at Taff Merthyr, circumneutral mine water emerges with iron concentrations of 5–15 mg $L^{-1}$. After an experimental reactor (600 L) showed removal rates of 95% after only 6 weeks, I had a containerised 24 $m^3$ tank constructed in 2016 that could treat all the mine water (Fig. 4.20). After 3 weeks, the iron removal rate had already increased to 55% at a flow rate of 5–30 L $min^{-1}$ and an average residence time of 11–13 h and has since reached up to 95%. The tank at Metsämonttu is the world's first containerised vertical flow reactor and the first that is able to treat the entire volume of mine water discharging.

**Fig. 4.20** Containerised vertical flow reactor (VFR) at the former Finnish polymetallic Metsämonttu mine with ventilation of the mine water (Sansox Oy OxTube). The outlet of the reactor with measuring weir for the flow is located at the rear end of the 20″-container (water volume: 24 $m^3$)

## 4.9 Passive Oxidation Systems (Cascades, "Trompe")

Most metal hydroxides are less soluble in higher oxidation states than in lower ones (Stumm and Morgan 1996). Oxygen is therefore one of the most important reagents in many water treatment systems (Wehrli 1990, p. 324), and therefore mine water should first be aerated before being treated in a treatment system. This is true for virtually all treatment plant systems except for anoxic limestone drains. The reason for this is that iron in the discharged mine water is often in the reduced ferrous ($Fe^{2+}$) species and must be oxidised to the oxidised ferric ($Fe^{3+}$) species (Glover 1975, p. 189; Gusek and Figueroa 2009, p. 90). The same is true for manganese. The supporting effect of oxidation allows mine water treatment systems to be operated more effectively with cost savings (see Sect. 3.2 for notes on mine water aeration in active treatment systems), which has long been known and studied for improving the quality of water bodies (Novak 1994). In addition to the oxygen concentration of the mine water, the reaction time of the oxidation of $Fe^{2+}$ to $Fe^{3+}$, the acid capacity, temperature, concentration of organics (Sharma 2001) and the residence time of the mine water in the settling pond play an important role in mine water treatment.

Leonardo da Vinci described very clearly what happens during oxidation (Codex A *f.* 59 *r* in the Bibliothèque de l'Institut de France in Paris):

> The water, when falling on other water, leaves its previous place in manifold, different, bifurcated and inwardly curved ramifications (it becomes more and more entangled and interwoven); these ramified parts rebound from the surface of the water, but the weight and the strike coming from said water are so strong that the air, because of the highest rapidity, does not have time to flee into its own element, but submerges in the manner described before (da Vinci and Schneider 2011).

Water can dissolve a maximum of about 10 mg $L^{-1}$ of oxygen. Therefore, to oxidise higher iron concentrations, the mine water must first be aerated due to the generally low oxygen concentrations in mine water. This can be achieved by active aeration, cascade aerators (Figs. 4.21 and 4.22), or chemical reagents (e.g., hydrogen peroxide, ozone; Younger et al. 2002, p. 278). While $Fe^{2+}$ requires a pH of about 8.5 to precipitate in sufficient quantity, a pH of about 7 is already sufficient for $Fe^{3+}$ – and the pH can certainly be adjusted to the desired level in passive treatment systems (Fig. 3.3; Table 3.6). Highly net alkaline mine water can even be passively treated with aeration and a settling basin only (Cravotta 2007), since iron oxidises rapidly at high pH values (Fig. 3.4). It is important to note that the hydrolysis of iron releases two protons per mole of iron (Eqs. 4.7 and 4.8). If the mine water is already highly acidic, then simple aeration will not achieve a positive effect, but rather lowers the pH (Fig. 4.23):

$$Fe^{2+} + \frac{1}{4}O_2 + \mathbf{H^+} \leftrightarrow Fe^{3+} + \frac{1}{2}H_2O \qquad (4.7)$$

$$Fe^{3+} + 3H_2O \leftrightarrow Fe(OH)_3 + \mathbf{3H^+} \qquad (4.8)$$

Since aeration experiments are time-consuming, this pH decrease to $pH_{Ox}$ by aeration can be approximately calculated using PHREEQC (Lausitzer und Mitteldeutsche Bergbau-Verwaltungsgesellschaft mbH 2019, p. 24, Annex VI-a). A mine water from Schlabendorf Nord, Germany, with an initial pH of 4.03, showed a $pH_{Ox}$ of 2.92 after the calculated aeration.

**Fig. 4.21** Oxidation cascade of the ventilator adit on the Leitzach river, Upper Bavaria, Germany. Flow rate about 2 $m^3$ $min^{-1}$; width of the channel about 1 m. In the meantime it was completely rebuilt

**Fig. 4.22** Oxidation cascade of the Neville Street passive mine water treatment plant, Nova Scotia, Canada, during a tracer test with uranine (sodium fluorescein). Flow rate about 1 $m^3$ $s^{-1}$; height of the individual cascade stages about 0.6 m

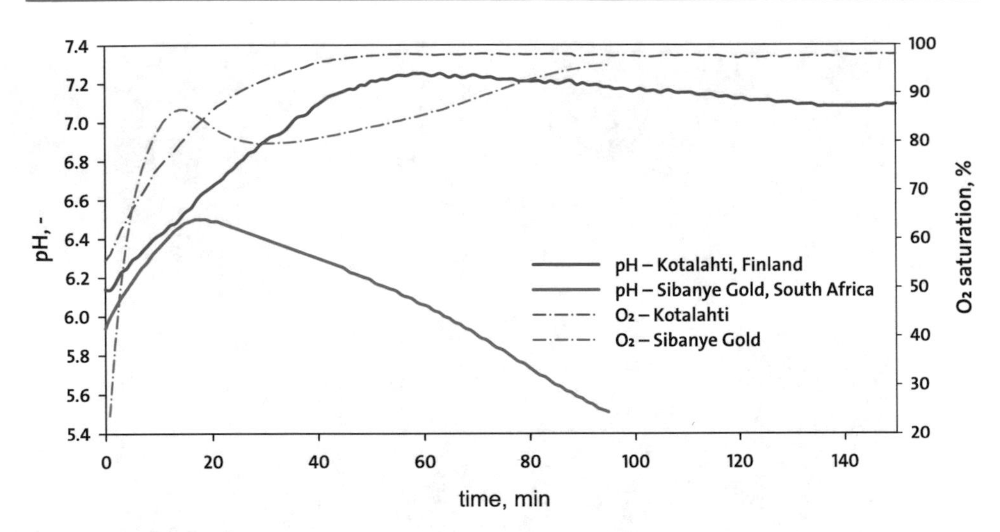


**Fig. 4.23** Results of two aeration experiments of mine water from Finland and South Africa. The mine water in Kotalahti, Finland, has sufficient buffer capacity to buffer the protons formed during iron hydrolysis, which is lacking in the mine water from the West Rand, South Africa, so that the pH value decreases

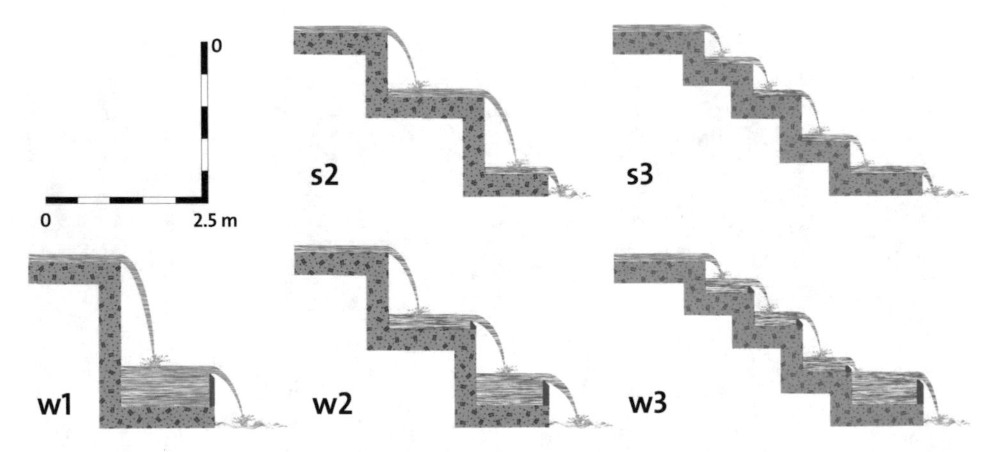


**Fig. 4.24** Five different stair- and weir-type aeration cascades. (Modified after Geroni 2011, Fig. 4.5)

Of the various systems used to aerate mine water, the cascade aerator is the most commonly used method, with the number of steps usually determining how much oxygen can be dissolved in the water. Two types of cascade aerators are used, a simpler one with steps only (stair type) and another one with weirs (weir type), often constructed as V-weirs (Fig. 4.24). Experience has shown that a cascade aerator can oxidise approximately 30–50 mg $L^{-1}$ of iron (National Coal Board – Mining Department 1982; Younger et al. 2002, p. 275). Trumm et al. (2009) were able to raise the oxygen concentration of a mine water from below 1 mg $L^{-1}$ to nearly 10 mg $L^{-1}$ using five cascade drops, with oxygen concentrations exceeding 9 mg $L^{-1}$ by the third cascade. This resulted in an increased oxidation rate of iron, leading to an overall decrease in iron concentration from more than 15 mg $L^{-1}$ to below 2 mg $L^{-1}$. Geroni (2011, pp. 89–93, 177–184) studied five different stair- and weir-type designs to determine which design was best for aerating mine water. Interestingly, she was unable to find much difference between the five types, with types w1 and w2 performing best for the mine water she studied. This result is interesting in that it differs from those of other studies, which generally conclude that type w3 should be used for aeration. Geroni (2011, p. 183) attributes this to the fact that the mine water she studied had relatively low iron concentrations and a comparatively high acid capacity.

Similar experiments to those carried out by Geroni (2011) were conducted by Oh et al. (2016) in mine water containing 25 mg $L^{-1}$ $Fe^{2+}$. They compared a stair- to a weir-type system, with each system having a total height of 4 m and a width of 10 cm, with steps 50 cm high each. In addition to the thin plated weirs (Fig. 4.24), the authors installed perforated plastic baffles about one-third of the way up the weirs, which increased the residence time on a step to 8 min. The extent to which a cascade with a width of only 10 cm can provide representative results due to edge effects is not discussed here. However, it could be shown that the weir-type cascades with baffles produced a slightly better oxidation of the mine water compared to the stair type and, as expected, that the overall height of the systems contributes substantially to the oxygen saturation and thus to the iron oxidation. The result of iron oxidation in the final settling basin ($1.2 \times 2.4 \times 1.2$ m) is interesting. There, the weir-type cascade shows better results than the stair type. While the 4 m stair type resulted in a decrease of the $Fe^{2+}$ concentration to 5 mg $L^{-1}$ after about 400 min, this value was already reached after about 100 min with the weir type.

Among others, Chanson and Toombes (2002) found that there is a strong relationship between the turbulence of the water and the trapped air. Novak (1994) recognised that the contact time between the air and the mine water exerts a decisive influence. Therefore, the following can be highlighted as an essential criterion for oxidation: The stronger the turbulence and greater the contact time between air bubbles and mine water, the more successful the aeration of the mine water (which, five centuries earlier, Leonardo da Vinci described in a very similar way).

Another patented passive oxidation method for mine water is a water-lime mixer in conjunction with a water drum ("Trompe": Leavitt 2011). This is an apparatus without moving parts in which air is drawn in from side openings by falling water and released as compressed air via the "drum" (Buff 1851; Veith 1871, p. 561). Water drums for mine water aeration are currently used in Pennsylvania, USA, in the two passive treatment systems Curley in Fayette County and the North Fork in Allegheny County (Leavitt et al. 2012, 2013).

An additional positive effect of aeration is the release of $CO_2$ and an associated increase in pH. Many mine waters contain substantial amounts of "dissolved" $CO_2$, and aeration alone can remove this and raise the pH (e.g., Cravotta 2007; Geroni 2010; Geroni et al. 2009, 2011; Janneck et al. 2007b; Jarvis 2006; Kirby et al. 2009; Petritz et al. 2009; Sapsford 2013; Younger et al. 2002, p. 279 f.).

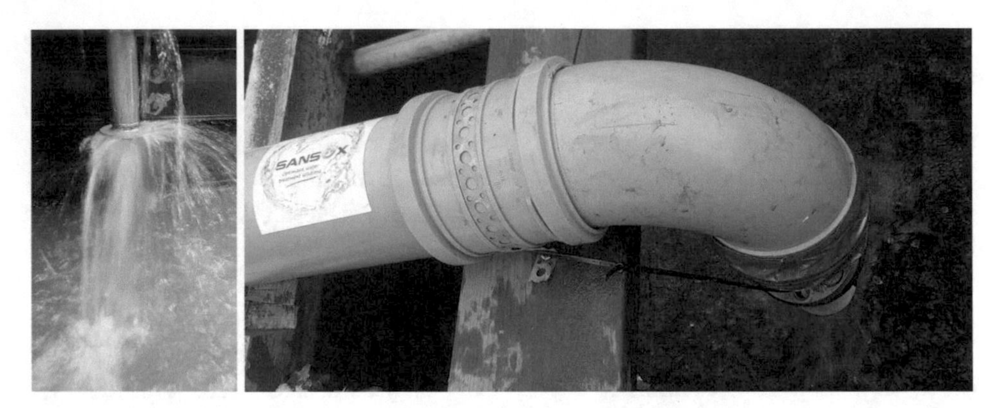


**Fig. 4.25** OxTube in 90° angle installation on the Metsämonttu vertical flow reactor. The oxygen saturation in the mine water could thus be raised from about 40 to 60%

Another passive treatment system for the oxidation of mine water is the OxTube developed in Finland (Fig. 3.11). It can be manufactured in various sizes, with 3-D printing being used for smaller ones. The operating principle of the OxTube is firstly an inlet, which functions according to the Venturi principle and through which the air is drawn in. This is followed shortly after by a diffuser, which finely distributes the sucked-in air and thus increases the oxygen saturation in the mine water. In the case of the Metsämonttu vertical flow reactor, the oxygen saturation could be raised from 40% to 60% in a first test, and the redox potential increased by about 50 mV. It is important to install the tube at a 45° angle at least, better still vertically to optimise the effect of the Venturi tube (Fig. 4.25).

At the abandoned Camphausen colliery, Germany, an OxTube system increased the oxygen saturation from 2% to 25%, the $S_2^-$ concentration decreased from 19 mg $L^{-1}$ to 5 mg $L^{-1}$, while a substantial decrease in the $H_2S$ smell was also observed.

## 4.10 ARUM (Acid Reduction Using Microbiology) Process

The ARUM process uses a combination of aerobic basins with downstream anaerobic water basins (polishing ponds) containing floating vegetation mats (macrophytes). These vegetation mats, once they completely cover the water surface, are supposed to stop the diffusion and mixing of oxygen and thus create reducing conditions in the entire water body. Further, the principle behind this is that the organic carbon is cycled through the vegetation, constantly renewing the organic substrate in the anaerobic pools. In this organic substrate, alkalinity is produced, and due to the reducing conditions, metal sulfides precipitate. In the ARUM system at the sludge ponds of the Copper Cliff mine in Sudbury, Canada (Makela), the anaerobic basins were supplemented with organic substrate from straw, potato waste and alfalfa and, in a system in Brazil, with waste from sugar cane as electron acceptors. These create a substrate with aerobic conditions that can be considered microbially active. After two years, 82–88% of the basins were covered with the vegetation mats (Johnson and Hallberg 2005; Kalin and Caetano Chaves 2003; essentially from MEND 2000, pp. 5–57). An aerobic system (Sect. 4.6) must be installed upstream of the anaerobic ARUM pools, because iron oxidation coats the surfaces of the sediment and the roots of the macrophytes with a layer of iron oxyhydrate, which prevents decomposition of the organic substrate. Thus, most of the iron is oxidised before it can precipitate in the ARUM process. This principle was first successfully tested at the Selminco coal mine tailings pile on Cape Breton Island, Canada (Kalin 1993), of which, however, there is only monitoring data for a very short period of 1–2 years; nothing remains of this system today, as it was unable to meet the expected treatment performance on a permanent basis (pers. comm. Joe Shea 2009). The ARUM process was developed by Boojum Research Ltd. in the late 1980s (Kalin et al. 1993; Kalin and Smith 1991) and has been steadily optimised since then, with mean residence times of 2–3 days and 1–1.5 m depths proving beneficial in the basins.

Physico-chemically, the ARUM process in the aerobic basins initially precipitates iron as iron oxyhydrate. In the anaerobic basins, the microbial processes create alkalinity, and an increase in pH is observed, and, due to the decreases of the redox potential and the microbially catalysed sulfate reduction, precipitation of metal sulfides.

Currently, a passive mine water treatment system based on the ARUM process is running successfully at the former Urgeiriça and Cunha Baixa uranium mines in Portugal (Carvalho et al. 2016). There, mine water is first passed through some aerobic basins until it reaches basins planted with macrophytes, which were not yet fully vegetated at the time of my visit in early 2019 (Fig. 4.26).

**Fig. 4.26** Macrophyte basin of an ARUM system at the former Cunha Baixa uranium mine, Portugal. Iron, uranium and radium are retained in the system

# 5 Alternative Methods for the Management of Mine Water

## 5.1 Thoughts on Alternative Treatment Methods and Their Application

Alternative methods for the treatment of mine water are methods that cannot be classified as either active or passive methods. Previous authors have often used the term "alternative methods" to refer to membrane technologies, ion exchange or electrochemical methods. However, I have decided to include these in the group of active methods. Therefore, which "alternative methods" remain? Strictly speaking, none, if we disregard the numerous interesting abatement strategies. These, however, are not the subject of this book. Therefore, the only remaining treatment strategy is "doing nothing". Whereby "doing nothing" is not quite correct, because two partners are involved here, one of whom "does" something while the other is an onlooker, a non-participating observer so to say: the natural processes and the human being, who monitors the processes with her possibilities. This brings us to "natural attenuation" or "monitored natural attenuation", which was first published in German by Max v. Pettenkofer in 1891 (if you are ever in Munich, you should not miss the MaxPett restaurant). I don't mean to recall the unhealthy discussion of the 1890s, nor that of the 1990s, where "natural attenuation" was occasionally understood as an invitation to sit back and wait until nature solves the problem itself – in fact, v. Pettenkofer's concept had also substantial negative effects on streams, as it was initially believed that the river would solve the problem anyway. On the contrary, it should mean that in certain areas we can let nature deal with our mine water itself; wastewater treatment has shown us for centuries, at times, when the loads from wastewater treatment plants were smaller, and there was hardly any industry, the receiving water was seen as a kind of bioreactor. Keywords include the Roman sewer system, *Cloaca Maxima,* or their septic drain fields (Hettersdorf 1908; Platner 1929). Today, Pettenkofer's approach will usually cause a conflict with existing laws, but closing our eyes (e.g. Schwefel adit [Schwefelstollen] in the



C. Wolkersdorfer, *Mine Water Treatment – Active and Passive Methods*,
https://doi.org/10.1007/978-3-662-65770-6_5

Selketal, Germany) or redefining mine water discharges ("naturally emerging source") does not help at this point either. However, instead of remediating every mine water contamination where a polluter can be found, the little money available should rather be spent on mine water discharges with elevated loads.

Monitored natural attenuation is therefore an alternative to the pump-and-treat or end-of-the-pipe methods commonly used today for the treatment of mine water. It depends on the individual case as well as on the legislative context and requires a comprehensive investigation of the respective discharge point based on generally accepted criteria, which have so far existed only to a limited extent for mine water discharges. Using a kind of checklist, it would then be possible to identify relatively quickly whether a mine water discharge needs to be comprehensively treated. However, all mine water discharges that are unlikely to cause lasting damage to the receiving watercourse or groundwater can then either be left to treat themselves naturally (Fig. 5.1) or treated using passive treatment technologies. From the juxtaposition of the two terms, you may see that "natural attenuation" is not the same as "passive treatment". Moreover, unlike the organic pollutants considered by v. Pettenkofer, the (semi)metals in the mine water are not removed from the material cycle – but merely fixed in a more harmless form or strongly diluted.

► Near Gernrode/Harz, Germany, there is a mine water discharge that has been polluting its receiving watercourse for some time (Anders 2002; Wolkersdorfer et al. 2003) and still does so despite extensive remediation measures. At one of the meetings with the authorities I attended, the owner of the mine water discharge wanted to know what limits he could expect when the water was treated and asked the competent authority representative sitting diagonally opposite him for a number. However, she said she could not give a figure until he applied for treatment, whereupon the owner asked the senior water authority representative sitting next to him for a number. The latter, however, pointed out that she was not entitled to set such figures, as this was a matter for the subordinate authority. Whereupon the owner sat back and said that he would not apply for a discharge permit and would simply let the water drain, because then it would not be officially known that mine water was flowing into the receiving watercourse – although he was prepared to install a passive

**Fig. 5.1** Natural attenuation using the example of a mine water adit in Bavaria, Germany (Phillipstollen on the Leitzach). Left: Outflow from the collapsed adit; middle: after a flow path of 5 m; right: after a flow path of 20 m. Shortly after the inflow into the Leitzach receiving watercourse, no more iron can be visually detected, as it has almost completely precipitated as iron oxyhydrate

system that would have improved the mine water's chemical composition, and the treated water would have caused less pollution of the receiving watercourse. From a legal and regulatory point of view, the discussion was correct.

Perhaps we should take an example from the USA, where instead of doing nothing at all, it is perfectly permissible to do at least a little optimisation (Kleinmann et al. 2021) – this means cooperation and discussion among all parties involved and the courage to tread less well-trodden paths (keyword: citizen participation; www.datashed.org). For example, there are hundreds of government-supported private volunteer groups in the USA that band together to passively and occasionally actively treat mine water (four random examples out of many are the Massachusetts Water Watch Partnership, a kind of umbrella organisation of volunteer water monitoring organisations, the U.S. Environmental Protection Agency's "Adopt Your Watershed" program, the Citizen's Volunteer Monitoring Program in Pennsylvania, or the Animas River Stakeholders Group; Fig. 5.2). Base of this protocol is the good Samaritian legislation, which, to a degree, also provides funding mechanisms. If the sentence "everything is different there" sparks your brain, then let's do it differently as well and look for an acceptable, common solution, specifically for your country.

In most countries, we do not as yet have a comprehensive inventory of mine water discharges, as we do in some watersheds in the United Kingdom, the USA, Australia or South Africa. In Germany, for example, North Rhine-Westphalia and Saxony are partly

**Fig. 5.2** William Simon (left, with walking stick), coordinator of the Animas River Stakeholders Group (Colorado, USA), explains the work of a group of volunteers in the remediation of water bodies contaminated by acid mine drainage from former gold mining operations to interested colleagues from the International Mine Water Association (IMWA)

well advanced in this respect, or the initiatives of individuals compiling data. In contrast to the database of North Rhine-Westphalia with several thousand entries, the unpublished mine adit database of Saxony contains only 78 entries. None of the entries in the Saxon database can be considered "problematic" (Table 5.1). Consequently, apart from North Rhine-Westphalia (Heitfeld et al. 2012), Germany does not yet have a prioritisation of mine water discharges that would allow us to identify and collectively solve the major problems (e.g. Jarvis et al. 2012). This is not at all about pointing fingers at the polluter, but about protecting our environment from avoidable negative effects. Without a comprehensive inventory of the potentially several thousand mine water discharges with a risk assessment, this will be a difficult task and we are likely to be less receptive to the alternative methods of mine water treatment that can be used there. A very similar demand was made as early as 1876 by the "Committee of the German Association for Public Health Care", which called for standards on the permissible contamination of surface streams based on v. Pettenkofer's findings on attenuation (*Brockhaus Enzyklopädie*, 14th edition, keyword "Health Care", p. 379 f.). It is true that in quite a few federal German states there are historical mining registers by mining associations, other private entities or Wiki-based, but as long as these are "afraid" of "moonlighters" who might destroy the mine workings or as long as historical mining enthusiasts do not agree with speleologists or bat conservationists, these private entities will not make their databases available to the authorities or to researchers. One of these private entities has compiled a list of 125 mine water seepages all over Germany, which can largely be regarded as problematic. Of these, 23 are in Saxony, and only six of them appear in the aforementioned Saxon mine adit database. Unfortunately, to date, it has not been published either.

Finally, let's investigate history to find out where the term attenuation originates from. In the 14th edition of the "*Brockhaus Enzyklopädie*" from 1894–1896, the keyword "river pollution" (*Flußverunreinigung*) is explained as follows (p. 945 f.):

"Not every river is polluted by the sewage it receives. There are rivers into which the refuse of an entire country, such as the Nile, has entered uninterruptedly for decades without ever causing river pollution. The Tiber has absorbed the dirty waters of the city of Rome for centuries without any visible pollution. This results from the fact that every river can process a large part of the waste, to treat itself, so to speak. For the river to clean itself, it must be long enough, have a sufficient volume of water, and be able to flow at a suffi-

**Table 5.1** Selected key parameters of the Saxon, Germany, mine adit database (updated to 2009): *n:* Number, $\bar{x}$: mean value. Number *n* refers to all individual analyses of the parameter, not to the number of sampling points; mean of the pH based on the proton activity (www.wolkersdorfer.info/pH_en)

| Parameter | Unit | $n$ | $\bar{x}$ | Histogram maxima |
|---|---|---|---|---|
| pH value (field) | – | 562 | 5.5 | 4.6; 6.6; 7.4 |
| Electrical conductivity | $\mu S\ cm^{-1}$ | 604 | 414 | 200; 400; 600; 850 |
| Sulfate | $mg\ L^{-1}$ | 579 | 107 | 40; 80; 120; 220 |
| Total iron, filtered | $mg\ L^{-1}$ | 335 | 0.99 | 0.1; 0.8; 2.3 |

cient rate. Incidentally, even a highly polluted river can, if it is given a certain time to clean itself, become completely clean again, as the Seine proves, which at Meulan, 70 km below Paris, has once more pure water. For some time now, efforts have been made to determine how much filth may be given to a river without exceeding its self-treatment capacity. Pettenkofer is of the opinion that river pollution must not be feared when the quantity of water in the river is at least fifteen times as great as the quantity of sewage, and further when the velocity of the current in the river is not less than that in the sewers, because otherwise there is opportunity for sedimentation and the formation of sludge. If the self-treatment capacity of the river is insufficient, if it is in danger of becoming permanently polluted because of too low a velocity or too small a quantity of water, etc., then the sewage may only be discharged in a treated state, after it has passed through filtering, clarifying and trickling plants. Von Pettenkofer's view has also been adopted by the German Association for Public Health Care, and the Reich Health Office has proceeded according to this principle in a number of cases where it was a matter of assessing the permissibility of direct discharge of sewage into public watercourses".

## 5.2 Natural and Monitored Natural Attenuation

### 5.2.1 Natural Attenuation

Natural attenuation is by far the most common alternative management method for mine water – although it is usually unplanned. Natural attenuation is not a treatment method *per se*, but rather natural processes that have been "used" for anthropogenic contamination since the dawn of mankind, as vividly described by Ercker v. Schreckenfels (1565). Stefan Wohnlich describes natural attenuation as follows: "'Natural Attenuation' does not represent a remediation measure in the proper sense, but is an assessment of whether the cleansing processes naturally present in the subsurface are sufficient to avert danger or eliminate damage" (Wohnlich 2001). In the process, starting from the pollutant source, physical, chemical and microbiological processes take place along the pollutant pathway, which usually result in a reduction of the pollutant load or concentration (Wiedemeier et al. 1999, p. 2 f.). This results in a decrease of (semi)metal concentrations and other potential pollutants in the mobile phase, e.g. water, and often in an enrichment or rearrangement in the immobile phase such as soil, sediment or rock. If the accumulation of (semi)metals in naturally occurring metal-rich waters is large and the available time long enough, this leads to the formation of ore deposits such as the uranium roll-front deposits or gold nuggets. However, metabolism can also transform organic or inorganic pollutants into substances that are more problematic than the parent substances (e.g. methylation of mercury; Blowes et al. 2014; Stumm and Morgan 1996, p. 627 f.; Wiedemeier et al. 1999). Furthermore, the (semi)metals are not "removed" from the mine water, but are fixed or diluted in other forms, e.g. as coprecipitates. This must be considered should mine water treatment be entirely left to nature.

The term first flush is often used in connection with natural attenuation of mine water (Sect. 1.2.8). However, this is not the attenuation of mine water after it reaches the receiving watercourses, but processes within the mine (Fig. 1.4). This process leads to a decrease in the concentration of almost all ions considered to be problematic and is essentially dependent on the infiltration of freshwater, the hydraulic conductivity of the mine workings and the geological and mineralogical conditions. The more advanced the process of the first flush in the mine behind the adit portal, the easier it is to implement measures downstream of the adit portal as monitored natural attenuation.

Cidu et al. (2011) show that natural attenuation is not always sufficiently effective to reduce pollutant loads from mine water to an acceptable level. Although the Montevecchio mine in Sardinia, Italy, had been flooded for 15 years at the beginning of the investigation, substantial contaminant loads occur, especially after intense rainfall. Even dilution and precipitation processes are not able to reduce the loads during such events. While, under normal runoff conditions, Fe, Al, and Pb show substantially lower concentrations already after 2–4 km and As is presumably reduced by coprecipitation, elevated concentrations and low pH values of other metals (Zn, Cd, Mn, Ni) can still be detected at a distance of 8 km from the adit portals. Only then does the pH value increase to over 6–7, and the concentrations of these metals decrease.

A study on the Denniston Plateau in New Zealand showed the limits of natural attenuation using two historic underground mines (Jewiss et al. 2020). Under normal flow conditions, the dilution of the mine drainage with stream water and the precipitation of schwertmannite improved the water quality, but within the first 10 h of rain events efflorescent salts are dissolved and flushed into the receiving watercourses.

In the Knappensee, covering an area of 52 ha and located near Wackersdorf, Germany, natural attenuation of the lake water has occurred, which the authors call "Natural Attenuating Pit lake Area (NAPA)" (Opitz et al. 2017, 2020), although support from discharged, treated mine water from the mine water treatment plant at a nearby lake may also have played a role. While the lake had a pH of 3.5 in 1999, it is now in the circumneutral range. Although iron and aluminium concentrations have decreased substantially since 2001, and the $k_{B3.4}$ value decreased to 0 mmol $L^{-1}$, sulfate concentrations do not yet indicate that sulfate reduction is occurring on a large scale, although disulfides have formed in the sediment. The water quality in the adjacent lake Steinberg to the south has also improved considerably in the meantime (see Sect. 6.3.1).

In cases such as the Montevecchio mine in Sardinia, monitored natural attenuation would not be a solution to protect the ecological balance in the long term. Yet, if after a certain period of time, characteristic initial flushing has established conditions such as those seen in Fig. 1.3 or Fig. 1.4, this approach may well be an ecologically acceptable solution. However, Metschies et al. (2018) highlight that often not all information is available to produce robust results at the laboratory and modelling scale to predict natural attenuation. However, they were able to show that it is possible to analyse and derive the general behaviour of the system within the known imprecise measurements.

### 5.2.2 Monitored Natural Attenuation

Monitored natural attenuation was first used in connection with the remediation of polluted groundwater. It showed that the physical, chemical and biological processes occurring in the subsurface can break down organic pollutants and convert inorganic pollutants into less mobile phases (Committee on Intrinsic Remediation et al. 2000; Wiedemeier et al. 1999). The method was then applied to the remediation of other contaminated sites or mining residues (Bekins et al. 2001), where buffering of pH, neutralisation of acids, precipitation reactions, sorption, and dilution are most relevant (Wilkin 2007). In the context of mine water, the term natural attenuation first appears in 1994 in the title of two papers by Webster et al. (1994) and Kwong and Van Stempvoort (1994). Previously, Kwong and Nordstrom (1989) had used the term mine water attenuation. Since then, the term has also become established in publications on mine water treatment, and the number of publications on this topic has steadily increased since 1994 (in the past decade, for example, in Breedveld et al. 2017; Johnson and Tutu 2016; Metschies et al. 2018; Sahoo et al. 2012). A comprehensive assessment and exposition of the method was provided by Wilkin (2008), and since then the U.S. Environmental Protection Agency has published three manuals on it (U.S. Environmental Protection Agency 2007a, b, 2010). These should be consulted in any case of monitored natural attenuation. To date, however, there are not many mine water discharges in Europe that are formally subject to monitored natural attenuation.

Monitored natural attenuation goes a step further than natural attenuation and thus actually represents a treatment method for mine water when the potential pollutant loads are relatively low (Fig. 2.18). Skinner (2009) has outlined this for South Africa, for example. He characterises the method as follows: "Routine monitoring to demonstrate that there is a managed impact on the water resources". This approach has already been elaborated and graphically illustrated by ERMITE Consortium et al. (2004; Fig. 5.3). Even if monitoring were to consist solely of visual inspection of the mine water discharge, it should not be disregarded in order to monitor the success of monitored natural attenuation. It is important to note that the decision-making process is iterative, as shown by the rejection arrows in the middle part of the graph. Monitored natural attenuation is consequently not a measure that is completed forever but can lead to active or passive treatment when the water chemistry no longer meets the quality requirements for the receiving watercourse or groundwater. There are regular examples of changes in the mine water chemistry, most notably discussed in detail by Younger et al. (1997), and which can only be detected by regular control measurements or walk-through observations (e.g. Renton et al. 1988; Skousen et al. 2002; Younger 1997; Younger and Banwart 2002). This demonstrates that monitored natural attenuation is more than simply sampling water a few times a year.

As can also be seen from the diagram (Fig. 5.3), natural attenuation must in no case be an argument for doing nothing. The first step is to carry out hydrogeological and hydrogeochemical investigations and, if necessary, numerical modelling to be able to demonstrate a reliable statement on the attenuation potential of the receiving watercourse or soil.

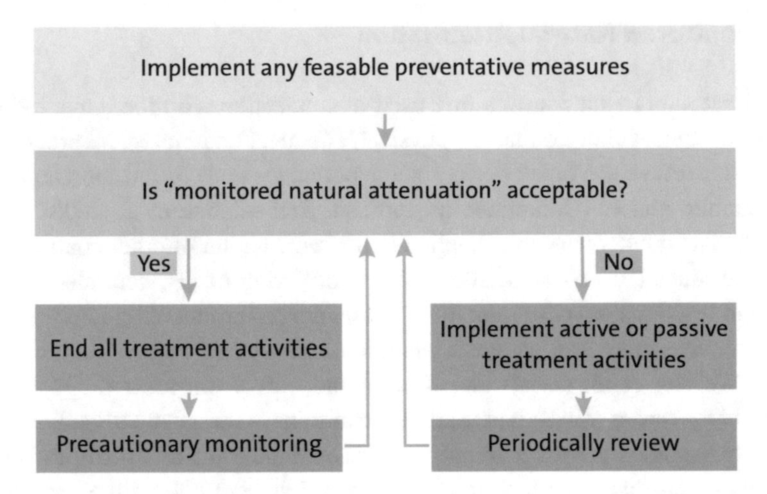


**Fig. 5.3** Decision tree for monitored natural attenuation. (After ERMITE Consortium et al. 2004)

The following – simplified – procedure would be useful and could be worked through in the manner of a checklist:

- Development of a conceptual model of mine water inflows and outflow mechanisms
- Hydrogeochemical inventory of the inflow
- Hydrogeochemical inventory upstream and downstream of the tributary after complete mixing
- Flow measurements, to allow load calculations
- Geological description of the surrounding area
- Deposit data
- History of the mine
- Hydrogeological conditions
- Description of the biocoenosis (biological community) upstream and downstream (ecology)
- Evaluation based on a generally accepted classification

Of course, the approach could be extended and complicated in any number of ways. For example, Turner et al. (2011) have proposed GIS-based assessment workflows, each related to watersheds and based on an extensive catalogue of criteria (watersheds in the United Kingdom are substantially smaller than those on the European continent). Such a measure proves to be suitable if, in addition to the assessment, an allocation for potential treatment methods is to be made, as shown for Wales, United Kingdom, by Rees et al. (2004).

It has been known since the late 1960s and early 1970s that the water quality of acidic open pit mine lakes can be improved through natural processes (King et al. 1974). Two main factors play a role: time, and the supply of organic material, for example in the form of leaves or grass. King et al. (1974) describe in a review-like article that the processes involved have been extensively studied in microcosms. It was shown that the natural treatment processes could be attributed to sulfate-reducing bacteria. The same processes took place in the microcosms as in the open pit mining lakes studied: an increase in the pH value with a simultaneous reduction in the sulfate and iron concentrations and the base capacity. Furthermore, studies showed that part of the base capacity decrease is due to hydrogen sulfide outgassing. Of the organic materials investigated, fresh green leaves proved to be less suitable, as conditions were similar to those in a swamp: brown water and pH values around 5. Substrates with low nitrogen and phosphorus concentrations were most suitable. The authors conclude that the naturally occurring process in a lake can be supported by the installation of micro-environments (enclosures) and the addition of organic material. In addition, the process could be further supported by the addition of "lime" – an approach that would usually be referred to as *enhanced natural attenuation* (details in Sect. 6.2.3).

Where do they find application in German-speaking countries? As far as I know, nowhere. I have explained the reasons for this above: As soon as an authority is officially aware that a problem exists, it would have to act and order an appropriate measure to treat the mine water. Since monitored natural attenuation is ultimately "controlled inaction" from an authority's point of view, it will not be tolerated as the sole measure in most cases. If there are conflicts with the water management objectives and if there is a status disturbance, "doing nothing" will probably not be tolerated. Even at the Burgfey adit in the Eifel, Germany's largest drainage adit by volume with up to 50 t of (semi)metals per day (Heitfeld et al. 2012; Mair 2002), or the Wilhelm-Erbstollen in Schwaz, Tyrol, Austria's highest antimony-rich mine water discharge (e.g. Millen 2003; Wolkersdorfer and Wackwitz 2004), no monitored natural attenuation of a *mine water discharge* is in place, but only a *naturally discharging source* is monitored.

## 5.3 Change in Mining Methods

As stated at the beginning of the chapter, changes in mining methods are not *per se* part of a mine water treatment method. However, a change can have a substantial effect on water quality – and ultimately leads to a reduction in mine water treatment costs. To date, however, the focus has been mainly on extracting the raw material by the cheapest route, and the treatment of the mine water has only been given attention at the end of the pumping strings or the dewatering tunnel. There are, however, other solutions for improving the mine water quality.

Mentz et al. (1975) studied how the mining method affects water quality after mining ceases. Using the Yorkshire and Shoff collieries (Clearfield County, Pennsylvania, USA) as an example, they were able to demonstrate that the water quality of mines that were operated down-dip was better than those operated up-dip. The same observation was reported by Commonwealth of Pennsylvania (1973, p. 124 f.). Similar results can be shown from the comparison of underground mines with shafts and adits. In areas of poor hydraulic contact between the shafts, the disulfide minerals are isolated from the atmospheric oxygen in the underground mine, and the weathering rates are disproportionately lower than in drift mines, where the disulfides are constantly in contact with atmospheric oxygen.

In their Mine 2030 concept (see end of Sect. 3.4.1), Bäckblom et al. (2010, p. 9) also conclude that "improvements in production … lead to even lower emissions to water and air". In addition, "innovative solutions … are being developed to mitigate the environmental impact of waste long after the mine has closed". We don't yet know with absolute certainty what such solutions might look like in 2030, but every step toward that goal is a step toward improving the environment.

And ultimately, overburden management also leads to an improvement in water quality if alkaline material is added to the pyrite-containing overburden or if overburden known to contain pyrite is stored in such a way that the environmental impact is minimised (e.g. Lengke et al. 2010; Merkel 2005; Miller et al. 1990; Wisotzky 2003). Similar concepts also exist at MIBRAG, in Germany, but publications on this subject have not yet appeared on a large scale (e.g. Friedrich et al. 2011).

## 5.4 Biometallurgy, Geobiotechnology, Biomimetics or Agro-metallurgy

This section combines the extraction of valuable materials using biomining and bioremediation and summarises results from both fields – the reason for this is that findings from one field are transferable to the other and there is a manageable number of publications on the subject. Phytomining, bioreactors or engineered (constructed) wetlands are already described in other sections of this book, as they do not extract metals *per se*, but remove these from mine water and immobilise them. Microorganisms have always played a substantial role in the context of mine water, and Rawlings (2005) has distinguished between the terms biooxidation for these and bioleaching for the extraction of metals. Biometallurgy (also called biohydrometallurgy) or geobiotechnology refers to any process that uses biotechnical methods to selectively extract metals from ores or mine water through bioleaching. I list these processes here because I can imagine using contaminated, metal-rich mine water in one of the aforementioned processes in order to extract the metals from the mine water and the process together – I am aware that the heap leaching on both Figs. 5.4 and 5.5 can alternatively be regarded as a "gigantic long-term source" with all the resulting problems.

One of the first, German-language publications on this subject appeared in 1961 (Marchlewitz et al. 1961) and investigated, on the one hand, the tolerance of *Acidithiobacillus* to iron, copper and zinc and, on the other hand, the copper ore bioleaching capability of three *Acidithiobacillus* strains. They were able to show that metal tolerance is dependent on the bacterial strain and that leaching of metals is enhanced in waters containing *Acidithiobacillus* compared to sterile samples. Ebner and Schwarz (1973) investigated whether it was possible to leach uranium ores using bacteria. In 1980, Bosecker wrote

**Fig. 5.4** Left: Collection trench of copper-containing solution from heap leaching; right: Preparation of a heap for microbiologically assisted acid leaching of copper-rich ores (Chilean El Salvador copper mine of the Corporación Nacional del Cobre de Chile – Codelco)

**Fig. 5.5** Heap leaching at the Finnish Terrafame nickel mine (formerly Talvivaara)

about "metal extraction with the aid of bacteria" (Bosecker 1980), and there were in situ experiments on the microbial leaching of ores at the Rammelsberg near Goslar, Germany, and at the Kleinkogel mine near Brixlegg in Tyrol, Austria (Beyer 1986; Brunner et al. 1993). An article published in 1985 in the newspaper *Die Zeit* presented the development with regard to metal extraction and desulfurisation of coals (Kemmer 1985). This newspaper article in particular attests to a high level of enthusiasm among the scientists and mine officials involved at the time. Although Lundgren and Silver (1980) wrote that bacteria would cause a revolution in metal mining in the future, the number of current applications has thus far fallen short of expectations. However, this research has contributed substantially to a better understanding of microbial processes on mineral surfaces than before (Kalin 2004b). For remediation of mineral processing residues, Duarte et al. (1990) conducted experiments, but these did not progress beyond the laboratory scale. Sand and Gehrke (2006) describe a plant in Ghana where gold is recovered by microbiological means prior to treatment with cyanides. If the studies of Paños and Bellini (1999) on the microbiological degradation of cyanide are added, a nearly self-contained biotechnological gold mine could well be designed using currently known methods. In zinc-rich synthetic mine waters, Nancucheo and Johnson (2012) succeeded in selective precipitation of transition elements at laboratory scale. However, they point out problems related to aluminium concentrations, which can lead to precipitation of gibbsite in the acidic environments of bioreactors. An interesting aspect from the field of bioleaching was investigated by Amin et al. (2018) using slags from the Mansfeld, Germany, copper shale mine and microorganisms from the mine water of the Kilian adit in Marsberg, Germany (Fig. 5.6). They found that *Acidithiobacillus ferrivorans* and *Leptospirillum* can remove copper, zinc, magnesium, lead and nickel from the slags and conclude that the slags of historical or heritage mines are a permanent source of pollutants, and that these microorganisms could be useful for bioleaching of these slags, provided that the microorganisms can develop according to the environment.

In the broadest sense, phytomining also falls into the category of this section, and indeed there are several hundred metal-accumulating plants, including many hyperaccumulating ones such as *Thlaspi calaminare* for zinc or *Thlaspi caerulescens* for cadmium (Brooks et al. 1998). Van der Ent et al. (2015) presented methods by which it might be possible to extract metals using agromining. However, as Wilfried Ernst (1996, p. 166), the founder of biogeochemistry summarised, "There is still a long way to go from the potential small-scale … to a realistic largescale approach".

**Fig. 5.6** Copper rich mine water with copper mineral precipitates from the Kilian adit, Marsberg, Germany (width of image 20 cm)

Without doubt, biometallurgical processes have great development potential (Rohwerder et al. 2003; Temporärer Arbeitskreis Geobiotechnologie in der Dechema e. V. 2013; van der Ent et al. 2015), as also evidenced by the EU projects BioMinE or BioHeap, the MINTEK projects in South Africa, or the BIOX™ process and Codelco plants in Chile (Fig. 5.4). Other commercial projects exist at the Terrafame nickel mine (formerly Talvivaara, Fig. 5.5) and Kittilä (gold) mines in Finland (Kauppila et al. 2013). They will gain further importance if mine water and mine waste can be turned into raw material instead of waste. The aim of research would have to be a large-scale plant, based on experience since the 1970s, in which mine water is deacidified and the metals in the mine water are selectively extracted.

# 6 In Situ and On-site Remediation Measures

## 6.1 Introductory Remark

During the preliminary research it seemed as if not many publications on in situ or on-site mine rehabilitation would be available. In fact, the first publications and experiments on in situ rehabilitation date back several decades: One of the oldest publications I could find on the subject dates back to 1936 (Anonymous 1936) and deals with the airtight sealing of mines to prevent pyrite oxidation. I see an immense potential for future mining and for research in the seven methods presented here. Some attempts seem exotic at first glance, while others have not progressed beyond laboratory investigations. In fact, a large playing field is opening for future research. In all cases, however, I consider it necessary to first gain a better understanding of the hydraulic conditions in open pits and underground mines. Tracer tests and numerical as well as analogue modelling are a basic prerequisite for this. During the discussions with colleagues it became apparent that there is also an interest in adit closures. I have therefore included the revised Sect. 5.7.4 from my earlier book here (Wolkersdorfer 2008, pp. 76–80). A good, partly in-depth overview of some of the measures presented in this section is given by Skousen et al. (2000) and, even if it goes back a bit further in time, Kleinmann (1990b). In principle, of course, it is most beneficial to keep the pyrite-rich material and the water as far apart as possible.

So let us embark on the playground together in the next seven sections.


C. Wolkersdorfer, *Mine Water Treatment – Active and Passive Methods*,
https://doi.org/10.1007/978-3-662-65770-6_6

## 6.2 In-lake Processes

### 6.2.1 Introduction

In-lake processes are defined as all processes in which the water quality of the lake (e.g. open pit residual lake) is improved within the lake itself. These include, for example, the liming of lakes, enclosures, the addition of caustic soda, the reactivation of sludge from mine water treatment that has been deposited into lakes, support for sulfate-reducing bacteria or electrochemical processes.

### 6.2.2 In-lake Liming

The liming of residual lakes from open pit mines for the neutralisation of lake water is sometimes presented as a process applied for the first time in Lusatia, Germany (e.g. Neumann et al. 2007). Actually, the process for liming natural lakes has been established in Scandinavia since the 1980s, as pointed out by Pust et al. (2010). The process was also successfully applied in the USA and Canada before it was used in Lusatia. I provide details on liming acidic lakes in Sect. 6.3, but include it here as a separate section because the "lime" is usually added to the lake from the outside and therefore, strictly speaking, it is not an in-lake process. But this definition and consequent separation of the sections may well be regarded as academic or even pedantic.

### 6.2.3 Stimulated Iron and Sulfate Reduction in Lakes

The aim of stimulated iron and sulfate reduction in lakes is to subject the entire lake to treatment. In the experimental stage, enclosures are used for this purpose. These microcosms or macrocosms are enclosures of varied sizes that are introduced into a surface water body and contain reactive material of different compositions (another term used is "in situ experimental facilities", pers. comm. Geller et al. 2013). In the enclosures, microbially catalysed sulfate reduction from sulfate to sulfide and eventual precipitation of metal sulfides occurs. For their metabolism, the microorganisms use organic substrate, which they oxidise, and the oxygen from the sulfate in the water. It is important that the enclosures are kept under anoxic conditions so that the expected reactions occur reliably and there is no re-oxidation of the precipitation products (Brugam et al. 1990). As the organic substrate is oxidised, protons are bound, resulting in an increase in the pH, and the resulting $S^{2-}$ can react with metal ions in the water, which are then fixed as metal sulfides in the lake substrate. The process is often referred to as a "passive" process. However, the definition given in the section on passive processes does not apply because the chemicals added must be renewed occasionally. Therefore, the very fact that a chemical must be added to allow the processes to run excludes the process from being classified as a passive process.

In 1988, for the treatment of mine water in Residual Lake 8 of the Will Scarlet coal mine (Illinois, USA), six cylindrical enclosures, each holding 9.4 $m^3$ and with the top end open to the environment, were installed in the lake with a pH of 3.1 (Brugam et al. 1990). The objective of this installation was to conduct sulfate reduction experiments within a well-defined lake area, with a controllable sequence of events within the experimental conditions. In addition to a control enclosure without treatment, the five other enclosures contained limestone (later replaced by calcium hydroxide), straw, sewage sludge, and combinations of limestone and straw and sewage sludge. During the experiment, the authors were able to demonstrate that the pH values of the water in the experiments with straw and lime had increased and that the sulfate concentration had decreased. However, they were not able to identify the extent to which these changes could be permanently adjusted. In other experiments, ethanol, carbolime or pyruvic acid (pyruvate) were used in addition to limestone or calcium hydroxide (Frommichen et al. 2003).

Enclosures (microcosms as well as macrocosms) were also installed in the Lusatian residual lake 111 (Plessa Lake, Germany) between 2001 and 2006 to test whether this method would be suitable for remediating the acid lake water (Fig. 6.1). It was shown that the smaller enclosures (approx. 20 $m^3$) were able to reduce the iron, whereas in the larger ones (approx. 4 500 $m^3$) iron sulfide was re-oxidised (Geller et al. 2013, p. 237 f.; Koschorreck et al. 2006, p. 59). In addition to carbolime, the authors also used ethanol as carbon sources. As the process has not been able to demonstrate sustainable deacidification of a residual lake in any of the trials to date, it must be regarded as an academic experiment that cannot currently be considered for long-term remediation of open pit lakes.

**Fig. 6.1** Enclosures ("macrocosms") of varying sizes in residual lake 111, Lusatia, Germany. (Photo: Peter Radke, LMBV)

### 6.2.4 Electrochemical and Electro-biochemical Treatment

Alternatively, electrons can be added to a system with the aid of electricity. The arrangement of cathodes and anodes then acts as electron donor and acceptor, respectively, and the microorganisms in the lake water as electron transporters (Fig. 4.11). Although not all of the processes involved in electron transfer and energy production in bacteria are fully understood, it is clear that there are many future uses for these processes (El-Naggar and Finkel 2013) – and not just in mine water treatment. The role of strains of the genus *Desulfobacter*, which can transport electrons, was already described by Sisler et al. (1977). A few examples, some of which do not even use the term electro-biochemistry because it had not yet been created, are described in this section.

Sisler et al. (1977) conducted laboratory studies on the electro-biochemical neutralisation of lake water. In their experiments, they supported biological sulfate reduction by passing current through iron or platinum electrodes. The basis of this idea is that the lake itself is an electrical half-cell built up from the organic lake substrate, sulfate ions in the water body, and microorganisms (in their case, *Desulfovibrio desulfuricans*). By incorporating an anode and a cathode, the half-cell could be converted into a galvanic cell that raised the pH of the laboratory lake water from 2.4 to 6.

Hilton et al. (1989) report on an experiment to electrochemically treat a 3 000 $m^2$ acidic water collection pond from the coal-rich tailings pile at Buck Lilly Colliery (Greenbrier County, West Virginia, USA). They installed a grid of electrodes in the pond and, after an initial failed attempt, observed air bubbles coming out of the water as if it were boiling. After a day or two, the sludge had to be cleared (it is not clear from the text whether the sludge was caused by the operation of the electrochemical system or by natural sedimentation). When the plant was uninstalled, it was damaged to such an extent that it could no longer be used in its original size. However, the operator of the Buck Lilly colliery used the system on a smaller scale together with neutralisation by sodium hydroxide.

Unfortunately, the authors give no details of the current or voltage they had used. However, since they write of dead amphibians, it may be assumed that the voltage and current were above those of an electrochemical method commonly used today (the article by Tiff Hilton is well worth reading – it is simply heart-breaking and inspires laughter). Today, a few volts and 1–2 amps of current are commonly used.

Adams and Peoples (2010) and Adams et al. (2012) claim to have found a completely novel method for treating mine water using a patented electro-biochemical method. They succeeded in removing the water constituents Se, As, and nitrate in a system exactly similar in principle to that of Sisler et al. (1977), using 3 V DC and a few microamperes of current. Opara et al. (2014) reported a drop in pH from 7.6 to 6.4, and *Desulfovibrio* sp., among others, could also be detected in this water. Unfortunately, the authors do not provide information on the location of their pilot plant. A University of Utah press release (January 5, 2011), the developer's website, and a presentation at the 2014 Mine Design, Operations & Closure Conference only indicate that the pilot plant was deployed at an abandoned gold mine (Landusky Mine, Montana) in 2010 and has since been evaluated at

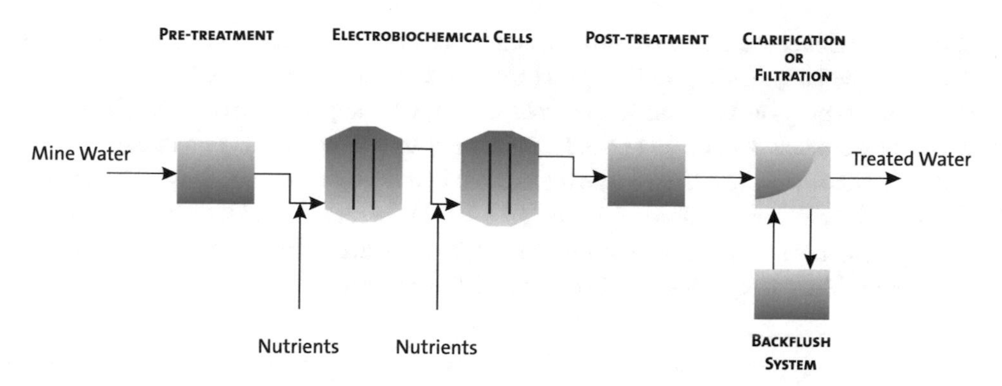


**Fig. 6.2** Principle of electro-biochemical water treatment (modified according to company brochure Inotec Inc., Salt Lake City, USA)

two other pilot plants (coal mine and presumably a porphyry copper deposit). Overall, the system consists of pre- and post-treatment, the addition of nutrients, and a settling pond or filtration (Fig. 6.2), although no details have been published about the individual steps.

## 6.3 Chemical Treatment Measures to Reduce Pollutants

### 6.3.1 Treatment of Acidic Lakes

Chemical treatment of acidic lakes involves either adding a chemical to the lake water to change its chemistry or applying calcium-containing sludge (ideally CaO) into the lake itself and spraying it therein (the latter known as the in-lake process). Injection of gaseous $CO_2$ to build up a buffer of hydrogencarbonate (bicarbonate) is also possible and has been successfully evaluated. In most cases, the aim of chemical treatment is to raise the pH of the lake, resulting in the processes described above under neutralisation. It is important that the ecological balance in the open pit lake is disturbed as little as possible. Consequently, highly active chemicals such as sodium hydroxide or slaked lime should only be used in exceptional cases (Geller et al. 2013). During remediation in Lusatia, Germany, it has been shown that monitoring of lake chemistry is important. This facilitates the controlled addition of the neutralisation agent and adjustment according to the parameters measured in each case (Benthaus et al. 2020; Luckner and Totsche 2017; Märten 2006; Merkel et al. 2010). In addition, this allows the continuous acquisition of data for a numerical lakewater model to reproduce the remediation progress. By means of a selection of case studies, I would like to introduce the procedure here, since – unlike other procedures – a generally applicable description of the procedure is not readily feasible and not every procedure is suitable for every lake.

The first chemical treatments of lakes took place in the 1970s in connection with the discussion on acid rain. Nearly 1 000 Swedish lakes were treated by liming between 1976 and 1979 (Bengtsson et al. 1980), and by 1991 this had increased to between 6300 and

8000 lakes (Geller et al. 2013; Henrikson and Brodin 1995; Lydersen et al. 2002). To raise the pH of the lake, 10–20 g $CaCO_3$ per cubic metre of lake volume was found to be suitable. For this purpose, between 50 and 75 kg $CaCO_3$ were applied to 10 000 $m^3$ of the lake, which lasted about 5 years and contained not only limestone but also dolomite, sodium hydroxide or olivine. The largest lake remediated in this way was the Swedish Unden lake (Edsån) with a volume of about $3 \cdot 10^9$ $m^3$ (Table 6.1). In Lusatia and the Central German mining area, more than a dozen lakes have since been treated with neutralisation or in-lake technology (Benthaus et al. 2020; Luckner and Totsche 2017).

**Table 6.1** Selection of acidic lakes whose water quality has been altered by the addition of chemicals or mine water treatment sludges

| Lake name | Volume $10^6$ $m^3$ | $k_{B4.3}$ mmol/L | Chemical | pH value change | Source |
|---|---|---|---|---|---|
| Lake Unden (Edsån)[a] | 3 000 | – | $Ca(OH)_2$ | – | Bengtsson et al. (1980) |
| Lake Orta[b] | 1 300 | 0.77 | $CaCO_3$ | 4.3 → 6.8 | Calderoni and Tartari (2000) |
| Iceland Copper Mine[c] | 241 | – | Fertiliser | Cu↓, Zn↓ | Fisher and Lawrence (2006) |
| Residual lake Skado[d] | 123 | 3.3 | CaO | 3.0 → 3.6 | Luckner and Totsche (2017) |
| Berkeley lake[e] | 160 | 28.97 | sludge | 2–3 → 4 4.5 | Gammons and Icopini (2020) |
| Residual lake Scheibe[f] | 110 | – | $CaCO_3 + CO_2$ | – | Strzodka et al. (2016) |
| Residual lake Koschen[g] | 82 | 1.6 | $CaCO_3$ | 3.0 → 3.5 | Benthaus and Uhlmann (2006) |
| Residual lake Lohsa[h] | 44 | 2.9 | $Ca(OH)_2$ | 2.7 → 5.4 | www.LMBV.de |
| Residual lake Burghammer[i] | 36 | 2.34 | $CaCO_3$, $Ca(OH)_2$ | 2.9 → 8 | Pust et al. (2010) |
| Residual lake Bockwitz[j] | 18 | 8.1 | $Na_2CO_3$ | 2.7 → 7.1 | Roenicke et al. (2010) |
| M4E pit lake[k] | 8,2 | – | $Ca(OH)_2$ | 3 → 6 | Gautama et al. (2014) |
| Lake Rävlidmyran[l] | 0.53 | – | $Ca(OH)_2$ | 3.7 → 7.5 | Lu (2004) |
| Anchor Hill Lake[m] | 0.3 | – | CaO, NaOH | 3 → 7 | Lewis et al. (2003) and Park et al. (2006) |
| Residual lake Sleeper[n] | – | – | CaO | 5 → 8 | Dowling et al. (2004) |

[a]Sweden, [b]Italy, [c]United States of America, [d]Germany: Partwitzer See, [e]Montana, USA, [f]Germany: Scheibe-See, [g]Germany: Geierswalder See, [h]Germany: Speicherbecken Lohsa II, [i]Germany: Bernsteinsee, [j]Germany, [k]Indonesia, [l]Sweden, [m]Australia, [n]United States of America: Liming during flooding of the residual lake

(Modified and supplemented from Geller et al. 2013)

In contrast to acidified natural lakes, the acid concentration in open pit mine lakes is disproportionately greater and characterised by elevated (semi-)metal concentrations, which are usually absent in natural lakes. Another difference between the lakes of Scandinavia and the open pit lakes is usually the continuous inflow of acidic groundwater from the pyrite-containing areas of the open pit mine dumps, which in principle would require a remediation strategy that starts directly at the dumps (Merkel 2005). Therefore, the method used in Scandinavia could only be applied to residual open pit mining lakes after adaptations (Pust et al. 2010). The first chemically remediated acidic open pit mining lake in Germany was the 176 ha lake Steinberg near Wackersdorf (Aalto et al. 2017; Hemm et al. 2002; Weilner 2013). In 1985, ash from the Schwandorf lignite-fired power plant was flushed into this lake, but without achieving a sustainable pH increase. After the pH was raised from the original 3.4–4 to a maximum of 10.4, it dropped again to 4.6 after 2 years, finally settling at 3.9 in 1999 and 6.5 in early 2012 through natural attenuation (Körtl 2012).

Uhlmann et al. (2001, p. 45 f.) used PHREEQC to model the development of pH in the Sedlitz residual lake (Sedlitzer See, Germany). As expected, they were able to show that sodium hydroxide and slaked lime caused the fastest neutralisation and limestone the slowest. To investigate the kinetics of the reactions, they carried out complementary laboratory experiments with sodium hydroxide, slaked lime and soda ash, which gave the same results as all other such experiments (see Sect. 2.5). The more important question of the sustainability of such chemical lake treatments could not be adequately modelled at the time, as too few data were available to act as input parameters for modelling. Instead, it was necessary to conduct in situ experiments on the lakes to investigate the method on a full scale and to accompany it with extensive monitoring (Benthaus and Uhlmann 2006). Nevertheless, the results of liming initially fell short of expectations. Merkel et al. (2010), Schipek (2011) and Schipek et al. (2011) conducted a research project to investigate the possibilities for liming lakes to show greater sustainability. To do this, they used numerical modelling and extensive monitoring in conjunction with laboratory and field investigations. They were able to show that the neutralisation of residual lakes can substantially be improved, for example, by using optimal wind conditions. A further improvement in remediation can be achieved by using $CO_2$ in addition to the neutralising agent. The $CO_2$ reacts with the CaO to form $CaCO_3$ (sum of Eqs. 6.1 and 6.2), which in turn dissolves in the mine water and thus contributes to the hydrogencarbonate alkalinity of the residual lake (Eq. 6.3). Watten and Schwartz (1996) already demonstrated that pre-treatment of acid mine drainage with $CO_2$ substantially increases the dissolution rate of limestone, thus optimising the neutralisation of the acid mine drainage:

$$CaO + 2\,\mathrm{H}^{+} \rightarrow Ca^{2+} + \mathrm{H_2O} \tag{6.1}$$

$$\mathrm{CO_{2(g)}} + \mathrm{H_2O} \rightarrow 2\,\mathrm{H}^{+} + \mathrm{CO_3^{2-}} \tag{6.2}$$

$$\mathrm{H}^{+} + \mathrm{CO_3^{2-}} \rightarrow \mathrm{HCO_3^{-}} \tag{6.3}$$

Strzodka et al. (2016) used this principle at Lake Scheibe, Germany, where a combination of $CO_2$ and lime was used (Fig. 6.3). The neutralisation potential of the lake could be increased from 0.1 mol $m^{-3}$ to 0.8 mol $m^{-3}$ after the experiment.

In principle, two different methods have been used thus far to raise the pH value of open pit mining lakes: external addition of alkaline material and distribution of alkaline lake sediment in the lake. To add external alkaline material to the lake, either specialised boats are used to distribute the alkaline material in the lake (e.g. Bernsteinsee; Lichtenauer See, Germany), or floating hose units with land-based injection stations, such as at Lake Scheibe (Figs. 6.3 and 6.4) or Lake Drehna (Koch and Mazur 2016). The alkaline lake sediment may be either sludge deposited in the lake from mine water treatment plants (Benthaus and Uhlmann 2006) or fly ash (e.g., Koch 2010; Koch et al. 2008; Loop et al. 2003; Schipek et al. 2006; Schipek et al. 2007; Werner et al. 2006).

An in-lake process was used for the first time in Saxony, Germany, in 2004, when Lake Bockwitz (formerly residual Lake Borna Ost) in the Central German mining district was treated with soda ash (Neumann et al. 2007). As a result, the pH of the lake was gradually raised from 2.7 to 7.1 between 2004 and 2008, and the base capacity was lowered from 8 mmol $L^{-1}$ to 0.5–1 mmol $L^{-1}$ (Roenicke et al. 2010). By 2011, the pH stabilised at 6.0–6.5, and then decreased to 5–6 by 2012 after soda addition ended (Heinrich et al. 2011; Ulrich et al. 2012). In total, the lake was treated with 15 kt of soda ash without

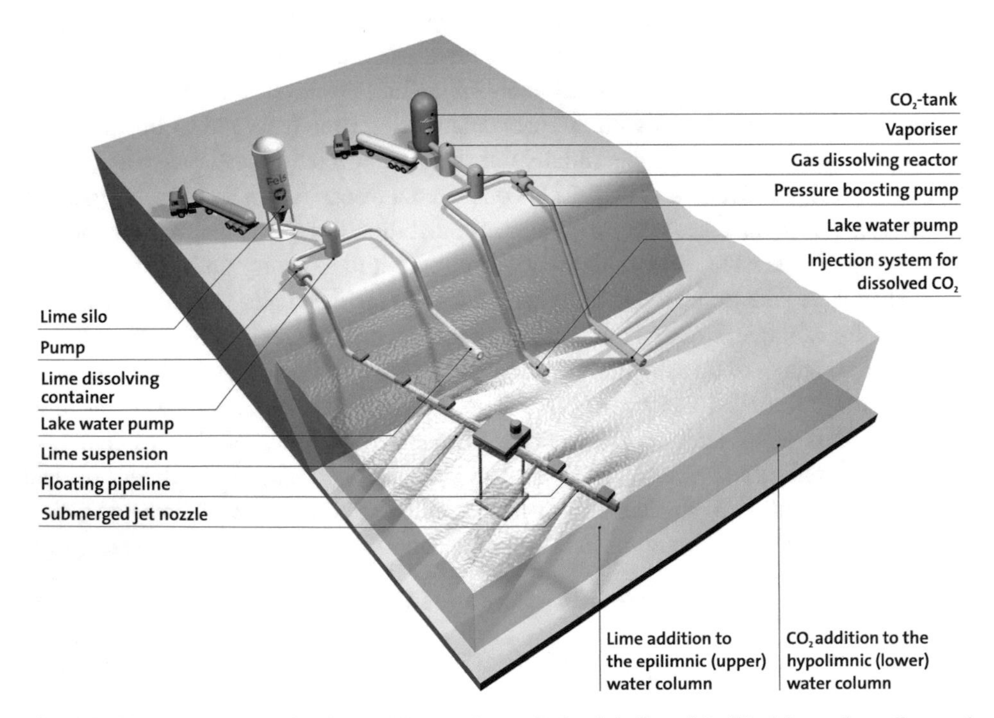


**Fig. 6.3** Neutralisation and $CO_2$ addition station at Lake Scheibe. (Modified from Strzodka et al. 2016, Figure: GMB GmbH)

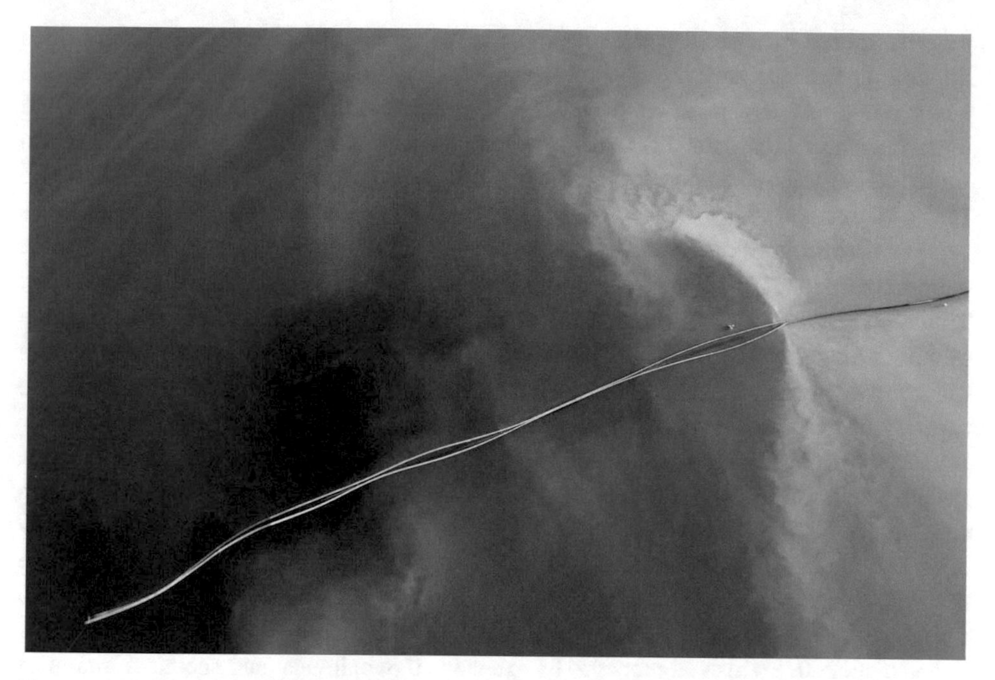

**Fig. 6.4** Injection of slaked lime into Lake Scheibe, Germany, in November 2010 via hose lines and a land-based injection station. (Image width about 500 m; Photo: Peter Radke, LMBV)

achieving the long-term neutralisation of lake water. Reasons for this may be higher acid loads from surrounding sediments, groundwater, or ion exchange reactions. Without doubt, the unexpected difficulties in neutralising the lake have contributed substantially to a better understanding of the mechanisms in this lake and to being able to transfer the process to other residual lakes.

In the residual lake Koschen (Geierswalder See, Germany), the alkaline sludge from the mine water treatment was circulated in 2004 and distributed in the lake by means of spray sprinklers. Chemical analysis showed that the sludge was composed of 85% calcite and portlandite. A total of 520 000 $m^3$ of sludge could be reprocessed using this technique. Apparently, a better distribution of the sludge suspension in the lake is obtained if the currents present in the lake are also used in a supportive manner. The method succeeded in raising the pH value by 0.5 from 3.0 to 3.5 (Benthaus and Uhlmann 2006).

In the residual lake Burghammer (Bernsteinsee) the lake sediment consists mainly of CaO-rich fly ash from a coal-fired power plant. A pilot test for lake water neutralisation took place there in 2007, in which the 4–8 m thick lake sediment was removed by dredging, $CO_2$ was added to support the neutralisation reaction, and the sediment was flushed back into the lake via a 150 m long hose line. However, as the evaluation of the Lake Bernstein pilot test indicates, substantial amounts of the fly ash would need to be reworked in the lake to permanently raise the lake pH to the circumneutral range (Koch 2010; Koch et al. 2008). Because the pilot tests did not produce sufficient neutralisation, Lake Bernstein was neutralised by spreading lime using specialised boats (Schipek et al. 2011). Since flooding of the lake began, pH values around 3 have been established (Schipek 2011), which could be raised to an initial pH of 8 by neutralisation with 10 kt of finely ground limestone and 1 kt of calcium hydroxide. However, inflowing acidic groundwater necessitated further neutralisation of the lake with about 0.5 kt of calcium hydroxide (Pust et al. 2011). With the water treatment vessel "Klara", the design of which is based on the many years of experience of the LMBV and its partners, the LMBV now has a vessel that enables almost continuous operation in the extended residual lake chain and in Lake Senftenberg. This vessel is capable of injecting about 40 kt of neutralisation agent into the lakes, with a turnaround time of about 1 h. One of the first lakes to be neutralised using "Klara" was Lake Partwitz, where 13 kt of quicklime raised the pH from 3.0 to 3.6 and lowered the base capacity from 3.3 to 0.4 mmol $L^{-1}$ (Benthaus et al. 2020; Luckner and Totsche 2017).

As part of the Big Gorilla project, the Big Gorilla residual lake (Schuylkill County, Pennsylvania, USA) was filled with fly ash and the pH was raised from 3.6 to 11.0 (Loop et al. 2003, 2007). The open pit residual lake had a volume of 45 000 $m^3$ and was connected to the underground anthracite mine, but without resulting in a pH increase in the mine water there (Loop 2004). To fill the lake with fly ash, approximately 3 Mt of fly ash from the nearby Northeastern Power Company plant was required between 1997 and 2004 (Loop et al. 2007). This was preceded by extensive mineralogical and chemical studies of the fly ash. In addition to the increase in pH, a decrease in the concentrations of potentially toxic water constituents was observed. In the receiving watercourse, however, the chemistry has changed since the fly ash was injected, with calcium and sulfate concentrations increasing in particular. It was not possible to determine whether this was due to the introduction of the fly ash or to the simultaneous remediation of the entire mining area. However, compared to the usual effects of acid mine drainage on receiving watercourses, the effect may be acceptable in the short term (See also Sect. 6.7).

### 6.3.2 On-site Chemical Treatment Measures

Although there have been numerous experiments to prevent the formation of acid mine drainage with chemical treatment measures on-site, it has not yet been possible to prevent it permanently. I describe some examples in Sect. 6.6 on "In situ remediation of uranium-containing mine and seepage water".

In the 1970s and 1980s, the former U.S. Bureau of Mines attempted to contain the microbial activity with bactericides. Since microbial catalysis accelerates pyrite weathering by a factor of one million, this would have the effect of substantially slowing down the kinetics of the process. This has been shown to be beneficial in preventing the formation of acid mine drainage in open pits, where disulfide-rich sections must be kept open for a period of time without producing excessive amounts of acid mine drainage (Skousen et al. 2000). As Kleinmann and Erickson (1982) have shown, anionic surfactants such as sodium lauryl sulfate ($C_{12}H_{25}NaO_4S$) prevent microbial pyrite oxidation because they denature the proteins of the microorganisms. They were able to show that pyrite oxidation in overburden from coal mines decreased substantially. In the long term, however, this method has limited ability to curb the formation of acid mine drainage. Yet, Gusek and Plocus (2015) used the example of the rehabilitated Fisher coal mine in Pennsylvania (USA) to show that the addition of bactericides accompanied by a comprehensive geological, chemical and geophysical characterisation of the rehabilitation object can still be successful after 15 years. Plaza-Cazón et al. (2021) used aqueous extracts of a South American plant of the aster tribe (*Parastrephia quadrangularis*) with antimicrobial characteristics to inhibit the growth of *Acidithiobacillus ferrooxidans*. They found that these plants grow in the vicinity of northwest Argentinian tailings dams and thus used them for their extracts.

Apatite dust applied to potentially acid-water-forming sections in a mine reduces the formation of acidity by up to 96%, with a ratio of 3:1 000 being sufficient (Kleinmann 1990b). Evangelou (1995, p. 221 ff.) was able to confirm and explain this effect in laboratory experiments.

## 6.4 Reinjection of Sludge, Treatment Residues or Lime

In contrast to the previous section, which dealt with the targeted addition of alkaline material to residual lakes for quality improvement, in this section I will show you case studies that have not *primarily* contributed to improving the mine water quality. As in the previous section, it is not possible to give a generally valid description, so I will again discuss selected case studies.

Low- or high density sludge may have alkaline material in concentrations that make it suitable for buffering acidic water in an underground mine or acidic residual lake. These residuals can therefore be injected into the underground mine workings, provided they are ecologically safe (Aubé et al. 2005; Coal Research Bureau 1971, p. 73 ff.; Hansen 1998; Kostenbader and Haines 1970; Wolkersdorfer and Baierer 2013). Jäger et al. (1990, 1991) have also made it unequivocally clear that underground disposal of residues should only be permitted if "the … pollutants contained can be kept away … from the biosphere with sufficient certainty over a sufficiently lengthy period of time. In principle, this can be achieved the deeper underground and thus the further away from the biosphere the disposal takes place, since the potential mass transport … proceeds correspondingly much more slowly" (Jäger et al. 1991, p. V/1).

The first report on the injection of sludge from a mine water treatment plant appeared in 1967 (Young and Steinman 1967, p. 478). In 1966, the Vesta-Shannopin Coal Division of the Jones and Laughlin Steel Corporation had begun to treat mine water by neutralisation at the Vesta No. 5 Colliery (Thompson Well, Pennsylvania, USA; between Marianna and Beallsville on PA-2011). This treatment resulted in large quantities of sludge being produced with only 6% solids concentration, which could not be deposited near the treatment plant due to space constraints. With official permission, this sludge was transported by truck to the nearby closed Vesta No. 6 colliery and pumped via a borehole into the flooded mine pool there. It is not known whether or not there was an improvement in the water quality at the Vesta No. 6 colliery.

Another test was conducted at the Driscoll No. 4 Mine in Pennsylvania, USA (Scott and Hays 1975, pp. 234–237; Stoddard 1973). After initial laboratory investigations had shown that flushing of alkaline sludge resulted in improved water quality, testing at the mine began in 1970. Although the pH increased to 11–12 during the first field test, two subsequent tests failed to adjust to the predicted pH of 11. Rather, the pH remained low at pH 3–5. Reasons that the sludge failed to improve the water quality are probably that the sludge settled in the flooded mine and there was thus no further contact with the discharging mine water. Unfortunately, neither Jones and Ruggeri (1969) nor Stoddard (1973) give details of the mine water chemistry during and after the experiment, so no conclusions can be drawn about improvements in water quality. In my opinion, this is another example of the fact that such tests should only be carried out if the hydrodynamic conditions in the mine have been explored well enough (Wolkersdorfer 2008, p. 268).

Other residues that have been reinjected into mines to raise the pH include fly ash in underground mines (Ashby 2001; Golden et al. 1996; Gray 1997, 1998; Gray et al. 1997; Loop et al. 2003) or open pits (Geller et al. 2013, p. 233 ff.; Uhlmann et al. 2001, pp. 45–65; Unger and Wolkersdorfer 2006). Generally, reinjection has been shown to produce a positive effect on mine water chemistry (Aljoe and Hawkins 1993, p. 36). Only in rare cases does the injection of alkaline material into an underground mine cause a deterioration of water chemistry, or no chemical effect is observed at all (Ziemkiewicz 2006, p. 65). Nevertheless, the issue is regularly discussed because the ashes themselves contain potentially toxic elements that are mobilised by the usual elution tests. In none of the international cases presented or which I am aware of, however, did relevant limits turn out to be exceeded. The reason for this is probably the same as that given in the last paragraph of this section.

The partially unreacted alkaline materials contained in the sludge allow some of the juvenile acidity to be buffered in the mine water. As a result, the pH of the mine water increases while still in the mine workings or in the open pit, and metals or semimetals precipitate or coprecipitate. Experience shows that the pH value only increases by a small amount, whereas there is a substantial increase in the acid capacity of the mine water.

There have been legal issues in the USA with the injection of residual substances not originating from mine water treatment, which were discussed at several conferences and eventually led to the development of guidelines (Vories and Harrington 2004, 2006). Some authors point out that from a scientific point of view there are no problems with the re-dissolution of potential pollutants, and they were not detectable. However, as soon as the interested public raises concerns, it is recommended to abandon or terminate a project. In the case of the Mettiki Colliery, USA, this led to the underground injection of alkaline material having to be stopped in 2002 after five years of injection, because the population feared that the slurries injected could pose an environmental hazard due to possible toxic components (Ziemkiewicz and Ashby 2007).

Since December 1990, residues from the mine water treatment plant at the Elbingerode mine in the Harz Mountains, Germany (Einheit mine) have been flushed into the flooded underground mine. Extensive laboratory investigations and chemical-thermodynamic modelling, as well as measurements in the mine water, did not reveal any negative effects on the mine water chemistry (Hansen 1998; Klinger et al. 2000). The flushing of residual materials into deeper areas of the mine workings can therefore be regarded as a suitable way of disposing of iron oxyhydrate sludge and removing it from the anthroposphere. Nevertheless, the flooding of the mine will not result in conditions that correspond to those of the pre-mining situation (Groß and Knolle 2011).

Wolkersdorfer and Baierer (2013) used the example of the German Straßberg/Harz fluorspar mine to show the positive effects on water quality of injecting sludges from mine water treatment. Based on laboratory tests, numerical modelling and measurements at the mine site, it was shown that even the smallest quantities of 0.5–1% of alkali-rich sludge can buffer the mine water's pH. The particulate sludge, on the other hand, remains in the mine workings because the effective velocities of the mine water are not sufficient to transport the deposited sludge again. However, it must be ensured that the sludge has sufficient time to settle in the mine workings before it reaches the vicinity of a main shaft. Regarding the amount of sludge required, Aubé et al. (2005) reached the same conclusions. Irrespective of the amount of sludge added, they were able to demonstrate in laboratory experiments that the relevant metal concentrations in the mine water were reduced.

Přikryl et al. (1999) used the Fluid Dynamics Analysis Package FIDAP 7.5 to numerically model the spread of the injected residuals in a mine shaft. Accordingly, the discharge of the residuals would take geological times. This finding, however, is entirely at odds with the results of my own tracer investigations in flooded mines. The reason for the contradiction arises from the authors' assumption that the main shaft is a single tube. However, from their Figure 1, it appears that the mine has two shafts that are interconnected. Numerous measurements of the concentration dispersion of water constituents also prove that the discharge of injected substances into a mine can proceed relatively quickly (cf. Wolkersdorfer 1996, 2008). Otherwise, it would not be possible to explain why, for example, in Straßberg/Harz the introduction of sludge has a positive effect on water chemistry (Wolkersdorfer and Baierer 2013). If the transport were slow, there would be no mixing of the sludge with the mine water and finally the positive effect on its quality.

Detailed hydrogeological investigations are necessary prior to an injection of alkaline material into an underground mine. In addition to the usual water chemistry parameters, these should consider the hydraulic conditions in the mine. Tests with natural or artificial tracers are suitable for this purpose (Wolkersdorfer 2008). Canty and Everett (1998) conducted a tracer test with Rhodamine WT and sodium chloride in the Howe-Wilburton coal province (Oklahoma, USA) to accompany a projected injection of fly ash. They succeeded in describing the hydrodynamic conditions of this mine. Aljoe and Hawkins (1993) also investigated the hydraulic conditions associated with the injection of alkaline material into an underground mine. They conclude that it is necessary to conduct such investigations or preparations for injection of alkaline material while the mine is still active, if possible. As it turned out, the access to already closed and flooded mines is problematic and the injection of alkaline material did not show any success either.

In addition to the aforementioned residual materials, lime or limestone can also be injected directly into a mine pool. Particularly in cases where mine closure is required – whether for safety reasons or as part of landscaping measures – it is a promising idea to fill the entrance to the mine with limestone. Burnett and Skousen (1996) discuss a case where limestone was brought into a mine and improved the quality of the discharging mine water.

Lake Velenje (*Velenjsko jezero*) west of Velenje (*Wöllan*) in the Slovenian Šalektal, serves as an example of the negative effects of uncontrolled flushing of alkaline material, as fly ash from the Šoštanj (*Schönstein*) lignite-fired power plant was flushed into Lake Velenje, which is of anthropogenic origin (Stropnik et al. 1991). Subsequently, the deposition of suspended fly ash led to an increase of the pH in Lake Velenje up to 12, and it was shown that this was primarily due to the high water content in the suspension. This high water content was necessary to transport the ashes by pipeline from the power plant to the lake located 2 km away. By changing the transport and deposition of the fly ash suspension, it was possible to gradually lower the pH so that it has fluctuated between 7 and 9 since 2004 (Šterbenk et al. 2011; Šterbenk and Ramšak 1999).

What are the reasons why the sludge from mine water treatment reinjected into the mine does not dissolve and the iron and the metals and semimetals that have also precipitated do not contaminate the mine water? Aubé et al. (2005) attribute this to the fact that the mine water is already in chemical equilibrium with the iron and consequently no dissolution occurs. Another reason could be that locally an alkaline environment forms in the alkaline sludge, which also counteracts further dissolution of the iron phases and coprecipitates. This is also indicated by the laboratory investigations and modelling of Wolkersdorfer and Baierer (2013). Consequently, there is probably nothing to prevent future underground disposal of mine water treatment sludges.

## 6.5 Remediation of Contaminated Watercourses

In addition to groundwater, surface waters are usually the ultimate recipient of mining influenced water. The type of inflow into the watercourse is usually surface inflow through adits or treatment plants. In rare cases, diffuse inflows of mining-influenced groundwater to a surface water body may occur (Fig. 6.5); these are generally difficult to control (Bezuidenhout 2009; Jarvis et al. 2007; Mayes et al. 2005, 2008). Recent examples include diffuse groundwater inflows into the River Spree, Germany, which led to ochre precipitates in the water body (Benthaus et al. 2015b; Grischek et al. 2016; Musche et al. 2016; Uhlig et al. 2016; Uhlmann et al. 2010), or mine water inflows in the Gessental near Ronneburg, Germany (Baacke et al. 2015).

To date, there are only a few examples of streams with diffuse inflow of acidic water that have been successfully remediated. In almost all cases, the remediation measures involved the addition of alkaline material to the watercourse (e.g. Maneval 1968; Menendez

**Fig. 6.5** Diffuse discharge of mining-influenced groundwater. Left: Cadegan Brook (Cape Breton Island, Nova Scotia, Canada; image width approx. 1 m); right: Gessenbach (Gessental near Ronneburg, Thuringia, Germany; image width approx. 2 m)

et al. 2000; Uhlig et al. 2016), remediation using the AquaFix system (Jenkins and Skousen 1993) or diversion of the contaminated mine water (Baacke et al. 2015). At the Shilbottle tailings pile (Northumberland, England, United Kingdom), diffuse leachates into a watercourse were successfully remediated using a reactive wall (Bowden et al. 2005). Olyphant and Harper (1998) describe anoxic limestone drains that were used for the remediation of such uncontrolled groundwater seeps.

Another remediation option could be electrocoagulation, in which the affected stream is treated directly (pers. comm. Philip G. Morgan 2013). This results in faster flocculation of the iron hydroxide and concomitant coprecipitation *s.l.* of other water constituents. To date, no research results are available on this subject.

## 6.6 In situ Remediation of Uranium-Containing Mine and Seepage Water

The in situ remediation of non-uranium mines, especially with alkaline material, is discussed in detail in Wolkersdorfer (2008). I will therefore refrain from a detailed repetition of the information already available in that book, but will supplement and extend the remarks made at that time to include in situ projects in flooded uranium mines, as much work has been done in the past decade. Unfortunately, we have not yet been allowed to dilute a uranium mine water with river water, as outlined by Merkel (2002, p. 269) – an experiment that would have its attractions. In the flooding experiments of Wismut GmbH in Königstein, Germany, only groundwater naturally percolating into the mine workings was used (Jenk et al. 2014a).

Zero-valent iron (ZVI; $Fe^0$) is known to be capable of chemically reducing mine water constituents (e.g., Fiedor et al. 1998; Klinger et al. 2002; Noubactep et al. 2002). Since reduced metal sulfide phases are generally less mobile than the associated oxidised species, this has the effect of improving water quality. This property has been used to treat uranium-containing mine water, seepage water and groundwater in the Bear Creek valley (Tennessee, USA), Erzgebirge (Saxony, Germany) or Pécs (Fünfkirchen, Hungary). In simple terms, oxidation of the zero-valent iron occurs at the surface, resulting in the reduction of relevant water constituents such as uranium, chromium, mercury, silver or technetium (Fiedor et al. 1998). In addition to the water constituents being reduced, however, the reduction of $H_2O$ by $Fe^0$ also occurs and hydrogen gas ($H_2$) is generated. Apart from iron granules, highly active nanoparticles of $Fe^0$ can also be used (Klímková and Cerník 2008; Klímková et al. 2011). However, exactly which mechanisms are responsible for the reduction at $Fe^0$ is subject to academic discussion (Noubactep 2010).

At the Niederschlag and Johanngeorgenstadt sites in the Ore Mountains, Germany, Schneider et al. (2001) used an Fe/Mn substrate from the flocculation stage of a waterworks as well as $Fe^0$ to treat uranium-containing mine water. At both sites, the latter generally showed better immobilisation rates than the Fe/Mn substrate. However, for the investigated parameters U, $^{226}Ra$, Pb, As, $SO_4$ and $NO_2$, there were substantial differences in immobilisation rates between the two mine waters and the two substrates, with uranium removal rates of 80 to 96% by $Fe^0$. Complementary to the laboratory and field experiments, numerical modelling of uranium and arsenic speciation was conducted using PHREEQC (Schneider et al. 2000).

Klinger et al. (2002) were able to improve the treatment properties of $Fe^0$ by adding coal to the iron. In addition to in situ tests in underground mines, tests with $Fe^0$ in reactive walls were also carried out (Biermann 2007; Roehl 2004). However, none of the methods could gain acceptance and were thus not permanently used for mine water treatment, although the experimental results generally showed good treatment performance. One of the major problems with the application of $Fe^0$ in reactive walls or iron beds is the reduction in permeability due to the precipitation products (e.g. Neitzel et al. 2000). In addition to hydrogen generation, this may be one of the reasons why there is currently no site where uranium-containing mine water is permanently treated using $Fe^0$.

Further in situ remediation experiments took place from 2001 to 2005 and from 2010 to 2011 at the former Königstein, Germany, uranium mine. In the first group of experiments, reactive minerals were to be immobilised with barium sulfate to lower the contaminant concentrations in the mine water during flooding. This was possible underground while the mine was open. Overall, after successful in situ tests, it was possible to inject over $1.1 \cdot 10^6$ $m^3$ of rock in phases with a chemical mixture based on $Ba(OH)_2 \cdot 8H_2O$, $Na_2SO_3$, $Na_2O$-$SiO_2$-$5H_2O$ and a precipitation inhibitor (Jenk et al. 2014b; Jenk et al. 2005; Ziegenbalg 1999). In principle, this artificially induces a process similar to crust formation on mine dumps (Regenspurg et al. 2005; Valente et al. 2019). The extent to which this measure will ultimately help to reduce contaminant concentrations in mine water will be difficult to determine with certainty, given the ratio of the total mine workings to the volume of the in situ measure (approximately 1:15–1:20). From the analogy of crust formation and in situ mobilisation, a positive effect can probably not be completely dismissed. However, it will also be difficult to verify whether these positive effects could perhaps be attributed to the aquifer processes calculated by Merkel (2002). If in the end there are positive effects for the receiving water or the aquifer, it is ultimately irrelevant which process played a role (although the criminologist in us always wants to know which preferential processes are taking place).

In the 2010/2011 in situ experiment, KOH and NaOH as well as $Na_2SO_3$ were pumped in three phases into the partially flooded underground area of the Königstein mine and the parameters essential for the process were subjected to continuous monitoring. It could be shown for the specific case that the injection of alkaline material had a beneficial effect on the mine water chemistry (Jenk et al. 2014b). Since the distances between the injection sites and the monitoring stations were relatively short, statements on the entire mine water pool can only be made to a limited extent. However, from the information given in the publication and the assumption of a flow rate of 100 $m^3\ h^{-1}$, it can be calculated that the recovery rate of the added alkalis is 3–4%, which corresponds to the also quite low rate of 5–6% at the Salem No. 2 mine (USA) (Aljoe and Hawkins 1994). This is primarily due to the complex hydraulic situation, which is difficult to predict in detail. In this context, Gerth et al. (2006, p. 320) point out that in Ronneburg, Germany, "in-situ liming … prior to initiation of flooding … however, did not prove to be sufficiently efficient" due to the pit geometry and underground flow conditions. As the above examples show, we still have much to learn in terms of optimising in situ remediation.

The Schwartzwalder uranium mine (Colorado, USA), which was closed in 2000 and sealed with two hydraulic dams in 2008, contains up to 26 mg $L^{-1}$ uranium in the mine water and is likely to require permanent treatment (Harrington 2016; Harrington et al. 2015; Sumi and Gestring 2013). To relieve ion exchange and reverse osmosis in the medium term, an in situ experiment to lower uranium concentrations in mine water was conducted in 2013. For this purpose, methanol and molasses were injected into the mine pool in several phases, and a decrease in uranium concentrations to 2 mg $L^{-1}$ was observed (Harrington et al. 2015). However, ten months after the end of the injection process, it was found that uranium concentrations had risen again to 7 mg $L^{-1}$, and remained within this range until the end of 2016; therefore the measure needs to be carried out regularly to keep concentrations permanently low. Details of the quantities injected, or the volume of the mine pool could not be found online or in the accessible literature. Yet, the mine is now a "Colorado Legacy Land" and a U.S. Superfund site, with Ensore Solutions being entrusted with identifying and installing treatment solutions. Currently, Reverse Osmosis as the primary treatment and ion exchange are the methods of choice for the pumped mine water (from company website, July 2022).

## 6.7 Mixing of Pyrite-containing Substrates with Alkaline Material

Since even the smallest amounts of disulfide can cause acidification of groundwater or surface water, it makes sense to mix the overburden, which is potentially susceptible to acidification, with buffering material. Wisotzky (2003) investigated this possibility in detail using the example of the Rhenish open pit lignite mines in Germany (Fig. 6.6). As part of the preliminary investigations, he conducted numerical calculations using PHREEQC, undertook extensive laboratory tests with the dump material, implemented his results at field scale and subjected the field tests to several years of monitoring (Wisotzky and Lenk 2006). His studies show that even tiny amounts of disulfide (0.04–0.26 mass % in the Rhenish mining area) in the overburden dump material can lead to acidification of groundwater or mine water with sulfate concentrations of up to 4 000 mg $L^{-1}$. He therefore recommended that 0.08 mass % limestone be added to the dumps by blending. Today, 0.2% limestone or a limestone-fly ash mixture is added to the acidification-prone substrate to ensure adequate safety (Wisotzky 2001, 2003). For the Helmstedt, Germany, lignite field, Simon et al. (2017) propose mixing the buffering Quaternary sediments (1 mass % carbonates) with the acidifying Tertiary sediments to avoid the formation of acid mine drainage. At the same time, it is proposed to keep the contact time of the acid-forming overburden material with the atmosphere as short as possible.

Similar investigations were conducted by Naumann and Wiram (1995) for waste rock in Sequatchie County, northwest of Dunlap, Tennessee, USA. They also recommended, based on intensive laboratory and field investigations, that limestone be added to the overburden, thereby preventing acidification of the waste rock dump seepage waters. Brant and

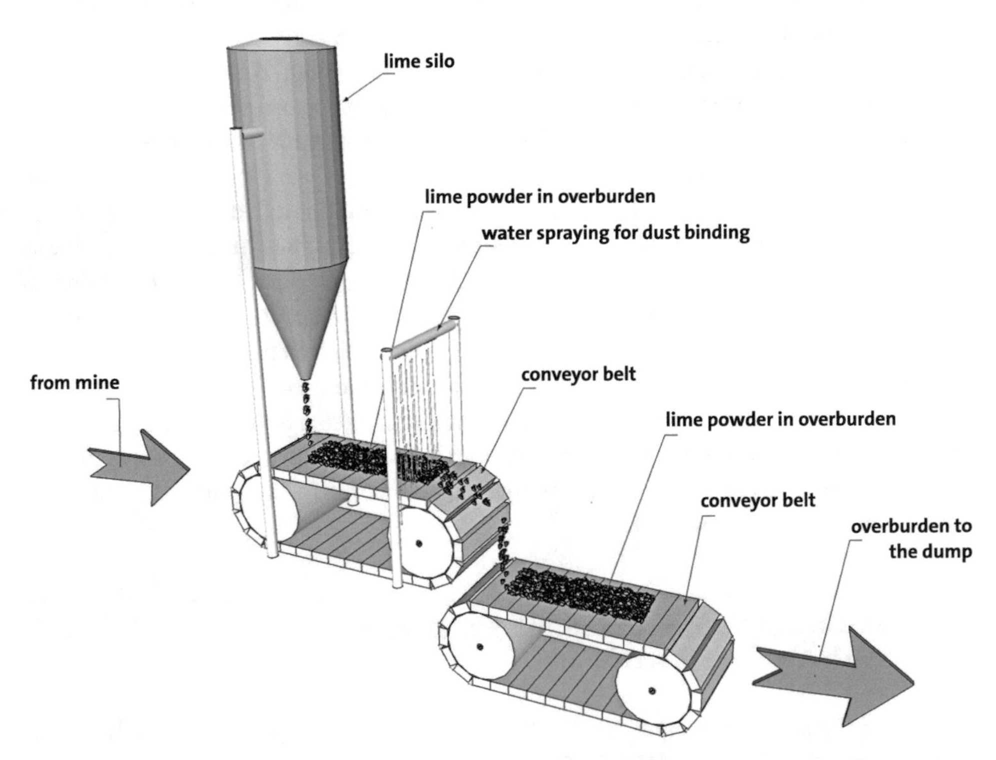


**Fig. 6.6** Mixing overburden and lime to prevent acid formation in the dumps. (Modified from Huisamen 2017; after Wisotzky 2003)

Ziemkiewicz (1997) demonstrated that buffering is particularly effective when the limestone is vigorously mixed with the dump substrate, as is also common practice in the Rhenish mining area. Extensive experiments with acid generating and neutralising waste rock material from the Gold Acres Mine, Nevada, USA, demonstrated that combining these rocks will result in circumneutral leachates (Davis et al. 2021).

The authorities in Pennsylvania and West Virginia, USA, took a different approach. There, alkaline fly ash was added to acid-producing overburden over a large area from 1986 onwards, without any discernible environmental damage being caused after 10 years of monitoring (Gray et al. 1997; Loop et al. 2003). At the McCloskey Colliery, over 37 $hm^2$ of overburden was covered with fly ash to prevent the seepage of acid mine drainage into the underlying and backfilled mine. This involved the use of over 263 kt of fly ash by 1998, and Hellier (1998) suggested that a further 58 kt would be required to cover all the overburden. He was able to demonstrate, after extensive monitoring, that the capping with fly ash had a positive effect on the chemistry of the water bodies (Loop et al. 2007). Nevertheless, they conclude that a case-by-case approach is essential and that further studies are needed.

At the Ronneburg site of Wismut GmbH in Thuringia, Germany, tailings leachate and mine water have been treated with power plant ash since the 1980s by trickling the water over the ash in the inner dump. In this way, a large proportion of the potentially toxic metals and sulfate could be removed from the leachate and mine water (Badstübner et al. 2010; Vogel et al. 1996).

The opinion that fly ash has a positive effect on water quality is not unanimously shared by everyone. Depending on the intent, results from fly ash applications are also interpreted negatively (e.g. Clean Air Task Force and Earthjustice 2007). However, the Pennsylvania Department of Environmental Protection has demonstrated on several occasions that the reports of the advocacy group Clean Air Task Force (CATF) contain substantial errors. In addition, the group appears to be pursuing an attempt to discredit mining and the use of fly ash as a way to treat acid mine drainage (e.g. Appendix "Pennsylvania Department of Environmental Protection Response to Clean Air Task Force Report: 'Impacts On Water Quality From Placement Of Coal Combustion Waste In Pennsylvania Coal Mines'" in Beadle et al. (2007)). In addition, the Clean Air Task Force report lacks a comparison between potential acid mine drainage effects and observed changes in chemistries in receiving waters or groundwater.

Since the method described here is not a mine water treatment *per se*, but a preventative measure, and thus does not exactly belong to the topic, I will refrain from a more detailed presentation. But again, allow me to make a comment: Do not repeat the experiments that researchers before you have done (Canty and Everett 1999; Yeheyis et al. 2009). Use the available evidence and implement it in field experiments. It is already well known that fly ash has alkaline properties and is capable of neutralising acid mine drainage or preventing its formation. If you want to work with fly ash, develop a general classification that makes it possible to evaluate fly ash according to how and whether it can be used for remediation or prevention of water contamination (and that goes beyond the classification used in the

USA), or develop a new method, such as for example, Musyoka et al. (2009) or Prasad et al. (2011) have done; they devised a method to convert fly ash into a more transportable and incorporable, and consequently higher value product so that less fly ash requires final disposal.

## 6.8 Closure of Drainage and Mine Adits

To prevent the uncontrolled discharge of contaminated mine water, complete or partial dams or barriers can be installed at the adit portal, as was done at the Schwartzwalder mine. A distinction should be made between dry dams, which only prevent oxygen ingress and consequently pyrite oxidation (Fig. 6.7), and wet dams, which prevent both oxygen ingress and water discharge to receiving waters (Fig. 6.8). They should not be used to impound the entire mine workings, which are above gallery level, in an uncontrolled manner to pre-mining water levels. What could become problematic, as the Parys Mountain, United Kingdom, example shows, is that damming the mine reactivates waterway features, such as karst or fault systems, which were dry during mining operations. This may cause mine water to leak in unpredictable locations and, as a result, would complicate mine water treatment. In summary, the purpose of dams is to prevent the following:

- Escape of contaminated mine water
- Disulfide oxidation (pyrite oxidation)
- Potentially toxic gases from escaping
- Unauthorised persons gaining access to the open mine workings (“moonlighters”, illegal mine explorers, Leupolt and Hocker 1999)
- Damage to cultural heritage

In the USA, investigations into the sealing of mine adits date back to the late 1920s. The aim was to reduce contamination of receiving waters, and by the end of the 1930s, over

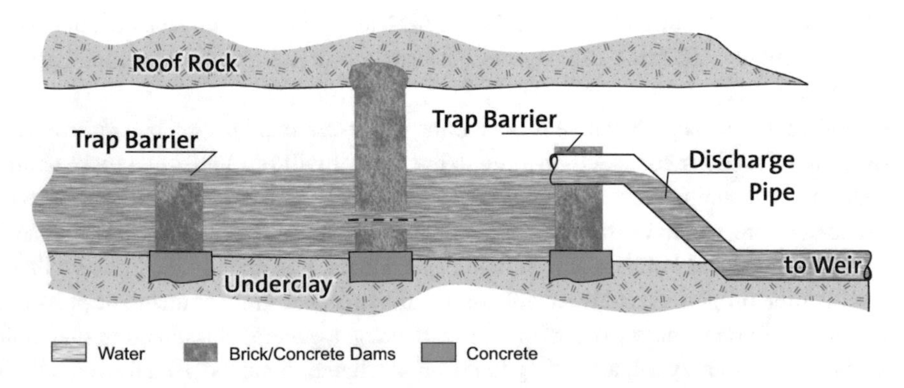


**Fig. 6.7** Partial damming of a drainage gallery. (After Foreman 1971)

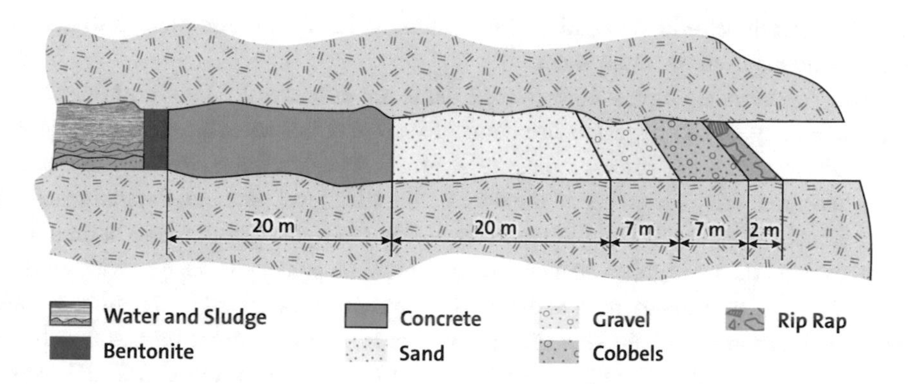


**Fig. 6.8** Example of an underground dam. (After Lang 1999)

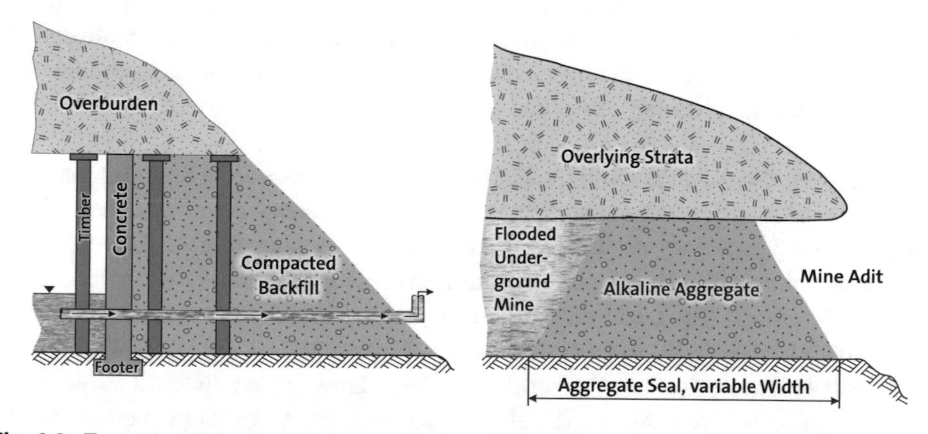


**Fig. 6.9** Examples of dammed drainage galleries. (Adapted from Halliburton Company 1970; Scott and Hays 1975)

1 000 coal mines had already been sealed. This resulted in a 50–90% reduction in loads (Fellows 1937; Leitch 1935). One of the most comprehensive descriptions of the prevention and control of contamination from abandoned mines was given by Scott and Hays (1975), although several of the methods described (Fig. 6.9) have since been optimised. As Foreman (1971) also pointed out, the layout of the mine is of critical importance to reliable damming, and accesses via adits should be treated differently from those through shafts or inclined shafts. In all cases, the mine should be cleared prior to damming, which means removing all debris such as wood, metal, cables and other material that could prevent effective damming. This also includes operating materials such as oils, PCBs (Bench 2008), chemicals or cleaning agents that could pose a risk to the aquatic environment. Many other materials used in mining are less mobile under anoxic conditions.

Another process, first described by Taylor and Waring (2001), prevents the formation of acid mine drainage by displacing oxygen with reducing gases, such as $CO_2$, $CH_4$ or $H_2S$, produced by anaerobic bacteria. This system, called GaRD (Gas Redox and Displacement), can reduce disulfide oxidation and acid formation and results in the precipitation of secondary minerals from the mine water.

As the literature examples show, not all dams are successful. Dam failures, although rare, are usually due to bypasses or chemical reactions between the dam and the mine water (this was the case, for example, at the Straßberg/Harz fluorspar mine in Germany where acid mine drainage corroded the dam on the fifth level between the main shaft and Brachmannsberg areas). It is important to note that the hydraulic pressure behind a dam can become markedly high, so special attention must be paid to its design. An example is the $1.1 \cdot 10^7$ Pa high-pressure dam in the Warndt coal mine, Saarland designed as a double cone, which separates the flooded French mine area, dammed more than 900 m higher, from the still unflooded German one (Ruhrkohle 2014; Sersch and Uhl 2006). This is 600 m higher than the current highest arch dam on earth, the *Jinping I* (锦屏一级水电站) in China. It should also be noted that the quality of the mine water often changes over time and can lead to chemical reactions between the mine water and the concrete. Most of the known cases of failure are due to inadequate hydrogeological or geotechnical investigations, which would ultimately have been much cheaper than the subsequent remediation measures!

Complete sealing of adits could conflict with species protection and should therefore be used with caution. Adits that are unlikely to release contaminated mine water in the future should not be permanently sealed. Such openings may serve as refuges for endangered animals such as bats. Grids can be installed to allow free drainage of water and free access for animals (Fig. 6.10). Local environmental groups, bat conservationists or mining archaeologists can help to preserve such sites permanently. However, bats usually only need small holes to enter the mine workings. Then, however, no effective reduction of pyrite oxidation will occur. Alternatively, the construction of dams or barriers can be moved several dozen meters into the mine adit if it is known that bats live in the mine.

A positive example of damming is the No.105 W underground mine near Buckhannon in West Virginia, USA. Two of the three mine portals had to be sealed with five dams to prevent the discharge of acid mine drainage. After damming (1984/1985), the mine was treated by in situ remediation. This involved introducing apatite dust into the area of potentially acid-forming mine sections above the mine water level and injecting alkaline substances into the flooded mine water body. Even after four water changes, the mine water, which originally had a pH of 3.9, retained a pH of about 7 (Hause and Willison 1986), and iron concentrations decreased from the original 144 mg $L^{-1}$ to 1–15 mg $L^{-1}$ by 1989 (Aljoe and Hawkins 1993). With respect to apatite dust, there were experiments as early as the 1980s that showed a 96% reduction in acidity (Kleinmann 1990b).

However, there are cases where damming has had a substantial negative effect on water quality at the mine portal and receiving watercourse – and as discussions with colleagues show, such unpublished cases are more common than previously thought. At Mynydd Parys mine in Anglesey, Wales (Parys Mountain, United Kingdom), for example, the receiving watercourse, Afon Goch ('Red River'), is heavily contaminated by acid mine drainage with pH values of 2–3 and elevated metal concentrations (e.g. Al 81 mg $L^{-1}$, Fe

**Fig. 6.10** "Königl. Verträglicher Gesellschaft Stolln" at the Rote Graben near Freiberg/Saxony, Germany, from the nineteenth century, closed with a grid. Part of the World Heritage Site *Montanregion Erzgebirge/Krušnohoří*

431 mg $L^{-1}$, Cu 39 mg $L^{-1}$, Zn 45 mg $L^{-1}$) (Boult 1996; Younger et al. 2004). In the early twentieth century, the Dyffryn Adda drainage adit was sealed with concrete, and copper was extracted ex situ from the highly copper-contaminated mine water; a process that is now being studied again in the context of circular economy (Sapsford et al. 2017). During and after this regular extraction of mine water, which was controlled by valves and pipes through the dam, a body of mine water containing approximately $270 \cdot 10^3$ $m^3$ of acid mine drainage and a hydrostatic pressure of $3.6 \cdot 10^5$ Pa built up behind the dam. One reason for the formation of this volume of water was the complete closure of the drainage adit after ex situ mining ceased in the mid-1950s. Since no data were available on the dam and its stability was not clear, $245 \cdot 10^3$ $m^3$ of acid mine drainage was pumped out to dismantle the dam. Following the removal of the dam, mine water quality initially improved slightly based on microbiological and geochemical analyses (Bryan et al. 2004; Coupland et al. 2004; Johnston 2004). Since then, however, the water quality of the Afon Goch South has improved substantially, while the Afon Goch North is as contaminated as before the remediation measure.

In principle, this is consistent with the results of an extensive study of 65 dammed mines in the eastern coal regions of the USA. There it was found that "the effectiveness of the mine closures with respect to the mine effluent quality by comparison with the preliminary mine effluent guidelines was observed to be usually less than 50 percent effective" (Bucek and Emel 1977, p. iv). In addition, the authors reported that local conditions, e.g., mining method and mining conditions, are of primary importance in terms of the effectiveness of the closure measure, whereas the damming technology is less relevant.

At the Upper Mining Authority in Saxony, Germany, interventions in the drainage system of a mine are always seen in view of the mine stability. Overall, the drainage capacity of the adits must remain guaranteed. In addition, there are concerns that complete mine sealing could cause the build-up of water behind the dam resulting in uncontrolled mine water discharges. These, in turn, could negatively affect the stability of the rock mass as well as rehabilitated or backfilled shafts.

In summary, the complete closure of a dewatering adit poses substantial risks. Let's make sure we never again have to deal with an incident like the 2015 Gold King mine mine water spill in Colorado, USA, where several million cubic metres of mine water contaminated the receiving watercourse, the Animas River (U.S. Department of the Interior – Bureau of Reclamation 2015).

# 7 Post-mining Usage of Mine Sites or Residues of the Treatment Process

## 7.1 Post-mining Usage of Remediated Sites

Mine water is not "sexy", said a former senior executive of a global engineering firm in Australia some time ago (and I'm not talking about Robin Strachan). So the question is, how can you make mine water attractive to people? Achieving this goal usually requires considerable financial resources, because – if we leave aside the Rio Tinto region of Spain (Fig. 1.8) – which non-specialist can possibly develop a passion for mine water? If we equate "sexy" with "confident; in touch; holistic; passionate; and conscious", then two things are necessary before we use our remediated sites: first the confidence that the remediation will succeed, and thereafter the conscious marketing of a holistic remediation project by getting involved with local residents. And if all of this is done with passion, then we can change the negative perceptions of mine water and improve the attractiveness of mine water in the future – and achieve greater acceptance of mining operations. Let's hope that soon we won't have to make a statement along the lines of Glover (1975, p. 179): "From the point of view of modern society, acidic and ferruginous mine drainages cannot be claimed to have any beneficial effects."

There are many examples of the after-use of abandoned mines. However, only those resulting from the use of mine water will be listed here. Primarily, abandoned mines, if they are put to an after-use, are prepared for tourist purposes. Some of the most beautiful examples of abandoned mine sites that turned into tourist attractions are the two German World Heritage Sites, Rammelsberg in Goslar/Harz and the *Montanregion Erzgebirge/Krušnohoří* (included in the World Heritage List since July 2019, Fig. 6.10), the former china clay pit, now the Eden Project in Cornwall, England, UK (Fig. 7.1), or the hotel complex built in an abandoned quarry, namely the InterContinental Shanghai Wonderland Hotel or Shimao Quarry Hotel in Songjiang, Shanghai, China (Fig. 7.2; Macadam and



C. Wolkersdorfer, *Mine Water Treatment – Active and Passive Methods*,
https://doi.org/10.1007/978-3-662-65770-6_7

**Fig. 7.1** Biosphere project "Eden" in a former china clay mine in Cornwall, United Kingdom. (Photo: Jürgen Matern; Wikimedia Commons, CC-BY-3.0)

**Fig. 7.2** The InterContinental Shanghai Wonderland Hotel in Songjiang, which opened in November 2018. Architects: JADE + QA (JADE + Quarry Associates), ECADI – East China Architectural Design & Research Institute and Atkins. (Photo: Martin Jochman, JADE + QA)

Shail 1999; Wolkersdorfer 2008). The trick fountain of the Steinhöfer waterfall in the *Bergpark Wilhelmshöhe* near Kassel, Germany, also a World Heritage Site, is another example; it is fed by mine water from the Herkules Colliery in the Habichtswald (Hessisches Ministerium für Wissenschaft und Kunst 2011, pp. 29, 39, "World Heritage List 2013" application supplement, p. 10).

At many places in Germany and worldwide, treated mine water is used as drinking water (e.g. Burbey et al. 2000; Stengel-Rutkowski 1993, 2002; Teaf et al. 2006; a more detailed list can be found in Wolkersdorfer 2008, p. 271). In Lusatia, Germany, treated mine water from the Schwarze Pumpe waterworks was made available as drinking water until 2018. Artificial groundwater recharge using mine water can also be a possibility to use mine water. Fernández-Rubio and Lorca Fernández (2010) give examples of this in Germany, the USA, Australia and Spain. Hobba Jr (1987) lists 70 water supply plants in Pennsylvania, USA, that provided drinking water from coal mines to nearly 82,000 people. The largest plant, Gary #2, supplied 5000 $m^3$ of drinking water daily to 2800 people at that time, and by 2003 it was supplying only 1500 $m^3$ to 1700 people (West Virginia Department of Health and Human Resources 2003).

Residual lakes or flooded mines could also be transformed into tourist attractions (mining tourism). There are hardly any limits to the possibilities, as long as the water quality is reasonably good: Think of surface or underground bathing lakes (Wackersdorf, Lusatia, both in Germany; Mina da Passagem de Mariana, Brazil), spas (Bad Gastein, Austria), diving facilities (I can't help but mention Ojamo in Finland, Bell Island in Newfoundland and Sala, Tuna Hästberg and Langban in Sweden here) or local recreation areas (St Aidan's, United Kingdom). For example, in Lusatia, Germany, some of the residual lakes in the Lusatian Lake District are used for tourism, and considerable future development potential is seen in this (Hunger et al. 2005; I. B. A. Internationale Bauausstellung Fürst-Pückler-Land 2010; Linke and Schiffer 2002; Pflug 1998). In addition, the Lohsa water reservoir, the Lake Dreiweibern and Lake Bernstein are to supply the Spreewald and the city of Berlin with the necessary water during dry periods, a water management system called *Speichersystem Lohsa II* (LMBV website).

The Knappensee Lake and the lake Steinberg near Wackersdorf, Germany, also provide an example of the advantages of using former open-pit mines for tourism. However, it should not be forgotten that for the development of the tourism infrastructure, local acceptance of the project should first be achieved, and extensive tourism marketing should be done. Marketing campaigns must ensure that opening and use of the residual lake will be released to the public promptly, whereby multilingual announcements are essential. At Lake Brombach in Franconia, Germany, foreign boating tourists arrived as early as 1983, because the lake was already indicated on many maps – although the lake was not completed until 1999 (personal experience in the summer of 1983 in Roth/Middle Franconia).

Mining lakes (pit lakes) or treated mine water can also be used in aquaculture. For example, Mallo et al. (2010) describe laboratory experiments for aquaculture in Batán (Buenos Aires, Argentina) and conclude that the mine water and the selected fish are suitable for aquaculture. Miller and D'Souza (2008) report experiences from West Virginia, USA, where open pit lakes have also been used for fish culture. In Minnesota, USA, where there are more than 4000 abandoned quarry pits, a small number of these pit lakes have been used for aquaculture. However, this caused environmental problems as the pit lakes accumulated nutrients and in turn required remediation (Axler et al. 1998). An interesting application exists in Collie, Western Australia. There, Premier Coal uses treated mine water to breed crayfish (*Cherax tenuimanus,* McCullough et al. 2009, 2020, McCullough and Lund 2011), which under natural conditions may only be caught by Aboriginal people and under strict legal conditions (Fig. 7.3). Hopefully, the farming of the crayfish will provide a continuous supply and a source of income for the Aboriginal people. On the other hand, limited survival of the crayfish was reported when stocking untreated surface mining lakes because they lacked nutrients (McCullough and Lund 2011). However, unlike in the aforementioned examples, the fish farms at the Jänschwalde and Schwarze Pumpe power plants in Lusatia, Germany, used cooling water and not mine water.

Because of the often higher temperatures compared to groundwater or surface water, mine water can in principle be used for geothermal purposes. Numerous such plants have been put into operation in the meantime. The problem, however, is still the clogging of the heat exchangers or the pipelines (one solution to this dilemma may be geothermal probes such as those used in Alsdorf, Germany – see below). For example, a Canadian city that

**Fig. 7.3** Premier Coal crayfish breeding in Western Australia. Left: Breeding tank; right: crayfish *(Cherax tenuimanus)*

**Fig. 7.4** Signage near the main geothermal boreholes of the flooded Springhill, Nova Scotia, Canada, coal mines

has long operated a geothermal system using mine water (Fig. 7.4) must replace pumps or pipelines on a regular basis, with the cost borne by the city and no apportionment to users. These aspects are usually not discussed in any detail when it comes to the geothermal use of mine water. They are one of the reasons why there have been only a marginal number of geothermal uses of mine water (Grab et al. 2018, p. 560), even though they were first systematically studied in 1978 (Lawson and Sonderegger 1978). Bernhard v. Cotta noted as early as 1853 (Cotta 1853, p. 140):

> "This elevated temperature, which is already of immense importance for humans and the conditions of their existence, could in the future possibly play a new important role among the resources of human life.
>
> Should once, in the future, on the increasingly populated earth, the forests be substantially thinned out everywhere and the coal deposits exhausted, it is conceivable that the inner heat of the earth may be made useful; that it may be conducted to the surface through special devices in shafts or boreholes and used for heating dwellings or even for heating machines. One will certainly not broadly and with advantage access this heat source, which probably will be costly in their application, until a serious shortage of fuel forces to do so; then however the warmth of the Mother Earth remains a safe last resort."

Since the EU funding for the Minewater Project Heerlen, in The Netherlands, ended, it has been comparatively quiet around the project, which once received much media attention. It distinguished itself mainly by the consciously or unconsciously false claims to be the "first project for mine water geothermal energy worldwide". Table 1 in the article by Ofner and Wieber (2008, p. 74) or Table 17.5 in Grab et al. (2018) should show with sufficient accuracy that this is not true: a mine water geothermal system was already installed in 1981 in Kingston, Pennsylvania, USA, and in 1984 at the Heinrich Colliery in Essen, Germany. An even earlier system went into operation in 1979 at the Midway Shopping Center near Wilkes-Barr, Pennsylvania, USA (Schubert and McDaniel 1982). So it can happen that a newspaper article like this appears: "Useful heat from mining has a future – … The oldest example is the Heinrich mine in Essen, where the 22 °C warm mine water … has been used since 1984. … In Heerlen near Aachen in the Netherlands, the world's first mine water power plant was commissioned in 2008" (Pasche 2013). This is unnecessarily confusing. A response to an e-mail query to the City of Heerlen asking about the status of geothermal extraction simply stated that the project ran from 2005 to 2008 (pers. comm. Ineke Lauscher, 8th March 2011). The project's website, which ran until about 2013, said "The first in the world", and the "energy savings meter" installed there was a simple meter built into a flash player and thus did not display actual performance data. Fortunately, this meter is now no longer displayed on the project's new website (www.mijnwater.com). Nevertheless, currently it is impossible to say with certainty how successfully the project is being run and what proportion of the heat actually comes from the geothermal use of mine water. Also the leaflet published in 2010 did not give any information about the operating data of the project. From a presentation by Jean Weijers at the Geothermal Energy International Meeting (31st August 2012) it became clear that the expectations of the project were far from being fulfilled and that figures were being glossed over (the table "Potential minewater energy demand" in the presentation lists all research projects and feasibility studies in addition to ongoing projects). In 2015, the project received the European Geothermal Innovation Award. This can be understood as a sign that the geothermal use of mine water has been successfully applied in the meantime. According to the project's website, a total of 27 users are to be supplied with mine water geothermal energy by the end of 2017 – at the time of translating this sentence in March 2022, there are nine objects installed.

It is not the only project that claims a unique position for itself. Similar presentations can be found for other mine water geothermal projects (e.g. on the Internet presentation of the WDR: Martin Teigeler: "Geothermal energy from the ex-coalmine – mine water heats up classrooms", as of 8th October 2012; or *Wochenspiegel – Zeitung für Freiberg und Umgebung*: "High-tech in the old adit – … Europe-wide unique project can begin", Ulbricht (2013)). In my opinion, such reporting does little to establish the technology, because it gives the impression that it is a completely new technology, which will tend to make potential financiers reluctant – a product does not become "sexy" by the fact that I assign it a unique selling point (and please do not use the term "potential literature" at this point). At least Jean Weijers' presentation ends with a note that applies to all mine water geothermal operators, but especially to the Heerlen project: we need "courage and perseverance" for such projects.

In Saxony, Germany, a mining operation and the Freiberg Mining Academy already distinguished themselves with a geothermal project shortly after the political turnaround: The Ehrenfriedersdorf tin mine, Germany, has operated a geothermal system intermittently since 1994 (Ofner and Wieber 2008), and the Freiberg Mining Academy has operated a geothermal system at the Reiche Zeche to heat this visitor mine since 1990 (Carsten Debes, unpublished presentation at the Geothermal Energy International Meeting, 30th August 2012) and to heat the Freudenstein Castle (Dillenardt and Kranz 2010; Kranz and Dillenardt 2010). Another project in Saxony is in progress in Zwickau (Felix et al. 2010); the well for this project was drilled in 2018, and a trial operation is underway (Grab et al. 2018) as part of the pilot project "Geothermal use of mine water from the mining cavities of the Zwickau coalfield". At the Eduard shaft near Alsdorf, a system has been in place since 2018 that heats the ENERGETICON Museum ("Experience energy – understand energy") using mine water geothermal energy. A prefabricated, 860 m long HDPE double-U geothermal probe DA75 is installed there (pers. comm. Kurt Schetelig 2018; Schetelig and Richter 2013). German pilot projects for energy recovery from mine water also exist at the Zollverein Coal Mine Industrial Complex in Essen, the Auguste Victoria and Prosper-Haniel collieries, the Robert Müser colliery in Bochum or the Wolf Mine in Siegerland (Hahn et al. 2018; Schetelig and Richter 2013; Wieber and Pohl 2008). There is no doubt that a rapid and most welcome development is emerging in this field, and the increasing number of plants will also lead to the development of procedures that will enable the plants to be operated more reliably than today. However, as experience from other areas of mining shows, in the medium and long term it will come about that mineral deposits (scaling, biofouling) can only be permanently prevented with the aid of chemical additives; or the plant will have to be cleaned regularly. One of the first larger mine water geothermal plants in the world, which is often used as a reference object, is only running "without problems", as described above, because the problems are only talked about behind closed doors. Without the enormous financial expenses of the operator to reactivate pumps or to rehabilitate ochre-clogged boreholes, none of the users would be willing to use the geothermal energy of the acidic and iron-rich mine water – especially since the users do not even have to pay a fee to the operator.

Another after-use of neutralised mine water is for irrigation in agriculture (du Plessis 1983). Voznjuk and Gorshkov (1983, p. 23) describe that soil salinisation occurs above a total mineralisation of 3 g $L^{-1}$, but no negative effects were apparent up to 2 g $L^{-1}$ (also van Zyl et al. 2001). Annandale (1998) and Annandale et al. (2001) dealt particularly extensively with the question of how "gypsum-containing" mine water could be used for irrigation (I have put the "gypsum-containing" in quotation marks, because, of course, mine water does not contain gypsum, but the ions $Ca^{2+}$ and $SO_4^{2-}$, which precipitate out of the mine water as gypsum when the gypsum solubility is exceeded). In almost all the cases described, where model calculations were also made over an irrigation period of 11 and 30 years (Annandale et al. 2009; Jovanovic et al. 2001), it was shown that no negative effects are to be expected from irrigation with mine water provided that the sodium concentration is not too high (van Zyl et al. 2001, p. 44). Future studies should show whether mine water can be used successfully in larger irrigation projects (Annandale et al. 2017, 2019). A negative example is known from the former GDR. There, the LPG Nöbdenitz as well as a nursery near Schmölln had pumped water from the Spree River to irrigate herbs and tea plants, for example. However, since SDAG Wismut discharged mine water from the Drosen and Beerwalde mining operations as well as from Paitzdorf, radioactive contamination of the plants has occurred (Beleites 1992, p. 54 f.).

In addition to the stated geothermal use, mine water can also be used to drive turbines. Where there are sufficient altitude differences and there is no danger of the mine water permanently damaging the turbine, larger hydroelectric power plants can be set up, but smart or micro-hydropower turbines in dewatering adits are also conceivable as a source of energy. The best known are probably the former hydroelectric power plants in the German Harz Mountains (Ottiliae-Schacht Clausthal; Hilfe Gottes Bad Grund) and in Freiberg, Saxony the cavern power plant Drei-Brüder-Schacht (Döring 1993; Galinsky et al. 2001). In Biberwier, Tyrol, Austria, a turbine has been in operation for almost 130 years, currently producing about 50 kW and feeding into the local power grid of Elektrizitätswerke Reutte (Wolkersdorfer 2008). A more comprehensive account, including historical data, of the use of mine water for electricity generation has been presented by Döring (1993). In Whillier (1977), it is discussed whether or not the water brought down in deep South African mine shafts for cooling purposes could be passed through turbines. On the one hand, this would supply energy, and on the other hand, it would cool the water again, so that a greater cooling capacity would be available underground.

It is also interesting to note that – at least in Lusatia, Germany, which is often characterised by extreme drought – the residual lakes of open-pit mines have been integrated into mine water management as reservoirs. Due to the storage lamellae in the residual lakes, which are connected with each other and with the watercourses there, the additional water at low-flow conditions can be useful for the Spreewald and the fishing industry and to support flood protection. In addition, it can help the flooding of residual lakes that have not yet been flooded.

In terms of the recovery of metals from mine water, we are still at the beginning of a hopeful development (Fosso-Kankeu et al. 2020). This can also be seen by the large number of research groups working on membrane processes for the recovery of commodities from liquid waste streams (Ali et al. 2021). It is sometimes possible to recover metals from mine water, as shown by the examples of Wismut GmbH in Königstein, Germany (Braun et al. 2008), Pécs in Hungary (Benkovics et al. 1997; Csővári et al. 2004) or the F-LLX process (Sect. 3.12). However, on a large scale, no recyclables are as yet produced from mine water anywhere. One of the problems is the higher reactivity of iron and aluminium compared to the priority materials, the rare earth elements (Nordstrom et al. 2017a). In this regard, there needs to be an optimisation of membrane processes or ion exchange technology in the coming years, which will help to recover the valuable or critical metals from mine water. The processes published thus far are still at the laboratory or pilot plant scale stage and therefore will not be discussed further here. Nordstrom et al. (2017a) point out a different challenge: Investors interested in recovering raw materials from abandoned mines could be held responsible for the "legacy" of former operators.

## 7.2 Treatment Residues as Recyclable Materials (Circular Economy)

To date, it has not been possible to use the residues emanating from mine water treatment universally as valuable materials, although many projects worldwide have tried and are trying to do so. These residues are generally sludges of varying composition, with carbonate or gypsiferous sludges predominating. Often the sludges contain higher concentrations of metals (e.g. copper, zinc, manganese, chromium) or semimetals (e.g. arsenic, antimony) that could theoretically be used as valuable materials (Dinardo et al. 1991). Mainly pure iron oxyhydrate sludges including coprecipitates are present in passive mine water treatment. In this regard, Georgaki et al. (2004, p. 305) state that "numerous attempts have been made by industry to reduce the environmental impact of sludge and, where possible, to market or reuse the waste or its components. Despite low contained values, limited success has been achieved".

▶ Here is a legal hint that the EU project "Re-Mining" stumbled upon and that can also be relevant for other projects that try to use mining residues. According to § 128 (old waste rock piles) of the German Federal Mining Act (BBergG) it has to be taken into account that, if necessary, a mining licence has to be obtained for those who carry out scientific investigations on waste rock piles: "Sections 39, 40, 42, 48, 50 to 74 and 77 to 104 and 106 apply *mutatis mutandis* to the exploration and extraction of mineral raw materials in waste rock piles if the mineral raw materials would fall under Section 3 (3) and (4) as mineral resources and originate from a previous exploration, extraction or processing of mineral resources". In addition to a notification of the investigations to the mining authority, an operating plan can even be demanded in the worst case. I praise Section 2 (5) (scope of application) of the Austrian Mineral Resources Act (MinRoG in the version of 15 November 2016): "This Federal Act does not apply to activities of the kind referred to in subsection 1 that serve exclusively scientific purposes, nor to the collection of minerals. Mining permits, however, shall be observed." With this provision, I have even been allowed to run a tracer test in Tyrol, Austria.

Although there are some good approaches, such as the production of pigments, we are still 15–25 years away from a technology that would help to generally recycle the residues of a mine water treatment plant. Taking a look at the practice of municipal waste disposal shows that as recently as between 1980 and 1990, mixed waste landfills were almost always used, and the mixed waste was mainly sent untreated to a single landfill. Today, municipal waste must be collected and reused separately in the German dual system, and the residual materials are largely sent for thermal disposal (Dehoust et al. 2005; Wacker-Theodorakopoulos 2000). All this was unthinkable before 1980–1990, and today, 30–40 years later, it has become a standard in many countries. This must also become our goal in mine water treatment! In principle, the following alternatives are available for this purpose:

- Avoid treatment residues by changing the technology
- Store the treatment residues or reinject them into the flooded mine pool
- Consider using alternative treatment plants
- Use sludge as a valuable material ("circular economy")
- Separate the treatment steps and use the individual stages

Currently, sludge is either stored in sludge tailings ponds, stored together with ore processing residues (tailings: processing tailings), used as a cover for waste rock piles, injected into mine workings, used partly for backfilling, or sent to a hazardous waste landfill (Zinck 2005, 2006). In some cases, the sludge has been used as a growth medium for plants or added to soil (e.g., Dudeney et al. 2004; Yeh and Jenkins 1971). However, in the strict sense, none of these processes represents a use as a valuable material. The following case studies are intended to illustrate the potential for the recovery of mine water treatment sludge.

Strictly speaking, the storage of sludge from mine water treatment in open pit lakes or underground mines cannot be regarded as recycling (see Sect. 6.4 for more details). However, as the alkalinity in the sludge can help to improve water quality, this process will also be presented in this section. Kostenbader and Haines (1970), Aubé et al. (2005), and Wolkersdorfer and Baierer (2013) describe the injection of sludges from mine water treatment. In all methods, sludges are collected from a low- or high density sludge plant and flushed back into the mine workings using either pipelines or trucks. Kostenbader and Haines (1970) do not provide details on whether the water quality was affected by the sludge injection into the underground mine. They state only that "the discharge of excess sludge via a borehole to an area adjacent to the underground mine-water pool proved to be a satisfactory disposal method" (p. 92). Aubé et al. (2005) conducted laboratory tests on high density sludge from a mine water treatment plant in Canada to determine whether the sludge was suitable for injection into mine workings. They concluded that the sludges could be readily injected into the mine workings with respect to the expected concentrations of iron and aluminium. For the metals cadmium, nickel and zinc, the results were less positive. Wolkersdorfer and Baierer (2013) investigated low density sludge from the Straßberg/Harz, Germany, mine water treatment plant and concluded through laboratory and field investigations that reinjecting the sludge into the flooded mine workings has a positive effect on mine water quality. They were able to show through chemical-thermodynamic modelling that the pH value of the mine water would have been lower if the low density sludge had not been reinjected.

In Montana, USA, sludge from a mine water treatment plant is discharged into the Berkeley open pit lake (Duaime and Metesh 2007; Geller et al. 2013, p. 365). Initially, no improvement in water quality occured because the volume of injected sludge was comparatively small in relation to the volume of the open pit lake, but after 15 years of operation, a substantial improvement in the pit water quality was observed, with iron concentrations decreasing from ± 1000 mg/L to < 5 mg/L and pH values increasing from 2–3 to 4–4.5 (Gammons and Icopini 2020). On the other hand, Naik et al. (1990), Ashby (2001), Schipek et al. (2006) or Schipek et al. (2007) discuss the positive effects of fly ash injection into open pit lakes or into underground mines.

In the Polish coal mines Dębieńsko and Budryk, high-salinity mine water (8–115 g $L^{-1}$ TDS) is treated by reverse osmosis or electrodialysis and distillation (Ericsson and Hallmans 1996; Sikora and Szyndler 2005; Turek 2004; Turek et al. 2005). The final products are drinking water and rock salt, which can be marketed. These two plants produce between 4400 and 8000 $m^3$ of mine water per day, which is treated in a multi-stage process. In this process, chemical treatment is first conducted prior to reverse osmosis and gypsum is separated from NaCl. In 2005, the energy consumption in the solution concentrator was 44 kWh $m^{-3}$ and in the brine crystalliser it was 66 kWh $m^{-3}$. As soon as the magnesium concentration in the brine exceeds 2110 meq $L^{-1}$, the crystalliser can no longer be operated optimally. Another limitation of the process is the high calcium and sulfate concentrations, which lead to uncontrolled gypsum precipitation. To avoid this, electrodialysis is used as a further pre-treatment step. In the recently completed plant, the following five process steps are included: Pre-treatment, reverse osmosis (desalination), brine concentration, crystallisation, and post-treatment (Fig. 7.5). Under optimum conditions, the

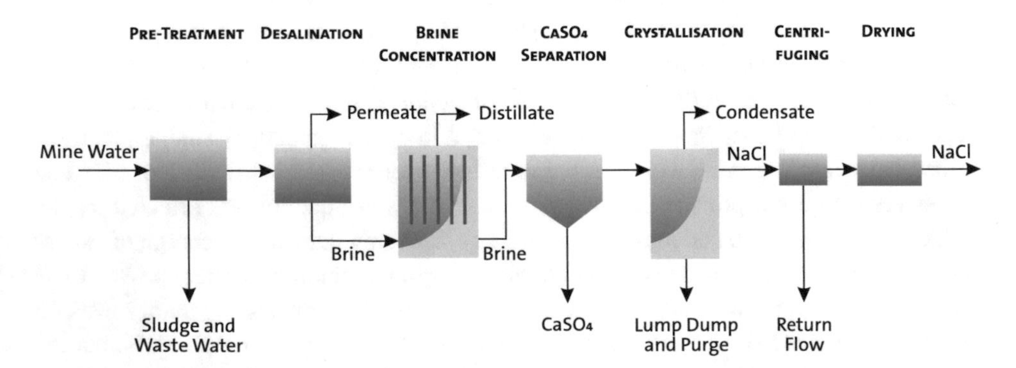


**Fig. 7.5** Mine water treatment process at the plant in Dębieńsko, Poland. (After Ericsson and Hallmans 1996)

following products are obtained: Distillate, rock salt, carnallite, magnesium chloride, iodine and bromine. A disadvantage of the method is the high energy demand, which is only worthwhile if high remediation targets are to be achieved and an inexpensive energy source is available.

Iron oxides are used in industry for a wide range of processes (Cornell and Schwertmann 2003, pp. 509–524), including the food industry (Voss et al. 2020). Although the sludges from mine water treatment consist for the most part of iron oxides or hydroxides, they are hardly used industrially. In Germany, the company Lanxess Deutschland (formerly Bayer) tried to use mine water treatment sludges for pigments and for the iron oxide Bayoxid E33 (Schlegel et al. 2005). Apparently, the coprecipitates from sludge precipitation are currently not suitable for industrial use (pers. comm. Andreas Schlegel). Hedin (1998, 2002, 2003) has successfully used passive mine water treatment sludges since the late 1990s to produce pigments for the brick industry and paints that are commercially available as EnvironOxide™. To achieve this, the sludge from passive mine water treatment plants is first extracted using excavators, then dewatered in a multi-stage process using tubes made of geotextiles (in the USA "Geotextile and Sludge Tubes", in Germany "Geotube® Geotextile Tubes") and thereafter the dewatered sludge is transferred to heaps (drying beds) until it is dry (Fig. 7.6a). Once dry, the material is burnt and sold on the market as EnvironOxide™ pigment (BR-832 red; BR-833 yellow; BR-834 black), depending on quality and colour (Fig. 7.6b). Approximately every seven to ten years, 2–2.5 kt of sludge can be recovered and marketed in this manner (pers. comm. Bob Hedin 2008). To operate this economically in the long term, seven to ten passive mine water treatment systems would be required in the vicinity of the production plant (Table 7.1).

Silva et al. (2011) took a different approach. In a four-step laboratory process, they were able to obtain goethite and hematite from mine water: selective iron precipitation → iron solution → crystallisation of goethite → production of hematite. They were able to determine a shade close to PANTONE® P 7–11 for the goethite and close to PANTONE® P 15–12 for the hematite powder. This makes them lighter than the EnvironOxide™ products which are burnt pigments. The product could then be used as a pigment for paints or concrete. However, industrial use has yet to be established.

**Fig. 7.6** Waste rock piles with ochre from a passive mine water treatment after dewatering and drying (**a**) and dried as well as burnt ochre (**b**) (Photos: Bob Hedin)

**Table 7.1** Cost comparison of a passive mine water treatment plant with marketing the recovered iron oxide and a conventional low density sludge plant (after pers. comm. Bob Hedin 2008). *Final storage costs if the sludge is not marketable

| | Unit | Passive FeOx | Conventional |
|---|---|---|---|
| Capital investment | € per (L $min^{-1}$) | 130 | 200…400 |
| Ongoing annual costs | € per 1000 $m^3$ | 2 | 10…20 |
| Total costs 25 years | € per 1000 $m^3$ | 14 (24*) | 56 |

Pigments have also been produced in Lusatia, Germany, from the residues of mine water treatment (including schwertmannite; Janneck et al. 2007a, pp. 100–109). In addition, the authors conducted extensive technical investigations on the pigments to verify their potential industrial suitability. The pigments called Nochten-Ochre and Nochten-Red were used to dye ceramics and concrete blocks and to produce pigment powders and paints. Overall, it was shown that the pigments are suitable for use in tile mass, concrete blocks, decorative chippings, bricks and clinkers, paints and varnishes, coloured plasters and wall paints.

Warkentin et al. (2010) describe laboratory experiments in which they succeeded in selectively removing metals (Cu, Zn, Co, Ni, Al, Fe, Mn) from mine water. To achieve this, they added different reagents to the mine water in four to six steps, depending on the experimental configuration. These included $H_2S$ from a bioreactor, $CaCO_3$, CaO and the residual solution from a bioreactor. In the past, methods have been repeatedly presented to selectively extract metals from (mine) water. Haber (1927), for example, wanted to produce gold from seawater, but he failed because of the tiny amounts contained in it. In the context of mine water, Valenzuela et al. (1999) proved that copper can be selectively removed from mine water. Menezes et al. (2009) prepared a ferric sulfate solution from which they were able to extract a product containing 12% iron and 1% aluminium. McCloskey et al. (2010) devised a method using metal-sensitive membranes to remove metals from mine water. Reddy et al. (2010) used eutectic freeze crystallisation and were able to recover sodium sulfate when crystallisation nuclei were added. Figuratively speaking, I share the sentiment expressed by the Nobel Prize winner Fritz Haber that "possibly, one day, somewhere [in the world's oceans] a kind of gold deposit will appear, where the precious metal particles regularly accumulate" (Haber 1927, p. 314) – although I am not entirely sure – perhaps he was just injecting a little irony here.

The eMalahleni mine water treatment plant in Gauteng Province, South Africa (Fig. 7.7), is the first plant to be described as a zero-discharge plant (Holtzhausen 2006a; Hutton et al. 2009), i.e. a plant that leaves no waste behind. However, this is only correct in a broader sense, since the highly concentrated brine from reverse osmosis is kept in a storage tank, as there is as yet no industrial use or treatment option for the brine. The HiPRo® process plant at Kromdraai (Optimum Coal Mine, South Africa), which started operation in 2012, is based on the same principle (Cogho 2012; Cogho and van Niekerk 2009; Karakatsanis and Cogho 2010). A multi-stage process is used in both plants in that

**Fig. 7.7** Mine water treatment plant eMalahleni. From top left to bottom right: Pre-treatment (leaching basin in the background on the left and coal mine on the right); reverse osmosis; building made of gypsum; mineral water produced in the plant. (Gauteng Province, South Africa)

the mine water is first treated conventionally, precipitating gypsum that can be reused (Fig. 7.8). The water is then subjected to ultrafiltration and reverse osmosis and can thereafter be discharged either to the local water supply or to the receiving watercourse. Mineral water can also be obtained from a partial stream. The water treatment is a four-step process developed by the company Keyplan Engineering Limited, and consists of the following steps (Cogho and van Niekerk 2009):

- Pre-treatment to remove solids and dissolved metals
- Filtering with a selective granular medium to remove manganese
- Ultrafiltration to remove suspended solids and colloids
- Reverse osmosis and nanofiltration for demineralisation of the residual water

In my opinion, this process is currently the optimum solution for treating mine water and complying with all the environmental specifications required today. If it should also be possible to use the brines (e.g. through freeze crystallisation), we will only be one small step away from "intelligent mine water treatment", namely its installation!

Earlier in the book I was a little critical of experiments with sorbents and phytoremediation for soil remediation. Let me now come back to sorbents – from a completely different perspective. If the earlier two sections on sorbents were about a critical examination of sorbents as a panacea for waterborne contamination, here I would like to discuss biosor-

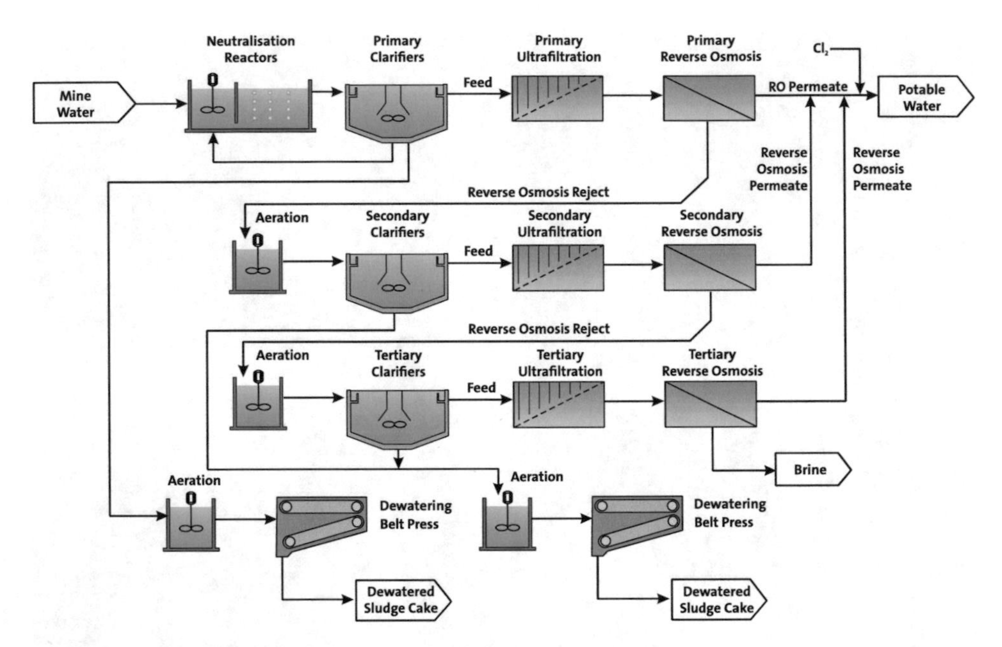


**Fig. 7.8** Flow chart of the eMalahleni and Kromdraai mine water treatment plants in Gauteng Province, South Africa. (Modified from Gunther and Mey 2008)

bents for the extraction of metals. Agricola was not only the first in modern literature to suggest and describe contamination of mine water or agricultural damage from mining, but he also described plants that can be used as ore indicators: "Likewise along a course where a vein extends, there grows a certain herb or fungus which is absent from the adjacent space, or sometimes even from the neighbourhood of the veins" (Agricola 1928 [1557], p. 30, Hoover & Hoover translation 1912). This is virtually identical to the findings on prospecting with plants that Cannon (1960) described. For a long time there has been uncertainty as to whether "heavy metal plants" (indicator plants, metallophytes) actually take up the metals or whether they are merely metal-tolerant. Since then, knowledge about indicator plants has expanded considerably, and it is now clear that there are indicator plants that take up (hyperaccumulate) metals on the one hand, and those that show tolerance to metals on the other (Ernst et al. 1975; Punz 2014; Punz and Mucina 1997). Hyperaccumulating metallophytes such as the genera *Thlaspi* or *Alyssum* can incorporate substantial amounts of different metals into their biomass (Brooks et al. 1998; van der Ent et al. 2015). If they are not suitable for phytoremediation (Blume et al. 2010, p. 468), they may be suitable as sorbents for different metals. In addition to plants, algae, fungi (think of the fungi still contaminated with $^{137}$Cs, especially in southern Bavaria, Germany), and bacteria are also suitable accumulators for metals, which could potentially make them important as biosorbents of metals or for bioaccumulation of metals in the future. A detailed summary of potential candidates with respect to mine water has been given by Dinardo et al. (1991) and a comprehensive account of the current state of knowledge by Phieler et al. (2015). A study from Poland detected high concentrations of As, Cr, Ni, U and rare earth elements in the Heiligkreuz Mountains in the Heiligkreuz Voivodeship (Gałuszka and Migaszewski 2018). I am convinced that further research will allow the use of bioreactors for the sorption of metals from mine water, just as bioreactors are used today for insulin production.

# 8 Finish

A journey ends here – for now!

The result of the literature review shows that mine water represents one of the largest waste streams by volume worldwide but is not contaminated in every case. Thousands of kilometres of water bodies, extensive areas in nature reserves and countless aquifers are contaminated or potentially endangered by acidic or (semi-)metal-enriched mine water. To keep the negative effects on the ecosphere and the anthroposphere as low as possible, it is therefore essential to treat contaminated mine water in such a way that the effects on the environment are kept to a minimum or are completely prevented. Wherever financial resources are available to treat mine water to an acceptable quality, usually specified by the authorities, treatment plants or passive systems are therefore being constructed. For this purpose I have presented many passive and active treatment methods.

What will be necessary in the future? Above all, cooperation between the research and development groups should be optimised so that duplicate research no longer occurs. To this end, the respective groups need to conduct more systematic literature reviews, because a great deal of duplicate research is due to insufficient knowledge of the literature and the results of other research or development groups. Furthermore, there is no need to conduct developments or research in areas where acceptable solutions already exist. Most importantly, research initiatives should fill the gaps of known methods rather than using the same research approaches of predecessors. Membrane-based methods are forward-looking in this respect, where integrative approaches and selective membranes are needed. It is not purposeful to show that a particular mine water can be treated using membrane-based technologies. Rather, the question is what general optimisation is needed to extend the lifespan of membranes for potential metal recovery from mine waters. Further research and development are needed in the fields of sensors, ion exchangers and electrochemical mine water treatment. Furthermore, there is a need for an intensification of research in


C. Wolkersdorfer, *Mine Water Treatment – Active and Passive Methods*,
https://doi.org/10.1007/978-3-662-65770-6_8

nano- and biotechnology, but without repeating the microbiological research of the 1960s to 1990s. The aim should be the valorisation of mine water and no longer a consideration from the point of view of treating the water. Finally, passive treatment methods should be given a higher priority, but this can only succeed if the differences between the individual methods are sufficiently internalised and if existing experience is systematically implemented. To this end, it is necessary to recognise the fundamental differences between constructed wetlands for municipal wastewater treatment and constructed wetlands for mine water treatment.

As I said at the beginning: This is not a cookbook with instructions on how you can build a mine water treatment plant. Rather, it was my intention to take you on a treatment journey to learn about the current methods to treat mine water. Just like a journey, you can get off at any point and look around using the literature I have listed for you. Write to me if you liked any of the stops or if you think I have failed to include particular aspects that you consider meaningful. You can reach me at the e-mail address minewater@Wolkersdorfer.info.

With a hearty "Glückauf" I thank you for accompanying me this far on our mine water treatment journey!

# Correction to: Introduction

## Correction to:

**Chapter 1 in: C. Wolkersdorfer, *Mine Water Treatment – Active and Passive Methods*, https://doi.org/10.1007/978-3-662-65770-6_1**

Table 1.6 on page 40 was mistakenly retained in German. This has now been updated in English version.

The updated original version for this chapter can be found at
https://doi.org/10.1007/978-3-662-65770-6_1


C. Wolkersdorfer, *Mine Water Treatment – Active and Passive Methods*,
https://doi.org/10.1007/978-3-662-65770-6_9

# Lagniappe – Bonus Chapter

## Explanation of Terms/Glossary/Abbreviations

### Introductory Remarks

In this chapter I explain briefly, rarely with the help of the online encyclopaedia Wikipedia, some important terms in this book. Commonly known facts, or terms that correspond to the level of a master's student, are largely excluded at this point. These include, for example, pH, acid, base, element, compound, redox potential, electrical conductivity, or other terms relevant in mine water treatment.

The list of abbreviations includes all abbreviations used in the text that usually cannot be found in a dictionary or that do not result from the context. Units which correspond to the International System of Units as well as to the coherent derived SI units (e.g. K, mol, S) or chemical symbols (e.g. K, V, Ca) are not included in the list of abbreviations. For the SI prefixes, see Table 1.2.

### Glossary

| | |
|---|---|
| Ad infinitum | Forever and ever |
| Alternative fact | Untruth, lie |
| Anthropoidea | Apes: Suborder of primates |
| Cape Breton Island | Île du Cap-Breton (fr.), Eilean Cheap Breatuinn (gaelic), U'namakika (mi'kmaq) |
| Chatham House Rule | When a meeting, or part thereof, is held under the Chatham House Rule, participants are free to use the information received, but neither the identity nor the affiliation of the speaker(s), nor that of any other participant, may be revealed. |
| Circa instans | A mid-twelfth-century medicinal pharmacology |
| Circumneutral | Water with pH values between ≈ 6 and 8 |



C. Wolkersdorfer, *Mine Water Treatment – Active and Passive Methods*,
https://doi.org/10.1007/978-3-662-65770-6

| | |
|---|---|
| Cloaca Maxima | Sewer system in ancient Rome |
| Coltan | Tantalum ore whose name is derived from the minerals of the columbite and tantalite solid solution series. Classified as a conflict mineral along with cassiterite, gold, diamond and wolframite. |
| Congener | One of several substances that are similar in origin, structure or function (https://doi.org/10.1351/goldbook.CT06819) |
| Deposit | An (ore) deposit is an accumulation of ore ("ore occurrence") that can be economically mined at a given time |
| Diadochy | Replacement of similar size elements or ions by another in the crystal lattice |
| Differential equation | System of signs and symbols, which seems like a book with seven seals to the normal mine water researcher and therefore is potentially excluded in this book as well |
| Eliot's sense | Thomas Stearns Eliot, awarded the Nobel Prize in Literature in 1948 and representative of modernist poetry, for whom the approximation to everyday language in poetry was important |
| Eutectic | Phase equilibrium within narrow limits of ambient conditions (e.g. pressure, temperature, concentration) |
| Fischer, Artur | German inventor of the Fischer expansion plug S, and holder of a further 1135 patents |
| Garimpeiro | Name for informal gold seekers in the Amazon region (word comes from Portuguese) |
| Heavy Metal | Erick Francisco Casas Ruiz is a Mexican professional wrestler currently working under the ring name Heavy Metal for AAA. He is the son of referee Pepe Casas and part of the Casas wrestling family; the brother of Negro Casas and Felino. |
| Hubbert curve | The Hubbert curve approximates the production rate of any finite resource over time |
| Hydrological year | A 12 month period, different from the calendar year, in which precipitation or hydrogeological parameters are measured. It runs, depending on the country, from October to September or November to October. Other terms are water year, discharge year or flow year |
| Juvenile acid | Newly formed acid in a mine by disulfide weathering, mainly pyrite and marcasite |
| Lagniappe | A term from the southern U.S. states of Louisiana and Mississippi, derived from *la ñapa* (Spanish), which |

| | |
|---|---|
| | there means "a small gift given by a merchant to a customer at the time of purchase" |
| hard landing on Mars | On the 23th September 1999, after the Trajectory Correction Manoeuvre 4, the Mars Climate Orbiter "disappeared" while on its way to Mars as a result of a unit mismatch of the total impulse where Newton seconds and pound seconds were confused |
| Meme | A term coined by Richard Dawkins, meaning the interconnectedness of consciousness content |
| Mine air | Air and ventilation in an underground mine (dull, toxic or nasty air, often summarised as "bad air") |
| Mineralisation | Total concentration of substances in water, usually expressed in mg $L^{-1}$ (TDS: total dissolved solids) |
| Nucleus (pl. nuclei) | Crystallisation nucleus |
| Oeuvre | The total creative output of an artist or author |
| Oulipo | L'*Ou*vroir de *Li*ttérature *Po*tentielle ("workshop of potential literature"), a group of authors, totally without formal constraint for talking and writing (look at: *A Void – La Disparition*) |
| Photo composite | A photo made up of several photos, often stitched together |
| Pinnacles | "mounds" within the excavation area, which thus far all excavators of Troia/Troja have left standing in order to give future generations of scientists the opportunity to verify old findings by means of modern methods |
| Potential literature | The art of creating works in compliance with certain self-selected rules |
| Reis, Philipp | German inventor of the first working telephone connection in 1861, which he called "das Telephon" |
| Rough (German: *ruppig*) affairs | Emil Rupp systematically falsified research results between 1926 and 1935 that fascinated even Albert Einstein. Based on this case, Charles Percy Snow wrote his book *The Affair.* |
| Semimetal | An element that (in simplified terms) stands between the metals and non-metals in terms of physico-chemical behaviour. |
| Silver bullet | A solution to a problem which contains all the necessary, even contradictory, characteristics |
| Sludge, slurry | Sludge from the mine water treatment plant |
| Statistically distributed | In the broadest sense: perfectly random distribution (there seems to be no specific definition for this, |

| | |
|---|---|
| | although the term is often used, especially in crystallography) |
| Stollentroll, the | The word cannot be explained, you'll have to read about it in: Walter Moers (1999) |
| Strachan, Robin | Author of a novel with remote mine water reference |
| Synthetic mine water | Synthetic mine water is produced artificially in the laboratory to enable experiments under controlled conditions or to prevent the flocculation of iron oxyhydrates. Results obtained in this way can only be transferred to real mine water to a limited extent or not at all. |
| Transition elements | Chemical elements of atomic numbers 21–30, 39 to 48, 57–80 and 89–112 |
| Turing Galaxy | A world in which networked computers deeply intervene in daily life. Derives from the name of the logician Alan Mathison Turing. |
| Ultrapure | Ultrapure chemicals are used for analysis in the ultra-trace range |
| Unobtainium | Neologism from the English word "unobtainable" and the end syllable "-ium" typical for elements |
| Valorisation | Any industrial reuse, recycling or composting of waste, useful products or energy sources (Kabongo 2013) |
| Water Make | Total volume of mine water inflow into a mine |

## Abbreviations

### Symbols and Units

| | |
|---|---|
| $A$ | Wetland area, $m^2$ |
| $c$ | Concentration, mg $L^{-1}$ |
| $c_{azi\text{-}eq}$ | Acidity equivalent concentration, mg $L^{-1}$ $CaCO_3$ |
| $c_i$ | Average daily concentration of the water constituent in the influent, mg $L^{-1}$ |
| $c_t$ | Average daily desired concentration of the water constituent in the effluent, mg $L^{-1}$ |
| $E_0$ | Redox potential of the standard hydrogen electrode, mV |
| $E_t$ | Measured voltage of the ORP electrode at temperature $T$, mV |
| $F$ | Conductivity coefficient, dag $m^{-2}$ $S^{-1}$ |
| $h$ | Water height above weir, cm |
| $I$ | Hydraulic gradient, – |
| $K$ | Conductivities of the mine cavities and their hydraulic connection |
| $k_B$ | Base capacity, mol $L^{-1}$ |
| $k_f$ | Hydraulic permeability, m $s^{-1}$ |
| $k_A$ | Acid capacity, mol $L^{-1}$ |
| $M$ | Load, g $s^{-1}$ |

| | |
|---|---|
| $Me_i$ | Concentration of species $i$ (e.g. Fe, Mn, Zn, Al, Co, Cu, Cd), g $L^{-1}$ |
| $M_i$ | Molar mass of species $i$, g $mol^{-1}$ |
| $n$ | Number, – |
| $n_e$ | Mean effective porosity of the bed in the limestone drain, – |
| pH | pH value, – |
| $Q$ | Flow rate (designation also $V$), L $min^{-1}$ |
| $Q_d$ | Average flow, $m^3$ $d^{-1}$ |
| $Q_e$ | Expected flow rate in the limestone drain, $m^3$ $s^{-1}$ |
| $R$ | Total mineralisation (total dissolved solids, TDS), mg $L^{-1}$ |
| $R_A$ | Area-adjusted treatment rates for the water constituent, g $m^{-2}$ $d^{-1}$ |
| $R_{GW}$ | Groundwater recharge |
| $r_w$ | Weathering rate of acidic (secondary) minerals, mol $s^{-1}$ |
| $T$ | Measured temperature, °C |
| $t$ | Time, s |
| $t_f$ | Duration of first flush, min |
| $t_r$ | Time elapsed to flood the mine workings (“rebound time”), min |
| $V$ | Volume, $m^3$ |
| $v_c$ | Critical particle velocity, m $s^{-1}$ |
| $v_f$ | Forward velocity, m $s^{-1}$ |
| $v_s$ | Settling velocity, m $s^{-1}$ |
| $\overline{x}$ | Mean value |
| $z_i$ | Charge of species $i$, – |
| $\delta_s$ | Density, g $cm^{-1}$ |
| $\gamma_{H_2O}$ | Activity of the mine water, – |
| $\Phi$ | Factor in the limestone drain, s |
| $\kappa$ | Electrical conductivity (sometimes also $\gamma$ or $K$), µS $cm^{-1}$ |
| $\tau$ | Residence time, s |

## Acronyms and Abbreviations

| | |
|---|---|
| ABA | Acid base accounting |
| ADCP | Acoustic Doppler current profiler |
| Adit Troll | see Walter Moers |
| ALD | Anoxic limestone drain |
| AMD | Acid mine drainage |
| AOP | Advanced oxidation process |
| ASTM | American Society for Testing and Materials (international standardisation organisation) |
| BAT | see BATEA |
| BATA | see BATEA |
| BATEA | Best available technology economically achievable; also abbreviated as BAT or BATA. |
| BBC | British Broadcasting Corporation |

| | |
|---|---|
| BBergG | Federal German Mining Act (Bundesberggesetz) |
| BTEX | Benzene, toluene, ethylbenzene and xylene(s) |
| CDT™ | Capacitive Deionisation Technology™ (capacitive deionisation) |
| COPEC | Contaminants of potential ecological concern |
| CPFM® | Colloid Polishing Filter Method® |
| DGGT | Deutsche Gesellschaft für Geotechnik e.V. (German Geotechnical Society) |
| DIN | Deutsches Institut für Normung e. V. (German Institute for Standardisation) |
| DVGW | Deutscher Verein des Gas- und Wasserfaches (German Technical and Scientific Association for Gas and Water) |
| DVWK | German Association for Water Management and Cultural Construction |
| ERMITE | Environmental Regulation of Mine Waters in the European Union (EU Project) |
| FTIR | Fourier transform infrared |
| GARD | Global Acid Rock Drainage Guide |
| GFP | Green fluorescent protein |
| GIS | Geographic information system |
| gpm | Gallons per minute (gallons per minute, can be US, UK imperial or metric gallons; 1 imp gal = 4.54609 L, 1 US liq. gal = 3.785411784 L, 1 met. gal = 4.00000 L) |
| HDS | High density sludge |
| IBC | Intermediate bulk container (1 $m^3$ plastic tank in a metal cage) |
| ICARD | International Conference of Acid Rock Drainage |
| imp gal | Imperial gallon (4.54609 L) |
| IMPI | Integrated Managed Passive Treatment Process (Pulles and Heath 2009) |
| IMWA | International Mine Water Association |
| INAP | International Network for Acid Prevention |
| IoMW | Internet of Mine Water |
| IPPC | Integrated Pollution Prevention and Control (European Union Agency) |
| ISI | Institute for Scientific Information |
| ISL | in situ leaching |
| ISO | International Organization for Standardization |
| ISR | in situ recovery |
| ISS | International Space Station |
| ISSN | International Standard Serial Number (journal and publication series number) |
| IUPAC | International Union of Pure and Applied Chemistry |
| LAWA | Bund/Länder-Arbeitsgemeinschaft Wasser (Interstate Working Group on Water, Germany) |
| LDO® | Luminescent Dissolved Oxygen Technology (oxygen sensor) |
| LDS | low density sludge |

| | |
|---|---|
| LfULG | Sächsisches Landesamt für Umwelt, Landwirtschaft und Geologie (Saxon State Office for Environment, Agriculture and Geology) |
| LLNL | Lawrence Livermore National Laboratories |
| LMBV | Lausitzer und Mitteldeutsche Bergbau-Verwaltungsgesellschaft mbH (Lausitz and Middle Germany Mining Administration Company) |
| MD | mine drainage |
| MEND | Mine Environment, Neutral Drainage Program |
| MIBRAG | Mitteldeutsche Braunkohlengesellschaft mbH (Central German Lignite Company) |
| MIFIM | Mine Water Filling Model |
| MiMi | Mitigation of the Environmental Impact from Mining Waste (Swedish research programme) |
| MIW | Mining-influenced water |
| ML | Mine leachate |
| MOC | Metals of concern |
| OLC | Open limestone channel |
| OLD | Open limestone drain |
| ORP | Oxidation reduction potential |
| PCB | Polychlorinated biphenyls |
| Pers. comm. | Personal communication |
| PFS | Polyferric sulfate |
| PIR | Passive in situ remediation |
| PIRAMID | Passive Remediation of Acidic Mine and Industrial Waters (EU Research Project) |
| PMFR | Perfectly mixed flow reactor |
| PTE | Potentially toxic element |
| PTM | Potentially toxic metal |
| PVC | Polyvinyl chloride |
| RAPS | Reducing and alkalinity producing system |
| RCRA | U.S. Resource Conservation and Recovery Act |
| RFID | Radio frequency identification |
| *s.l.* | *sensu lato* (in the broad sense) |
| *s.s.* | *sensu stricto* (in the strict sense) |
| SAPS | Successive alkalinity producing system |
| SCI | Science Citation Index |
| SCOOFI | Surface-catalysed oxidation of ferrous iron ($Fe^{2+}$) |
| SEM | Scanning electron microscope |
| SHE | Standard hydrogen electrode |
| SRO | Seeded reverse osmosis |
| SSSP | Saudi Strategic Storage Program (Saudi Arabia's military program that employed up to 12,000 people between 1988 and 2008 to build underground fuel storage facilities). |

| | |
|---|---|
| TDS | Total dissolved solids |
| TIC | Total inorganic carbon |
| TOC | Total organic carbon |
| Troglotroll | see Walter Moers |
| U.S. liq. gal | US gallon (3.785411784 L) |
| UAV | Unoccupied Aerial Vehicle (Joyce et al. 2021), Drone |
| UNEXMIN | Underwater Explorer for Flooded Mines (EU research project) |
| USGS | U.S. Geological Survey |
| USSR | Union of Soviet Socialist Republics |
| UTM | Universal Transverse Mercator System |
| VDI | Verein Deutscher Ingenieure (Association of German Engineers) |
| VFR | Vertical flow reactor |
| VKTA | Verein für Kernverfahrenstechnik und Analytik Rossendorf e.V. (Rossendorf Association for Nuclear Process Technology and Analysis) |
| VODAMIN | "VODA" (water) and "MIN" (mining) (EU research project) |
| Walter Moers | *see* Adit Troll or Troglotroll |
| WDR | West German Broadcasting, Cologne |
| WDS | World Data System |
| WGS84 | World Geodetic System 1984 |
| WQA | Water quality assessment |
| XRD | X-ray diffraction |
| ZVI | Zero-valent iron |

# References

7. Senat des Bundesverwaltungsgerichts (2014) Urteil des Bundesverwaltungsgerichts vom 18. Dezember 2014 (7 C 22.12) – Verpflichtung eines Bergwerksunternehmers zur Vorlage eines Sonderbetriebsplans zur Grubenwasserreinigung [Judgment of the Federal Administrative Court of 18 December 2014 (7 C 22.12) – Obligation of a mine operator to submit a special operating plan for mine water treatment] (ECLI:DE:BVerwG:2014:181214U7C22.12.0)

Aalto J-M, Kankkunen J, Vepsäläinen J (2017) Sulfate removal from water streams down to ppm level by using recyclable biopolymer. In: Proceedings IMWA 2017 – mine water & circular economy. Lappeenranta III, pp 1356–1361

Adams DJ, Peoples M (2010) New electrobiochemical reactor and water gas mixing technology for removal of metals, nitrate, and BOD. In: Wolkersdorfer C, Freund A (eds) Mine water and innovative thinking – international mine water association symposium. Cape Breton University Press, Sydney, pp 99–103

Adams DJ, Peoples M, Opara A (2012) A new electro-biochemical reactor for treatment of wastewater. In: Drelich J, Hwang J-Y, Adams J, Nagaraj DR, Sun X, Xu Z (eds) Water in mineral processing. SME, Littleton, pp 143–154

Adams N, Carroll D, Madalinski K, Rock S, Wilson T, Pivetz B, Anderson T, Chappell J, Huling S, Palmiotti J, Sayre P (2000) Introduction to phytoremediation, EPA-600/R-99/107. U. S. Environmental Protection Agency, Cincinnati, 72 p.

Adeniyi A, Maree J, Mbaya R, Popoola P (2013) HybridICE™ filter design in freeze desalination of mine water. In: Brown A, Figueroa L, Wolkersdorfer C (eds) Reliable mine water technology. International Mine Water Association, Golden, pp 507–512

Adeniyi A, Mbaya RKK, Onyango MS, Popoola API, Maree JP (2016) Efficient suspension freeze desalination of mine wastewaters to separate clean water and salts. Environ Chem Lett 14(4):449–454. https://doi.org/10.1007/s10311-016-0562-6

AG "PCB-Monitoring" (2018) Bericht der AG "PCB-Monitoring" nach Beendigung des Sondermessprogramms 2016–2017 – Ergebnisse der chemischen und ökotoxikologischen Analytik von Grubenwässern und von Grubenwasser beaufschlagten kleineren Bächen rechts der Saar, report no. PCB-2018-01. Saarländisches Landesamt für Umwelt- und Arbeitsschutz, Saarbrücken, 89 p.

Agnew AF, Corbett DM (1969) Hydrology and chemistry of coal-mine drainage in Indiana. Am Chem Soc Div Fuel Chem Prepr 13(2):137–149



C. Wolkersdorfer, *Mine Water Treatment – Active and Passive Methods*,
https://doi.org/10.1007/978-3-662-65770-6

Agricola G (1928) Zwölf Bücher vom Berg- und Hüttenwesen (new translation) (ed: Matschoss C). VDI-Verlag, Berlin, 564 p. (First publication 1557)
Agricola G (1974) Bergbau und Hüttenkunde, 12 Bücher. Ausgewählte Werke, VIII. VEB Deutscher Verlag der Wissenschaften, Berlin, pp 53–723 (First publication 1557)
Agricola G, Hoover HC, Hoover LH (1912) De re metallica – translated from the First Latin Edition of 1556 (trans: Hoover HC, Hoover LH). The Mining Magazine, London
Ahrens RH (2013) Bakterien holen Industriemetalle aus der Giftbrühe. VDI Nachrichten, 22. März 2013, p 19
Alcolea A, Vázquez M, Caparros A, Ibarra I, García C, Linares R, Rodríguez R (2012) Heavy metal removal of intermittent acid mine drainage with an open limestone channel. Miner Eng 26:86–98. https://doi.org/10.1016/j.mineng.2011.11.006
Ali A, Quist-Jensen CA, Jørgensen MK, Siekierka A, Christensen ML, Bryjak M, Helix-Nielsen C, Drioli E (2021) A review of membrane crystallization, forward osmosis and membrane capacitive deionization for liquid mining. Resour Conserv Recycl 168:33. https://doi.org/10.1016/j.resconrec.2020.105273
Aljoe WW, Hawkins JW (1993) Neutralization of acidic discharges from abandoned underground coal mines by alkaline injection. Bur Mines Rep Invest 9468:1–37
Aljoe WW, Hawkins JW (1994) Application of aquifer testing in surface and underground coal mines. In: Proceedings, 5th international mine water congress, Nottingham, vol. 1, pp 3–21
Al-Zoubi HS, Al-Thyabat SS (2012) Treatment of a Jordanian phosphate mine wastewater by hybrid dissolved air flotation and nanofiltration. Mine Water Environ 31(3):214–224. https://doi.org/10.1007/s10230-012-0197-1
Am J, Zhiwei Z, Kang Kug L, Chong HA, Paul LB (2011) State-of-the-art lab chip sensors for environmental water monitoring. Meas Sci Technol 22(3):032001. https://doi.org/10.1088/0957-0233/22/3/032001
American Public Health Association, American Water Works Association, Water Environment Federation (2012) Standard methods for the examination of water and wastewater, 22nd edn. American Public Health Association, Washington, 1496 p.
Amin N, Schneider D, Hoppert M (2018) Bioleaching potential of bacterial communities in historic mine waste areas. Environ Earth Sci 77(14):542. https://doi.org/10.1007/s12665-018-7714-x
Anders D (2002) Zeitbombe tickt im Kupferberg. Mitteldeutsche Zeitung – Quedlinburger Harz-Bote. 4 Nov 2002
Annandale JG (1998) Irrigating with gypsiferous mine water. S Afr J Sci 94(7):359–360
Annandale JG, Beletse YG, Stirzaker RJ, Bristow KL, Aken ME (2009) Is irrigation with coal-mine water sustainable? In: Water Institute of Southern Africa, International Mine Water Association (eds) Proceedings, International mine water conference. Document Transformation Technologies, Pretoria, pp 337–342
Annandale JG, Jovanovic NZ, Pretorius JJB, Lorentz SA, Rethman NFG, Tanner PD (2001) Gypsiferous mine water use in irrigation on rehabilitated open-cast mine land: crop production, soil water and salt balance. Ecol Eng 17(2–3):153–164. https://doi.org/10.1016/S0925-8574(00)00155-5
Annandale JG, Tanner PD, Burgess J (2017) Where there's muck there's brass: irrigated agriculture with mine impacted waters. In: Wolkersdorfer C, Sartz L, Sillanpää M, Häkkinen A (eds) IMWA 2017 – mine water & circular economy. II. Lappeenranta University of Technology, Lappeenranta, pp 915–922
Annandale JG, Tanner PD, du Plessis HM, Burgess J, Ronquest ZD, Heuer S (2019) Irrigation with mine affected waters: a demonstration with untreated colliery water In South Africa. In: Proceedings mine water – technological and ecological challenges (IMWA 2019), Perm, pp 71–76

Anonymous (1859) Entsäuerung der Grubenwasser zum Speisen der Dampfkessel. Polytech J 152:74

Anonymous (1934) The Henry process for the clarification of polluted water. Engineering 138(8, 9):213–215, 293–295

Anonymous (1935) Wallsend – at their rising sun pit the Wallsend & Hebburn Coal Co. Ltd. have erected what is probably the most complete coal preparation plant in the world. Colliery Eng 12(144), London

Anonymous (1936) Mine-sealing program to reduce acid pollution in streams. Eng News Rec 2:42–43

Anonymous (1995) Urteil des Bundesverwaltungsgerichts vom 9. November 1995 (4 C 25.94) – Unternehmerverantwortung bei der teilweisen Betriebseinstellung eines Bergwerks für die Abwehr gemeinschädlicher Einwirkungen – Beseitigung von Sauerwasser im Erzbergwerk Rammelsberg/Harz. Z f Bergrecht 136:290–301

Anonymous (2006) Heavy Metal' does not necessarily mean 'Toxic. Nickel 21(4):6

Anonymous (2008) Liquid loop cleans up coal mining – 2008 R&D 100 Winner. R&D Magazine – Online

Anthony JW, Bideaux RA, Bladh KW, Nichols MC (2003) Handbook of mineralogy – borates, carbonates, sulfates, V. Mineralogical Society of America, Chantilly, p 791

Arbeitsausschuss Markscheidewesen im Normenausschuss Bergbau im DIN (2003) Hydrologie – Ergebnisdarstellungen unter Tage. Empfehlung des Arbeitsausschusses Markscheidewesen. FABERG Normenausschuss Bergbau im DIN e. V., Herne, 3 p.

Arbeitskreis 4.6 "Altbergbau" der Fachsektion Ingenieurgeologie in der DGGT (2013) Empfehlung "Geotechnisch-markscheiderische Bewertung und Sanierung von altbergbaulich beeinflussten Gebieten hinsichtlich ihrer baulichen Nachnutzung". In: Meier G (eds) 13. Altbergbau-Kolloquium 7. bis 9.11.2013 in Freiberg. Wagner, Freiberg, pp 1–15 (Anhang)

Ariza LM (1998) River of vitriol. Sci Am 279(3):26–27

Arous O, Amara M, Kerdjoudj H (2010) Selective transport of metal ions using polymer inclusion membranes containing crown ethers and cryptands. Arab J Sci Eng A 35(2A):79–93

Ashby JC (2001) Injecting alkaline lime sludge and FGD material into underground mines for acid abatement. In: Proceedings, West Virginia surface mine drainage task force symposium, vol. 22, pp 1–8

ASTM D1125 (2014) Standard test methods for electrical conductivity and resistivity of water. In: ASTM International (eds) West Conshohocken. https://doi.org/10.1520/D1125-14

Atiba-Oyewoa O, Onyango MS, Wolkersdorfer C (2016) Application of banana peels nanosorbent for the removal of radioactive minerals from real mine water. J Environ Radioact 164:369–376. https://doi.org/10.1016/j.jenvrad.2016.08.014

Atkins PW (2006) Physical chemistry, 8th edn. Oxford University Press, Oxford, p 1097

Attri P, Arora B, Bhatia R, Venkatesu P, Choi EH (2014) Plasma technology: a new remediation for water treatment with or without nanoparticles. In: Mishra AK (ed) Application of nanotechnology in water research. Scrivener, Salem, pp 63–77. https://doi.org/10.1002/9781118939314.ch4

Aubé BC, Clyburn B, Zinck JM (2005) Sludge disposal in mine workings at Cape Breton Development Corporation. In: Proceedings securing the future 2005, Skellefteå

Aubé BC, Payant SC (1997) The Geco process: a new high density sludge treatment for acid mine drainage, I. In: Proceedings 4th international conference on acid rock drainage, Vancouver, pp 165–180

Aubé BC, Zinck JM (2003) Lime treatment of acid mine drainage in Canada. In: Proceedings, Brazil-Canada seminar on mine rehabilitation – technological innovations. Desktop Publishing, Florianópolis, pp 89–105

Aubé BC, Zinck JMM (1999) Comparison of AMD treatment processes and their impact on sludge characteristics. In: Proceedings Sudbury '99 – mining and the environment II. Sudbury, vol. 1, pp 261–270

Axler R, Yokom S, Tikkanen C, McDonald M, Runke H, Wilcox D, Cady B (1998) Restoration of a mine pit lake from aquacultural nutrient enrichment. Restor Ecol 6(1):1–19. https://doi.org/10.1046/j.1526-100x.1998.00612.x

Baacke D, Snagowski S, Jahn S (2015) Entwicklung der Grundwasserbeschaffenheit im Flutungsraum des Grubengebäudes Ronneburg. In: Paul M (ed) Sanierte Bergbaustandorte im Spannungsfeld zwischen Nachsorge und Nachnutzung – WISSYM 2015. Wismut GmbH, Chemnitz, pp 117–123

Baas Becking LGM, Kaplan IR, Moore D (1960) Limits of the natural environment in terms of pH and oxidation-reduction potentials. J Geol 68(3):243–284. https://doi.org/10.1086/626659

Bäckblom G, Forssberg E, Haugen S, Johansson J, Naarttijärvi T, Öhlander B (2010) Smart mine of the future conceptual study 2009–2010. Nordic Rock Tech Centre, Luleå, p 34

Badstübner L, Bauer E, Baumann R, Bergner W, Beyer H-E, Bommhardt KH, Böttcher J, Böttcher R, Brandt R, Brettschneider G, Damm H, Damm W, Decker J, Demmler K, Dicke HJ, Dietel R, Ducke G, Dusemond D, Felber B, Fischer K, Fleischer E, Fleischer P, Freitag W, Freyhoff G, Friedrich K, Fritsch E, Gahler G, Geilert D, Gottsmann J, Götze D, Gräbner R, Gressel G, Grimmer K, Groth P, Hähne R, Hambeck L, Hänsler E, Harlaß E, Hartmann L, Hiller A, Hommel M, Janke H, Kahlenbach H, Kalisch S, Keller K, Kempe E, Kießlich H, Klier H, Klose G, Knoch-Weber J, Kohlisch R, Kröber P, Krull E, Lange G, Lange H, Lange RC, Lange RG, Lötzsch F, Luderer J, Luft W, Lutze H, Malech S, Markstein H, Matthees G, Matthes W, Mehlhorn E, Meyer S, Motz H, Müller H, Müller H, Neldner D, Neubert H, Neubert W, Neugebauer H, Neumann H, Oczadly V, Otto G, Päßler D, Prokop R, Radon H, Reinhold P, Reinisch A, Renneberg L, Riedel L, Riedel P, Rochhausen D, Rosmej D, Rothe K, Sachse K, Schauer M, Schenke G, Schierz W, Schiffner W, Schilk M, Schirmer E, Schmiermund B, Schöfer M, Scholz A, Schönherr W, Schuppan W, Seidel G, Stübner J, Thiele R, Tonndorf H, Vater A, Viehweg M, Vogel H, Wagner G, Wagner P, Waskowiak R, Weber K, Weigelt H, Wicht E, Wild E, Wildner G, Winkler G, Wolf E (2010) Chronik der Wismut – Mit erweitertem Sanierungsteil (1998–2010). Chemnitz, Wismut GmbH, 3134 p. (unpubl.)

Baglikow V (2012) Grubenwasseranstieg in Steinkohlegebieten – Auswirkungen auf die Oberfläche. bergbau, vol. 1, pp 16–23

Ball JW, Nordstrom DK (1991) User's manual for WATEQ4F, with revised thermodynamic data base and test cases for calculating speciation of major, trace and redox elements in natural waters. Open-File Rep US Geol Surv Of 91–183:1–189

Banks D (2001) A variable-volume, head-dependent mine water filling model. Ground Water 39(3):362–365. https://doi.org/10.1111/j.1745-6584.2001.tb02319.x

Banks D (2004) Geochemical processes controlling minewater pollution. In: Prokop G, Younger PL, Roehl KE (eds) Conference papers 35. Umweltbundesamt, Wien, pp 17–44

Banks D, Frolik A, Gzyl G, Rogoż M (2010) Modeling and monitoring of mine water rebound in an abandoned coal mine complex: Siersza Mine, Upper Silesian Coal Basin. Poland Hydrogeol J 18(2):519–534. https://doi.org/10.1007/s10040-009-0534-z

Barbier E, Logette S, Coste M, Cook RG, Blumenschein CD (2008) Precipitation processes in the treatment of industrial inorganic wastewaters: impact of reactor and process configuration. In: Proceedings, 1st International conference hazardous waste management:CD-ROM B07.1_1–8

Barnes A (2008) The rates and mechanisms of Fe(II) oxidation in a passive vertical flow reactor for the treatment of ferruginous mine water. Unpubl PhD thesis. Cardiff University, Cardiff, p 341

Barnes HL, Romberger SB (1968) Chemical aspects of acid mine drainage. J Water Pollut Control Fed 40(3):371–384

Batty LC (2003) Wetland plants – more than just a pretty face? Land Contam Reclam 11(2):173–180. https://doi.org/10.2462/09670513.812

Batty LC, Younger PL (2002) Critical role of macrophytes in achieving low iron concentrations in mine water treatment wetlands. Environ Sci Technol 36(18):3997–4002. https://doi.org/10.1021/es020033+

Bayless ER, Olyphant GA (1993) Acid-generating salts and their relationship to the chemistry of groundwater and storm runoff at an abandoned mine site in southwestern Indiana, U.S.A. J Contam Hydrol 12(4):313–328. https://doi.org/10.1016/0169-7722(93)90003-B

Beadle JF, Berger J, Brady KBC, Dalberto AD, Hill SA, Hellier W, Hornberger RJ, Houtz NA, Kania T, Killian TD, Koury DJ, LaBuz L, Levitz KJ, Laslow KA, Loop C, Menghini MJ, Owen T, Pounds WF, Scheetz BE, Schueck JH, Stehouwer R, Strock GN, Tarantino JM, Walters SE, White WB, Williams DL (2007) Coal ash beneficial use in mine reclamation and mine drainage remediation in Pennsylvania. The Pennsylvania Department of Environmental Protection, Harrisburg, pp 369–379

Beckmann M (2006) Grenzen der Zumutbarkeit der Nachsorgeverantwortung eines Bergwerksunternehmens? Z f Umweltrecht Das Forum für Umwelt und Recht 17(6):295–300

Behum PT, Hause DR, Stacy MA, Branam TD (2008) Passive treatment of acid mine drainage – the enos reclamation project, Indiana: preliminary results. New opportunities to apply our science. In: Proceedings National Meeting of the American Society of Mining and Reclamation, Richmond, pp 129–147. https://doi.org/10.21000/JASMR08010129

Bekins B, Rittmann BE, MacDonald JA (2001) Natural attenuation strategy for groundwater cleanup focuses on demonstrating cause and effect. EOS Trans Am Geophys Union 82(5):53, 57–58. https://doi.org/10.1029/01EO00028

Beleites M (1992) Altlast Wismut – Ausnahmezustand, Umweltkatastrophe und das Sanierungsproblem im deutschen Uranbergbau, 1st edn. Brandes & Apsel, Frankfurt, 174 p.

Bench DW (2000) PCBs, mining, and water pollution. Eng Min J 201(8):36–38

Bench DW (2008) Identification, management, and proper disposal of PCB-Containing Electrical Equipment used in mines. U.S. Environmental Protection Agency, p. 11.

Bengtsson B, Dickson W, Nyberg P (1980) Liming acid lakes in Sweden. Ambio 9(1):34–36

Benkovics I, Csicsák J, Csövári M, Lendvai Z, Molnár J (1997) Mine water treatment – anion-exchange and membrane process. In: Proceedings, 6th International Mine Water Association Congress, Bled, vol. 1, pp 149–157

Benner SG, Blowes DW, Ptacek CJ, Mayer KU (2002) Rates of sulfate reduction and metal sulfide precipitation in a permeable reactive barrier. Appl Geochem 17(3):301–320. https://doi.org/10.1016/S0883-2927(01)00084-1

Benthaus FC, Totsche O, Luckner L (2020) In-lake neutralization of East German lignite pit lakes: technical history and new approaches from LMBV. Mine Water Environ 39(3):603–617. https://doi.org/10.1007/s10230-020-00707-5

Benthaus F-C, Totsche O, Radigk S (2015a) Der Grundwasserwiederanstieg in den bergbaubeeinflussten Gebieten des Braunkohlenbergbaus und Konzepte zur Minderung der Auswirkungen. In: Paul M (ed) Sanierte Bergbaustandorte im Spannungsfeld zwischen Nachsorge und Nachnutzung – WISSYM 2015. Wismut GmbH, Chemnitz, pp 151–160

Benthaus F-C, Uhlmann W (2006) Die chemische Behandlung saurer Tagebauseen in der Lausitz – Erfahrungen zur Kalkschlammresuspension im Tagebausee Koschen. Wiss Mitt Inst Geol 31:85–96

Benthaus F-C, Uhlmann W, Totsche O (2015b) Investigation and strategy on iron removal from water courses in mining induced areas. In: Brown A et al (eds) Agreeing on solutions for more sustainable mine water management. Gecamin, Santiago, pp. 1–11 (Electronic document)

Berner RA (1972) Sulfate reduction, pyrite formation, and the oceanic sulfur budget. In: Proceedings the changing chemistry of the oceans – nobel symposium 20, Aspenäsgarden, pp 347–361

Berner RA (1984) Sedimentary pyrite formation: an update. Geochim Cosmochim Acta 48(4):605–615. https://doi.org/10.1016/0016-7037(84)90089-9

Berner RA (1985) Sulfate reduction, organic matter decomposition and pyrite formation. Philos Trans R Soc Lond 315(1531):A25–A38. https://doi.org/10.1098/rsta.1985.0027

Bertrand S (1997) Performance of a nanofiltration plant on hard and highly sulfated water during two years of operation. Desalination 113(2–3):277–281. https://doi.org/10.1016/S0011-9164(97)00141-0

Best GA, Aikman DI (1983) The treatment of ferruginous ground water from an abandoned colliery. Wat Pollut Control 82(4):557–566

Bethke CM (2008) Geochemical and biogeochemical reaction modeling, 2nd edn. Cambridge University Press, New York, p 564

Beyer W (1986) Zur mikrobiellen In-situ-Laugung komplexer Sulfiderze des Rammelsbergs bei Goslar. Wissenschaft und Forschung, Bd 1. Weidler, Berlin, 118 p.

Bezuidenhout Y (2009) Implementing an efficient mine water monitoring and management program: key considerations. In: Water Institute of Southern Africa, International Mine Water Association (eds) Proceedings, international mine water conference. Document Transformation Technologies, Pretoria, pp 958–966

Bhattacharya J, Ji S, Lee H, Cheong Y, Yim G, Min J, Choi Y (2008) Treatment of acidic coal mine drainage: design and operational challenges of successive alkalinity producing systems. Mine Water Environ 27(1):12–19. https://doi.org/10.1007/s10230-007-0022-4

Biagini B, Mack B, Pascal P, Davis TA, Cappelle M (2012) Zero discharge desalination technology – achieving maximum water recovery. In: International mine water association symposium. Edith Cowan University, Bunbury, pp 577–584

Biermann V (2007) Langzeitverhalten von elementarem Eisen und Hydroxylapatit zur Uranrückhaltung in permeablen reaktiven Wänden bei der Grundwassersanierung. BAM-Dissertationsreihe 25:201. https://doi.org/10.14279/depositonce-1520

Bigham JM, Schwertmann U, Carlson L, Murad E (1990) A poorly crystallized oxyhydroxysulfate of iron formed by bacterial oxidation of Fe(II) in acid mine drainages. Geochim Cosmochim Acta 54(10):2743–2758. https://doi.org/10.1016/0016-7037(90)90009-A

Bijmans MFM, Dopson M, Peeters TWT, Lens PNL, Buisman CJN (2009) Sulfate reduction at pH 5 in a high-rate membrane bioreactor: reactor performance and microbial community analyses. J Microbiol Biotechnol 19(7):698–708. https://doi.org/10.4014/jmb.0809.502

Bilek F (2012) Reinigungsverfahren von Grundwasser und Oberflächengewässern – Bearbeitet im Rahmen des Ziel-3-Projektes VODAMIN – Arbeitspaket 4: Problemlagen des obertägigen Braunkohlenbergbaus Teilprojekt P 04. Dresdner Grundwasserforschungszentrum e. V., Dresden, 79 p.

Bilek F (2013) Grubenwassergenese und -behandlung – Beiträge zur Modell- und Technologie-Entwicklung. In: Proceedings des Dresdner Grundwasserforschungszentrums e, vol. 48, pp 1–234

Bilek F, Koch C, Bücker J, Luckner L (2013) Bergrechtlich bestimmter Umgang mit den in den Folgegebieten des Braunkohlenbergbaus anfallenden Eisenhydroxidschlämmen in Süd-Brandenburg. Grundwasser-Zentrum Dresden, Dresden, 53 p.

Bilek F, Wagner S, Pelzel C (2007) Technikumsversuch zur Eisen- und Sulfatabscheidung durch autotrophe Sulfatreduktion im in-situ Reaktor – bisherige Ergebnisse. Wiss Mitt Inst Geol 35:49–56

Blachère A, Metz M, Rengers R, Eckart M, Klinger C, Unland W (2005) Evaluation of mine water quality dynamics in complex large coal mine fields. Mine water 2005 – mine closure. University of Oviedo, Oviedo, pp 551–557

Blowes DW, Bain JG, Smyth DJ, Ptacek CJ (2003) Treatment of mine drainage using permeable reactive materials. In: Jambor JL, Blowes DW, Ritchie AIM (eds) Environmental aspects of mine wastes, Bd 31. Mineralogical Association of Canada, Waterloo, pp 361–376

Blowes DW, Ptacek CJ, Bain JG, Waybrant KR, Robertson WD (1995) Treatment of mine drainage water using *in situ* permeable reactive walls. In: Proceedings Sudbury '95 – mining and the environment. Sudbury, vol. 3, pp 979–987

Blowes DW, Ptacek CJ, Benner SG, McRae CWT, Bennett TA, Puls RW (2000) Treatment of inorganic contaminants using permeable reactive barriers. J Contam Hydrol 45(1–2):123–137. https://doi.org/10.1016/S0169-7722(00)00122-4

Blowes DW, Ptacek CJ, Jambor JL, Weisener CG, Paktunc D, Gould WD, Johnson DB (2014) The geochemistry of acid mine drainage. In: Turekian HD, Holland KK (eds) Treatise on geochemistry, 2nd edn. Elsevier, Oxford, pp 131–190. https://doi.org/10.1016/B978-0-08-095975-7.00905-0

Blume H-P, Brümmer G, Horn R, Kandeler E, Kögel-Knabner I, Kretzschmar R, Stahr K, Wilke B-M (2010) Lehrbuch der Bodenkunde ("Scheffer/Schachtschabel"), 16th edn. Spektrum, Heidelberg, 569 p. https://doi.org/10.1007/978-3-662-49960-3

Bogner JE, Doehler RW (1984) Ettringite formation following chemical neutralization treatment of acidic coal mine drainage. Int J Mine Water 3(1):13–24. https://doi.org/10.1007/BF02504589

Böhler M, Siegrist H (2004) Partial ozonation of activated sludge to reduce excess sludge, improve denitrification and control scumming and bulking. Water Sci Technol 49(10):41–49. https://doi.org/10.2166/wst.2004.0604

Bollmann K (2000) Ettringitbildung in nicht wärmebehandelten Betonen. Unpublished dissertation, Bauhaus-Universität Weimar, Weimar, 197 p. https://doi.org/10.25643/bauhaus-universitaet.61

Bolzicco J, Ayora C, Rötting T, Carrera J (2003) Performance of the Aznalcóllar permeable reactive barrier. In: Nel PJL (ed) Mine water environment. Proceedings 8th International Mine Water Association Congress, Johannesburg, pp 287–299

Bomberg M, Arnold M, Kinnunen P (2015) Characterization of the bacterial and sulfate reducing community in the alkaline and constantly cold water of the closed Kotalahti mine. Minerals 5(3):452–472. https://doi.org/10.3390/min5030452

Bosecker K (1980) Bakterielles Leaching – Metallgewinnung mit Hilfe von Bakterien. Metall 34(1):36–40

Botha M, Bester L, Hardwick E (2009) Removal of uranium from mine water using ion exchange at driefontein mine. In: Water Institute of Southern Africa, International Mine Water Association (eds) Proceedings, international mine water conference. Document Transformation Technologies, Pretoria, pp 382–391

Boult S (1996) Fluvial metal transport near sources of acid mine-drainage: relationships of soluble, suspended and deposited metal. Mineral Mag 60(399):325–335. https://doi.org/10.1180/minmag.1996.060.399.07

Bowden LI, Jarvis A, Orme P, Moustafa M, Younger PL (2005) Construction of a novel Permeable Reactive Barrier (PRB) at Shilbottle, Northumberland, UK: engineering design considerations and preliminary performance assessment. Mine water 2005 – mine closure. University of Oviedo, Oviedo, pp 375–381

Bowell RJ (2004) A review of sulfate removal options for mine waters. In: Jarvis AP, Dudgeon BA, Younger PL (eds) Mine water 2004 – proceedings International Mine Water Association Symposium. 2. University of Newcastle, Newcastle upon Tyne, pp 75–91

Brant DL, Ziemkiewicz PF (1997) Alkaline foundation drains and alkaline amendments for amd control in coal refuse piles. In: Proceedings, West Virginia surface mine drainage task force symposium 18 (Electronic document)

Brassington R (2017) Field hydrogeology, 4th edn. Wiley, Chichester, p 294. https://doi.org/10.1002/9781118397367

Brauer C (2001) Bergleute verändern das Gesicht einer Landschaft – Der Bergbau und die Topographie des Rammelsberges. Der Rammelsberg – Tausend Jahre Mensch-Natur-Technik, Bd 2. Goslarsche Zeitung, Goslar, pp 358–375

Braun L, Märten H, Raschke R, Richter A, Sommer K, Zimmermann U (2008) Flood water treatment at the former uranium mine site Königstein – a field report. In: Proceedings, 10th International Mine Water Association Congress, pp 189–192

Breedveld GD, Klimpel F, Kvennås M, Okkenhaug G (2017) Stimulation of natural attenuation of metals in acid mine drainage through water and sediment management at abandoned copper mines. In: Wolkersdorfer C, Sartz L, Sillanpää M, Häkkinen A (eds) IMWA 2017 – mine water & circular economy. II. Lappeenranta University of Technology, Lappeenranta, pp 1133–1137

British Coal Corporation (1997) Monitoring and utilization of drained mines gas. Technical coal research – mining operations, EUR 17393 EN. Office for Official Publications of the European Comunities, Luxembourg, p 89

Brix H (1987a) The applicability of the wastewater treatment plant in Othfresen as scientific documentation of the root-zone method. Water Sci Technol 19(10):19–24. https://doi.org/10.2166/wst.1987.0093

Brix H (1987b) Treatment of wastewater in the rhizosphere of wetland plants – the root-zone method. Water Sci Technol 19(1–2):107–118. https://doi.org/10.2166/wst.1987.0193

Brodie GA, Britt CR, Tomaszewski TM, Taylor HN (1991) Anoxic limestone drains to enhance performance of aerobic acid drainage treatment wetlands – experiences of the tennessee valley authority. Wetland design for mining operations. In: Proceedings international conference on constructed wetlands for water quality improvement, Pensacola, pp 17–34

Brookins DG (1988) Eh-pH diagrams for geochemistry. Springer, Berlin, p 176

Brooks RR, Chambers MF, Nicks LJ, Robinson BH (1998) Phytomining. Trends Plant Sci 3(9):359–362. https://doi.org/10.1016/S1360-1385(98)01283-7

Brown M, Barley B, Wood H (2002) Minewater treatment – technology, application and policy. IWA Publishing, London, 500 p. https://doi.org/10.2166/9781780402185

Brugam R, Gastineau J, Ratcliffe E (1990) Tests of organic matter additions for the neutralization of acid mine drainage influenced lakes. In: Proceedings, West Virginia surface mine drainage task force symposium 11 (Electronic document). https://doi.org/10.21000/JASMR90010425

Brunner H, Strasser H, Pümpel T, Schinner F (1993) Mikrobielle Laugung von Kupfer und Zink aus Tetraedrit-haltigem Baryt (Region Kleinkogel/Tirol) durch *Thiobacillus ferrooxidans*. Arch f Lagerstforsch Geol B-A 16:5–12

Brusseau ML (1996) Evaluation of simple methods for estimating contaminant removal by flushing. Ground Water 34(1):19–22. https://doi.org/10.1111/j.1745-6584.1996.tb01860.x

Bryan CG, Hallberg KB, Johnson DB (2004) Microbial populations in surface spoil at the abandoned Mynydd Parys copper mines. In: Jarvis AP, Dudgeon BA, Younger PL (eds) Mine water 2004 – proceedings international mine water association symposium. 1. University of Newcastle, Newcastle Upon Tyne, pp 107–112

Bucek MF, Emel JL (1977) Long-term environmental effectiveness of close down procedures – eastern underground coal mines, EPA-600/7-77-083. U.S. Environmental Protection Agency – Office of Research and Development, Cincinnati, 139 p.

Buck RP, Rondinini S, Covington AK, Baucke FGK, Brett CMA, Camões MF, Milton MJT, Mussini T, Naumann R, Pratt KW, Spitzer P, Wilson GS (2002) Measurement of pH – definition, standards, and procedures (IUPAC Recommendations 2002). Pure Appl Chem 74(11):2169–2200. https://doi.org/10.1351/pac200274112169

Bucksteeg K (1986) Sumpfpflanzenkläranlage – Verfahrensvarianten, Betriebserfahrungen, Problem Bodenhydraulik. GWF, Gas- Wasserfach: Wasser/Abwasser 127(9):429–434

Buff H (1851) Ueber das Wassertrommelgebläse. Ann Chem Pharm 79(3):249–255. https://doi.org/10.1002/jlac.18510790302

Buhrmann F, van der Waldt M, Hanekom D, Finlayson F (1999) Treatment of industrial wastewater for reuse. Desalination 124(1–3):263–269

Bunce NJ, Chartrand M, Keech P (2001) Electrochemical treatment of acidic aqueous ferrous sulfate and copper sulfate as models for acid mine drainage. Water Res 35(18):4410–4416. https://doi.org/10.1016/S0043-1354(01)00170-1

Bundesministerium der Justiz (2016) Verordnung zum Schutz der Oberflächengewässer (Oberflächengewässerverordnung – OGewV). Bundesgesetzblatt I:1373–1443

Burbey TJ, Younos T, Anderson ET (2000) Hydrologic analysis of discharge sustainability from an abandoned underground coal mine. J Am Water Resour Assoc 36(5):1161–1172. https://doi.org/10.1111/j.1752-1688.2000.tb05718.x

Bureau International des Poids et Mesures (2019) Le Système international d'unités – the international system of units, 9th edn. Organisation Intergouvernementale de la Convention du Mètre, Sèvres Cedex, p 216

Burghardt D, Coldewey WG, Melchers C, Meßer J, Paul M, Walter T, Wesche D, Westermann S, Wieber G, Wisotzky F, Wolkersdorfer C (2017) Glossar Bergmännische Wasserwirtschaft, 1st edn. Fachsektion Hydrogeologie in der DGGV, Neustadt/Wstr., p 79

Burke SP, Banwart SA (2002) Using Fe(II) adsorption and surface-catalysed oxidation for Fe(II) removal from predicted mine water discharges, vol 72. IAHS Publication, pp 271–275

Burnett M, Skousen JG (1996) Injection of limestone into underground mines for AMD control. In: Proceedings, West Virginia surface mine drainage task force symposium 17 (Electronic document)

Buzzi DC, Viegas LS, Cianga Silvas FP, Romano Espinosa DC, Siqueira Rodrigues MA, Bernardes AM, Tenório JA (2011a) The use of microfiltration and electrodialysis for treatment of acid mine drainage. In: Mine water – managing the challenges – 11th International Mine Water Association Congress. RWTH Aachen University, Aachen, pp 287–292

Buzzi DC, Viegas LS, Cianga Silvas FP, Siqueira Rodrigues MA, Romano Espinosa DC, Homrich Schneider IA, Bernardes AM, Soares Tenório JA (2011b) Treatment of acid mine drainage by electrodialysis. In: Supplemental proceedings: materials processing and energy materials, vol 1. Wiley, Hoboken, pp 277–284. https://doi.org/10.1002/9781118062111.ch29

Calderoni A, Tartari GA (2000) Evolution of the water chemistry of Lake Orta after liming. J Limnol 60(Supp 1):69–78. https://doi.org/10.4081/jlimnol.2001.69

Cannon HL (1960) Botanical prospecting for ore deposits. Science 132(3427):591–598. https://doi.org/10.1126/science.132.3427.591

Cánovas CR, Hubbard C, Olías M, Nieto JM, Black S, Coleman M (2005) Water quality variations in the Tinto River during high flood events. Mine water 2005 – mine closure. University of Oviedo, Oviedo, pp 133–139

Canty GA, Everett J (1998) Using tracers to understand the hydrology of an abandoned underground coal mine. In: Proceedings 15th annual national meeting – American Society for Surface Mining and Reclamation, Princeton, pp 62–72. https://doi.org/10.21000/JASMR98010062

Canty GA, Everett J (1999) Remediation of underground mine areas through treatment with fly ash, Report No. OCC Tasks 50 FY 1993 319(h) Task #600 EPA Grant C9-996100-0. Oklahoma Conservation Commission, Oklahoma City, p 167

Carayannis EG, Del Giudice M, Soto-Acosta P (2018) Disruptive technological change within knowledge-driven economies: the future of the Internet of Things (IoT). Technol Forecast Soc Change 136:265–267. https://doi.org/10.1016/j.techfore.2018.09.001

Cartwright PS (2012) Application of membrane separation technologies to wastewater reclamation and reuse. In: Drelich J, Hwang J-Y, Adams J, Nagaraj DR, Sun X, Xu Z (eds) Water in mineral processing. SME, Littleton, pp 115–129

Carvalho E, Diamantino C, Pinto R (2016) Environmental remediation of abandoned mines in Portugal – balance of 15 years of activity and new perspectives. In: Drebenstedt C, Paul M (eds) IMWA 2016 – mining meets water – conflicts and solutions. TU Bergakademie Freiberg, Freiberg, pp 554–561

Castendyk D, Hill B, Filiatreault P, Straight B, Alangari A, Cote P, Leishman W (2018) Experiences with autonomous sampling of Pit Lakes in North America using drone aircraft and drone boats. Paper presented at the IMWA 2018 – Risk to Opportunity, Pretoria, vol. II, pp 1036–1041.

Castendyk D, Voorhis J, Kucera B (2020) A validated method for pit lake water sampling using aerial drones and sampling devices. Mine Water Environ 39(3):440–454. https://doi.org/10.1007/s10230-020-00673-y

Cath TY, Childress AE, Elimelech M (2006) Forward osmosis: principles, applications, and recent developments. J Membr Sci 281(1–2):70–87. https://doi.org/10.1016/j.memsci.2006.05.048

Chałupnik S, Wysocka M (2000) Removal of radium from mine waters – the experience from the coal mine. In Proceedings, 7th International Mine Water Association Congress, pp 352–362

Chamberlain S, Moorhouse AM (2016) Baffle Curtain installation to enhance treatment efficiency for operational coal mine water treatment schemes. In: Drebenstedt C, Paul M (eds) IMWA 2016 – mining meets water – conflicts and solutions. TU Bergakademie Freiberg, Freiberg, pp 812–819

Chang N-B, Houmann C, Lin K-S, Wanielista M (2016) Fate and transport with material response characterization of green sorption media for copper removal via desorption process. Chemosphere 154:444–453. https://doi.org/10.1016/j.chemosphere.2016.03.130

Chanson H, Toombes L (2002) Air–water flows down stepped chutes: turbulence and flow structure observations. Int J Multiphase Flow 28(11):1737–1761. https://doi.org/10.1016/S0301-9322(02)00089-7

Chapman PM (2012) "Heavy metal" – cacophony, not symphony. Integr Environ Assess Manag 8(2):216. https://doi.org/10.1002/ieam.1289

Charmbury HB, Maneval DR, Girard L, III (1967) Operation yellowboy-design and economics of a lime neutralization mine drainage treatment plant. Society of Mining Engineers at AIME, preprint 67F35

Chartrand MMG, Bunce NJ (2003) Electrochemical remediation of acid mine drainage. J Appl Electrochem 33:259–264. https://doi.org/10.1023/A:1024139304342

Chen G (2004) Electrochemical technologies in wastewater treatment. Sep Purif Technol 38(1):11–41. https://doi.org/10.1016/j.seppur.2003.10.006

Cheremisinoff NP (2002) Handbook of water and wastewater treatment technologies. Butterworth-Heinemann, Woburn, p 636

Cherian MG, Goyer RA (1978) Role of metallothioneins in disease. Ann Clin Lab Sci 8(2):91–94

Chesters S, Morton P, Fazel M (2016) Membranes and minewater – waste or revenue stream. In: Drebenstedt C, Paul M (eds) IMWA 2016 – mining meets water – conflicts and solutions. TU Bergakademie Freiberg, Freiberg, pp 1310–1322

Chilingar GV (1956) Durov's classification of natural waters and chemical composition of atmospheric precipitation in USSR: a review. Trans Am Geophys Union 37(2):193–196. https://doi.org/10.1029/TR037i002p00193

Cidu R, Frau F, Da Pelo S (2011) Drainage at abandoned mine sites: natural attenuation of contaminants in different seasons. Mine Water Environ 30(2):113–126. https://doi.org/10.1007/s10230-011-0146-4

Clark BT, McCormick P (2021) Gold metal waters – The Animas River and the Gold King Mine spill. University Press of Colorado, Louisville, p 277.
Clark JP (2010) Treatment of mine site runoff containing suspended solids using sedimentation ponds – optimizing flocculant addition to ensure discharge compliance. In: Wolkersdorfer C, Freund A (eds) Mine water and innovative thinking – international mine water association symposium. Cape Breton University Press, Sydney, pp 217–221
Clark LC Jr, Wolf R, Granger D, Taylor Z (1953) Continuous recording of blood oxygen tensions by polarography. J Appl Physiol 6(3):189–193. https://doi.org/10.1152/jappl.1953.6.3.189
Clark S, Muhlbauer R (2010) Mine closure water management: choosing the right alternative within a changing environment. In: Wolkersdorfer C, Freund A (eds) Mine water and innovative thinking – international mine water association symposium. Cape Breton University Press, Sydney, pp 463–467
Clean Air Task Force, Earthjustice (2007) The impacts on water quality from placement of coal combustion waste in Pennsylvania Coal Mines. Fact Sheet, pp 1–4
Coal Research Bureau (1971) Dewatering of mine drainage sludge. Wat Poll Contr Res Ser 14010 FJX 12(71):1–90
Cogho VE (2012) Optimum coal mine – striving towards a 'zero effluent' mine. J S Afr Inst Min Metall 112(2):119–126
Cogho VE, van Niekerk AM (2009) Optimum coal mine water reclamation project. In: Water Institute of Southern Africa, International Mine Water Association (eds) Proceedings, international mine water conference. Document Transformation Technologies, Pretoria, pp 130–140
Cohen ER, Cvitas T, Frey JG, Holmström B, Kuchitsu K, Marquardt R, Mills I, Pavese F, Quack M, Stohner J, Strauss HL, Takami M, Thor AJ (2008) Quantities, units and symbols in physical chemistry, 3rd edn. International Union of Pure and Applied Chemistry, Cambridge, p 233. https://doi.org/10.1039/9781847557889
Colorado Department of Public Health and Environment – Water Quality Control Commission (2013) The basic standards and methodologies for surface water (Regulation No. 31) – (last amendment 2012-11-09, effective 2013-01-31), 5 CCR 1002-31. Department of Public Health and Environment, Denver, p 213
Committee on Intrinsic Remediation, Water Science and Technology Board, Board on Radioactive Waste Management, National Research Council (2000) Natural attenuation for groundwater remediation. National Academy Press, Washington, p 292
Commonwealth of Pennsylvania (1973) Scarlift report – Clearfield Creek Moshannon Watershed, SL 1-73:1–101.7. Commonwealth of Pennsylvania, Harrisburg, p 333
Comninellis C, Chen G (2010) Electrochemistry for the environment. Springer, Heidelberg. https://doi.org/10.1007/978-0-387-68318-8
Cook RB, Kelly CA, Schindler DW, Turner MA (1986) Mechanisms of hydrogen-ion neutralization in an experimentally acidified lake. Limnol Oceanogr 31(1):134–148. https://doi.org/10.4319/lo.1986.31.1.0134
Cornell RM, Schwertmann U (2003) The iron oxides – structure, properties, reactions, occurrences and uses. Wiley, Weinheim, p 664. https://doi.org/10.1002/3527602097
Costello C (2003) Acid mine drainage – innovative treatment technologies. National Network for Environmental Management Studies Program, Washington, p 47
Cotta B (1853) Das Zentralfeuer. Unterhaltungen am häuslichen Herd 1(46):727–730
Coulton R, Bullen C, Dolan J, Hallett C, Wright J, Marsden C (2003a) Wheal Jane mine water active treatment plant – design, construction and operation. Land Contam Reclam 11(2):245–252. https://doi.org/10.2462/09670513.821
Coulton R, Bullen C, Hallett C (2003b) The design and optimisation of active mine water treatment plants. Land Contam Reclam 11(2):273–280. https://doi.org/10.2462/09670513.825

Coulton R, Bullen C, Williams C, Williams K (2004a) The formation of high density sludge from minewater with low iron concentrations. In: Jarvis AP, Dudgeon BA, Younger PL (eds) Mine water 2004 – Proceedings International Mine Water Association symposium. 1. University of Newcastle, Newcastle upon Tyne, pp 24–30

Coulton R, Bullen C, Williams K, Dey M, Jarvis A (2004b) Active treatment of high salinity mine water. In: Jarvis AP, Dudgeon BA, Younger PL (eds) Mine water 2004 – Proceedings International Mine Water Association symposium 2. University of Newcastle, Newcastle upon Tyne, pp 119–124

Coupland K, Rowe O, Hallberg KB, Johnson DB (2004) Biogeochemistry of a subterranean acidic mine water body at an abandoned copper mine. In: Jarvis AP, Dudgeon BA, Younger PL (eds) Mine water 2004 – Proceedings International Mine Water Association symposium. 1. University of Newcastle, Newcastle upon Tyne, pp 113–119

Cravotta CA III (2007) Passive aerobic treatment of net-alkaline, iron-laden drainage from a flooded underground anthracite mine, Pennsylvania, USA. Mine Water Environ 26(3):128–149. https://doi.org/10.1007/s10230-007-0002-8

Cravotta CA III (2021) Interactive PHREEQ-N-AMDTreat water-quality modeling tools to evaluate performance and design of treatment systems for acid mine drainage. Appl Geochem 126:104845. https://doi.org/10.1016/j.apgeochem.2020.104845

Cravotta CA III, Hilgar GM (2000) Considerations for chemical monitoring at coal mines. In: Kleinmann RLP (ed) Prediction of water quality at surface coal mines. National Mine Land Reclamation Center, Morgantown, pp 195–218

Cravotta CA III, Parkhurst DL, Means BP, McKenzie RM, Arthur W (2010) Using the computer program AMDTreat with a PHREEQC titration module to compute caustic quantity, effluent quality, and sludge volume. In: Wolkersdorfer C, Freund A (eds) Mine water and innovative thinking – International Mine Water Association symposium. Cape Breton University Press, Sydney, pp 111–115

Cravotta CA III, Trahan MK (1999) Limestone drains to increase pH and remove dissolved metals from acidic mine drainage. Appl Geochem 14:581–606. https://doi.org/10.1016/S0883-2927(98)00066-3

Cravotta CA III, Ward S, Hammarstrom J (2008) Downflow limestone beds for treatment of net-acidic, oxic, iron-laden drainage from a flooded anthracite mine, Pennsylvania, USA: 2 laboratory evaluation. Mine Water Environ 27(2):86–99. https://doi.org/10.1007/s10230-008-0031-y

Cravotta CA, III (1998) Oxic limestone drains for treatment of dilute, acidic mine drainage. In: Proceedings, West Virginia surface mine drainage task force symposium 19 (Electronic document)

Cravotta CA III (2008) Dissolved metals and associated constituents in abandoned coal-mine discharges, Pennsylvania, USA. Part 2: Geochemical controls on constituent concentrations. Appl Geochem 23(2):203–226. https://doi.org/10.1016/j.apgeochem.2007.10.003

Cravotta CA III (2010) Abandoned mine drainage in the Swatara Creek Basin, Southern Anthracite Coalfield, USA: 2. Performance of treatment systems. Mine Water Environ 29(3):200–216. https://doi.org/10.1007/s10230-010-0113-5

Cravotta CA III, Means BP, Arthur W, McKenzie RM, Parkhurst DL (2015) AMDTreat 5.0+ with PHREEQC titration module to compute caustic chemical quantity, effluent quality, and sludge volume. Mine Water Environ 34(2):136–152. https://doi.org/10.1007/s10230-014-0292-6

Cravotta CA III, Watzlaf GR (2002) Design and performance of limestone drains to increase pH and remove metals from acidic mine drainage. In: Naftz D, Morrison SJ, Davis JA, Fuller CC (eds) Handbook of groundwater remediation using permeable reactive barriers – applications to radionuclides, trace metals, and nutrients. Academic Press, Amsterdam, pp 19–66. https://doi.org/10.1016/B978-012513563-4/50006-2

Crittenden JC, Trussell RR, Hand DW, Howe KJ, Tchobanoglous G (2012) MWH's water treatment: principles and design, 3rd edn. Wiley, Hoboken., 1920 S. https://doi.org/10.1002/9781118131473
Crooks J, Thorn P (2016) A sustainable approach to managing the treatment of mine waters associated with historic mining. In: Drebenstedt C, Paul M (eds) IMWA 2016 – mining meets water – conflicts and solutions. TU Bergakademie Freiberg, Freiberg, pp 1303–1309
Croxford SJ, England A, Jarvis AP (2004) Application of the PHREEQC geochemical computer model during the design and operation of UK mine water treatment schemes. In: Jarvis AP, Dudgeon BA, Younger PL (eds) Mine water 2004 – Proceedings International Mine Water Association symposium. 2. University of Newcastle, Newcastle upon Tyne, pp 125–134
Csővári M, Berta Z, Csicsák J, Hideg J, Varhegyi A (2004) Case study: remediation of a former uranium mining/processing site in Hungary. In: Prokop G, Younger PL, Roehl KE (eds) Conference papers, vol 35. Umweltbundesamt, Wien, pp 81–100
Cullimore DR (1999) Microbiology of well biofouling (PDF-Version). Sustainable well series, Bd 3. Lewis, Boca Raton, p 456
DIN Deutsches Institut für Normung e.V. (1984) DIN 38404-C6 Deutsche Einheitsverfahren zur Wasser-, Abwasser- und Schlammuntersuchung – Physikalische und physikalisch-chemische Kenngrößen – Bestimmung der Redox-Spannung, 8 p.
DIN Deutsches Institut für Normung e.V. (1993) DIN EN 27888 Wasserbeschaffenheit – Bestimmung der elektrischen Leitfähigkeit (ISO 7888:1985); Deutsche Fassung EN 27888:1993, 8 p.
DIN Deutsches Institut für Normung e.V. (2009) DIN 38404-C5 Deutsche Einheitsverfahren zur Wasser-, Abwasser- und Schlammuntersuchung – Physikalische und physikalisch-chemische Kenngrößen – Bestimmung des pH-Werts, 24 p.
DIN Deutsches Institut für Normung e.V. (2013) DIN 38402-62 Deutsche Einheitsverfahren zur Wasser-, Abwasser- und Schlammuntersuchung – Allgemeine Angaben (Gruppe A) – Plausibilitätskontrolle von Analysendaten durch Ionenbilanzierung, 10 p.
da Vinci L, Schneider M (2011) Das Wasserbuch – Schriften und Zeichnungen. Schirmer/Mosel, München, 103 p.
Damons RE, Petersen FW (2002) An Aspen model for the treatment of acid mine drainage. Eur J Miner Process Environ Prot 2(2):69–81
Davies M, Figueroa L, Wildeman T, Bucknam C (2016) The oxidative precipitation of thallium at alkaline pH for treatment of mining influenced water. Mine Water Environ 35(1):77–85. https://doi.org/10.1007/s10230-015-0349-1
Davies T, Long P, Dunn R (2012) Horden passive mine water treatment scheme – a collaborative partnership delivering a sustainable solution to the legacy of mine closures on the north east coastline. Lancaster, UK Water Projects, pp 203–205
Davis A, Whitehead C, Sims N, Collord J, Lengke M (2021) Random mixing of acid generating and neutralizing waste rock as a management strategy at the Gold Acres Mine, NV USA. Mine Water Environ 40(3):752–772. https://doi.org/10.1007/s10230-021-00790-2
Davison J (1991) Bio-carb and wetlands – passive, affordable acid mine drainage treatment. In: Proceedings, West Virginia surface mine drainage task force symposium 12 (Electronic document)
Davison J, Jones S (1990) Mine drainage bioremediation – the evolution of the technology from microbes to bio-carb. In: Proceedings, West Virginia surface mine drainage task force symposium 11 (Electronic document)
de la Vergne J (2003) Hard rock miner's handbook, 3rd edn. McIntosh Engineering, Tempe, p 314
de Vries AH (1989) The overall accuracy of the measurement of flow. Discharge measurement structures. International Institute for Land Reclamation and Improvement, Wageningen, pp 356–367
Dehoust G, Wiegmann K, Fritsche U, Stahl H, Jenseit W, Herold A, Cames M, Gebhardt P (2005) Beitrag der Abfallwirtschaft zum Klimaschutz – Umweltstudie, Forschungsbericht 205 33 314 UBA-FB III. RT Reprotechnik, Berlin, 65 p.

Demers I, Benzaazoua M, Mbonimpa M, Bouda M, Bois D, Gagnon M (2015) Valorisation of acid mine drainage treatment sludge as remediation component to control acid generation from mine wastes, part 1: Material characterization and laboratory kinetic testing. Miner Eng 76:109–116. https://doi.org/10.1016/j.mineng.2014.10.015

Dempsey BA, Jeon B-H (2001) Characteristics of sludge produced from passive treatment of mine drainage. Geochem-Explor Environ Anal 1:89–94. https://doi.org/10.1144/geochem.1.1.89

Deul M, Mihok EA (1967) Mine water research – neutralization. Bur Mines Rep Invest 6987:1–24

Deutscher Verband für Wasserwirtschaft und Kulturbau e. V. (1992) DVWK Schriften – Anwendung hydrogeochemischer Modelle, Bd 100. Parey, Hamburg, 344 p.

Dey M, Sadler PJK, Williams KP (2003) A novel approach to mine water treatment. Land Contam Reclam 11(2):253–258. https://doi.org/10.2462/09670513.822

Dill S, Cowan J, Wood A, Bowell RJ (1998) A review of sulfate removal options from mine waters. Mine water and environmental impacts, 2. In: Proceedings international mine water association symposium, Johannesburg, pp 329–342

Dill S, Du Preez L, Graff M, Maree J (1994) Biological sulfate removal from acid mine drainage utilizing producer gas as carbon- and energy source – process limitations and their resolution. In: Proceedings, 5th International mine water congress, Nottingham, vol. 2, pp 631–641

Dillenardt J, Kranz K (2010) Geologisch-Hydrogeologische Kartierung im Alten Tiefen Fürstenstolln zur geothermischen Grubenwassernutzung im Schloss Freudenstein. TU Bergakademie Freiberg, Freiberg, p 56

Dinardo O, Kondos PD, MacKinnon DJ, McCready RGL, Riveros PA, Skaff M (1991) Study on metals recovery/recycling from acid mine drainage phase Ia: literature survey, MEND 3.21.1a. The Mine Environment Neutral Drainage [MEND] Program, Ottawa, pp 48–11

Dold B (2015) Pre-mining characterization of ore deposits – what information do we need to increase sustainability of the mining process? In: Brown A et al (eds) Agreeing on solutions for more sustainable mine water management. Gecamin, Santiago, pp 1–7 (Electronic document)

Dold B (2017) Acid rock drainage prediction: a critical review. J Geochem Explor 172:120–132. https://doi.org/10.1016/j.gexplo.2016.09.014

Dörfler H-D (2002) Grenzflächen und kolloid-disperse Systeme: Physik und Chemie. Springer, Heidelberg, p 989

Döring M (1993) Unterirdische Wasserkraftwerke im Bergbau. Wasserwirtschaft 83(5):272–278

Double ML, Bissonnette GK (1980) Enumeration of coliforms from streams containing acid mine drainage. J Water Pollut Control Fed 52(7):1947–1952

Dow – Water and Process Solutions (2013) FILMTEC™ reverse osmosis membranes – technical manual, Form No. 609-00071-1009. Dow, Midland, p 181

Dowling J, Atkin S, Beale G, Alexander G (2004) Development of the sleeper pit lake. Mine Water Environ 23(1):2–11. https://doi.org/10.1007/s10230-004-0038-y

Drioli E, Macedonio F (2012) Membrane operations in water treatment and reuse. In: Drelich J, Hwang J-Y, Adams J, Nagaraj DR, Sun X, Xu Z (eds) Water in mineral processing. SME, Littleton, pp 105–113

Drury WJ (1999) Treatment of acid mine drainage with anaerobic solid-substrate reactors. Water Environ Res 71(6):1244–1250. https://doi.org/10.2175/106143096X122375

du Plessis HM (1983) Using lime treated acid-mine water for irrigation. Water Sci Technol 15(2):145–154. https://doi.org/10.2166/wst.1983.0030

Du Preez LA, Maree JP, Jackson-Moss CA (1991) Biological sulfate removal from mining effluents utilizing producer gas as energy source. In: Proceedings, 4th international mine water association Congress, vol. 2, pp 255–264

Duaime TE, Metesh JJ (2007) 2005 Consent Decree update, water-level monitoring and waterquality sampling, Butte underground mines and Berkeley Pit, Butte, Montana, 1982–2005. Montana Bureau of mines and geology open file report 549, p 120

Duarte JC, Estrada P, Beaumont H, Sitima M, Pereira P (1990) Biotreatment of tailings for metal recovery. Int J Mine Water 9(1–4):193–206. https://doi.org/10.1007/BF02503692

Dudeney AWL, Neville KJ, Tarasova I, Heath AOD, Smith SR (2004) Utilisation of ochreous sludge as a soil amendment. In: Jarvis AP, Dudgeon BA, Younger PL (eds) Mine water 2004 – proceedings International Mine Water Association symposium. 1. University of Newcastle, Newcastle upon Tyne, pp 31–40

Duffus JH (2002) "Heavy Metals" – a meaningless term? Pure Appl Chem 74(5):793–807. https://doi.org/10.1351/pac200274050793

Duren SM, McKnight DM (2013) Wetland photochemistry as a major control on the transport of metals in an acid mine drainage impacted watershed. In: Brown A, Figueroa L, Wolkersdorfer C (eds) Reliable mine water technology. International Mine Water Association, Golden, pp 973–975

Durov SA (1948) Klassifikacija prinodnych vodi graficeskoje izobrazenie ich sostava [Classification of natural waters and graphic representation of their composition]. Doklady Akad Nauk SSSR 59(1):87–90

Dvorácek J, Vidlár J, Šterba J, Heviánková S, Vanek M, Bartak P (2012) Economics of mine water treatment. J S Afr Inst Min Metall 112:157–159

Dzombak DA, Morel FMM (1990) Surface complexation modeling – hydrous ferric oxide. Wiley, New York, p 393. https://doi.org/10.1016/S0016-7037(97)81467-6

Earthworks (2012) U.S. copper porphyry mines – the track record of water quality impacts resulting from pipeline spills, tailings failures and water collection and treatment failures. Earthworks, Washington, p 33

Earthworks, Oxfam America (2004) Dirty metals – mining, communities and the environment. Earthworks and Oxfam America, Washington/Boston, p 33

Ebeling W (1986) Betriebserfahrungen mit dem Wurzelraumverfahren. Schriftenr WAR 26:31–49

Ebner HG, Schwarz W (1973) Untersuchungen über die Laugung von Uranerzen mit Hilfe von Bakterien. Erzmetall 26(10):484–490

Edwards RW, Stoner JH (1990) Acid waters in wales. Monographiae Biologicae, Bd 66. Kluwer, Dordrecht, p 337. https://doi.org/10.1007/978-94-009-1894-8

El-Naggar MY, Finkel SE (2013) Live wires – discoveries of microbial communities that transfer electrons between cells and across relatively long distances are launching a new field of microbiology. The Scientist 27(5):38–43

Eloff E, Greben HA, Maree JP, Radebe BV, Gomes RE (2003) Biological sulfate removal using hydrogen as the energy source. In: Nel PJL (ed) Mine water environ. Proceedings 8th International Mine Water Association Congress, Johannesburg, pp 99–108

Eppink FV, Trumm D, Weber P, Olds W, Pope J, Cavanagh JE (2020) Economic performance of active and passive AMD treatment systems under uncertainty: case studies from the Brunner Coal Measures in New Zealand. Mine Water Environ 39(4):785–796. https://doi.org/10.1007/s10230-020-00710-w

Ercker v. Schreckenfels L (1565) Vom Rämelsbergk vnd desselbigen Berckwergks ein kurtzer bericht. Durch einen wohl erfahrnen vnd Vorsuchten desselbigen Berckwergks etlichen seinen guten Freunden vnd Liebhabern der Berckwerge zu ehren vnd nutz gestellet [VOm Rãmels=||bergk vnd desselbigen || Berckwergks/ ein kurtz||er bericht.|| Durch einen wohl er=||fahrnen vnd Vorsuchten desselbigen || Berckwergks/ etlichen seinen guten || Freunden vnd Liebhabern der || Berckwerge zu ehren vnd || nutz gestellet.||] [A short report on the Rammelsberg and its mine. By a well-experienced and well-intentioned person of the same mine, to honour and benefit some of his good friends and lovers of the mines]. Georg Baumann d.Ä., Erfurt, no page numbers, 56 p.

Ericsson B, Hallmans B (1996) Treatment of saline wastewater for zero discharge at the Debiensko coal mines in Poland. Desalination 105(1–2):115–123. https://doi.org/10.1016/0011-9164(96)00065-3
ERMITE Consortium, Younger PL, Wolkersdorfer C (2004) Mining impacts on the fresh water environment: technical and managerial guidelines for catchment scale management. Mine Water Environ 23(Supplement 1):S2–S80. https://doi.org/10.1007/s10230-004-0028-0
Ernst W, Mathys W, Janiesch P (1975) Physiologische Grundlagen der Schwermetallresistenz – Enzymaktivitäten und organische Säuren. Forschungsberichte des Landes Nordrhein-Westfalen 2496:3–38
Ernst WHO (1996) Bioavailability of heavy metals and decontamination of soils by plants. Appl Geochem 11(1–2):163–167. https://doi.org/10.1016/0883-2927(95)00040-2
European Commission (2000) Directive 2000/60/EC of the European Parliament and of the Council of 23 October 2000 establishing a framework for community action in the field of water policy. Off J Eur Communities L 372:1–72
European Commission (2006) Directive 2006/21/EC of the European Parliament and of the Council of 15 March 2006 on the management of waste from extractive industries and amending directive 2004/35/EC. Off J Eur Communities L 102(49):15–34
European Commission (2009) Management of tailings and waste-rock in mining activities. European Commission, Luxembourg, p 511
European Innovation Partnership on Raw Materials (2016) Raw materials scoreboard. Publications Office of the European Union, Luxembourg, p 104. https://doi.org/10.2973/686373
Evangelou VP (1995) Pyrite oxidation and its control – solution chemistry, surface chemistry, Acid Mine Drainage (AMD), molecular oxidation mechanisms, microbial role, kinetics, control, ameliorites and limitations, microencapsulation. CRC Press, Boca Raton, p 293
Everett DJ, du Plessis J, Gussman HW (1993) The treatment of underground mine waters for the removal of calcium and sulfates by a GYP-CIX Process. In: Proceedings First African symposium on mine drainage and environment protection from mine wastewater disposal, Chililabombwe, pp 463–491
Expert Team of the Inter-Ministerial Committee (2010) Mine water management in the witwatersrand gold fields with special emphasis on acid mine drainage – report to the inter-ministerial committee on acid mine drainage. Council for Geoscience, Pretoria, p 128
Fabian D, Jarvis AP, Younger PL, Harries ND (2006) A Reducing and Alkalinity Producing System (RAPS) for passive treatment of acidic, aluminium rich mine waters. CL:AIRE Technology Demonstration Project Report, TDP5, p 38
Farmer JC, Fix DV, Mack GV, Poco JF, Nielsen JK, Pekala RW, Richardson JH (1995) Capacitive deionization of seawater. In: Proceedings Pacific Rim environmental conference (LLNL preprint) UCRL-JC-121958, pp 1–12. https://doi.org/10.2172/125000
Faulkner BB, Wyatt EG, Chermak JA, Miller FK (2005) The largest acid mine drainage treatment plant in the world. In: Proceedings, West Virginia surface mine drainage task force symposium, vol. 26, pp 1–10
Felix M, Möllmann G, Wagner S, Görne S (2010) Zur geothermischen Nutzbarkeit des Grubenwassers im Bergbaurevier Lugau/Oelsnitz. Geoprofil 13:169–175
Fellows PA (1937) Sealing projects – sharply reduce stream pollution from abandoned mines. Coal Age 42:158–161
Feng D, Aldrich C, Tan H (2000) Treatment of acid mine drainage by use of heavy metal precipitation and ion exchange. Miner Eng 13(6):623–642. https://doi.org/10.1016/S0892-6875(00)00045-5
Fernández-Rubio R, Lorca Fernández D (2010) Artificial recharge of groundwater in mining. In: Wolkersdorfer C, Freund A (eds) Mine water and innovative thinking – International Mine Water Association symposium. Cape Breton University Press, Sydney, pp 77–81

Fernández-Torres MJ, Randall DG, Melamu R, von Blottnitz H (2012) A comparative life cycle assessment of eutectic freeze crystallisation and evaporative crystallisation for the treatment of saline wastewater. Desalination 306:17–23. https://doi.org/10.1016/j.desal.2012.08.022

Fernando WAM, Ilankoon IMSK, Syed TH, Yellishetty M (2018) Challenges and opportunities in the removal of sulfate ions in contaminated mine water: a review. Miner Eng 117:74–90. https://doi.org/10.1016/j.mineng.2017.12.004

Ficklin WH, Mosier EL (1999) Field methods for sampling and analysis of environmental samples for unstable and selected stable constituents. In: The environmental geochemistry of mineral deposits, 6A. Society of Economic Geologists, Littleton, pp 249–264. https://doi.org/10.5382/Rev.06.12

Ficklin WH, Plumlee GS, Smith KS, McHugh JB (1992) Geochemical classification of mine drainages and natural drainages in mineralized areas. In: Proceedings, international symposium on water-rock interaction, vol. 7, pp 381–384

Fiedor JN, Bostick WD, Jarabeck RJ, Farrell J (1998) Understanding the mechanism of uranium removal from groundwater by zero-valent iron using X-ray photoelectron spectroscopy. Environ Sci Technol 32(10):1466–1743. https://doi.org/10.1021/es970385u

Fischer P (2016) Ende des Steinkohlenbergbaus im Ruhrrevier – Zeitplan und Herausforderungen. Veröff Dt Bergbau-Mus Bochum 217:173–179

Fiscor S (2008) Battelle develops VEP to remediate acid rock drainage. Eng Min J 209(7):96–100

Fisher TSR, Lawrence GA (2006) Treatment of acid rock drainage in a meromictic mine pit lake. J Environ Eng 132:515–526. https://doi.org/10.1061/(ASCE)0733-9372(2006)132:4(515)

Florence K (2014) Mechanisms of the removal of metals from acid and neutral mine water under varying redox systems. PhD, Cardiff University, p 203

Florence K, Sapsford DJ, Johnson DB, Kay CM, Wolkersdorfer C (2016) Iron mineral accretion from acid mine drainage and its application in passive treatment. Environ Technol 37(11):1428–1440. https://doi.org/10.1080/09593330.2015.1118558

Foreman JW (1971) Deep mine sealing. In: Proceedings, 1st acid mine drainage workshop, vol. 1, pp 19–45

Fosso-Kankeu E, Wolkersdorfer C, Burgess J (2020) Recovery of byproducts from acid mine drainage treatment. Scrivener, Salem, p 462

Frau F, Cidu R (2010) Diel changes in water chemistry in the Baccu Locci stream (Sardinia, Italy) affected by past mining. In: Wolkersdorfer C, Freund A (eds) Mine water and innovative thinking – International Mine Water Association symposium. Cape Breton University Press, Sydney, pp 339–343

Friedrich A, Guse M, Jolas P, Simon A, Hoth N, Rascher J (2011) Results of studies performed for reduced impact of material conversion processes in mixed soil dumps. In: Mine water – managing the challenges – 11th International Mine Water Association Congress. RWTH Aachen University, Aachen, pp 17–22

Friedrich H-J (2016) Membrane electrolysis – a promising technology for mine water treatment, radionuclide separation and extraction of valuable metals. In: Drebenstedt C, Paul M (eds) IMWA 2016 – mining meets water – conflicts and solutions. TU Bergakademie Freiberg, Freiberg, pp 1280–1286

Friedrich HJ, Zaruba A, Meyer S, Knappik R, Stolp W, Kiefer R, Benthaus F-C (2007) Verfahren und kleintechnische Anlage zur Aufbereitung schwefelsauerer Grubenwässer (RODOSAN®-Verfahren). In: Proceedings 11. Dresdner Grundwasserforschungstage, Dresden, pp 205–210

Frommichen R, Kellner S, Friese K (2003) Sediment conditioning with organic and/or inorganic carbon sources as a first step in alkalinity generation of acid mine pit lake water (pH 2–3). Environ Sci Technol 37(7):1414–1421. https://doi.org/10.1021/es026131c

Fu F, Wang Q (2011) Removal of heavy metal ions from wastewaters – a review. J Environ Manage 92(3):407–418. https://doi.org/10.1016/j.jenvman.2010.11.011

Galinsky G, Leistner J, Scheuermann G, Ebert S, Förderverein Drei-Brüder-Schacht e. V. (2001) Kavernenkraftwerk Drei-Brüder-Schacht – Geschichte und Überlegungen zur Rekonstruktion, 2nd edn. Saxonia Standortentwicklungs- und -verwaltungsgesellschaft, Freiberg, 72 p.

Gałuszka A, Migaszewski ZM (2018) Extremely high levels of trace elements in aerial parts of plants naturally growing in the Wiśniówka acid mine drainage area (South-Central Poland). In: Proceedings IMWA 2018 – risk to opportunity. Pretoria II, pp 598–603

Gammons CH, Icopini GA (2020) Improvements to the water quality of the acidic Berkeley pit lake due to copper recovery and sludge disposal. Mine Water Environ 39(3):427–439. https://doi.org/10.1007/s10230-019-00648-8

Gan WY, Selomulya C, Tapsell G, Amal R (2005) Densification of iron(III) sludge in neutralization. Int J Miner Process 76:149–162. https://doi.org/10.1016/j.minpro.2004.12.008

Gandy CJ, Jarvis AP (2012) The influence of engineering scale and environmental conditions on the performance of compost bioreactors for the remediation of zinc in mine water discharges. Mine Water Environ 31(2):82–91. https://doi.org/10.1007/s10230-012-0177-5

Garrels RM, Christ CL (1965) Solutions, minerals, and equilibria. Harper & Row, New York, p 450. https://doi.org/10.1017/S001675680005411X

Gautama RS, Novianti YS, Supringgo E (2014) Review on in-pit treatment of acidic pit lake in Jorong Coal Mine, South Kalimantan, Indonesia. In: Sui W, Sun Y, Wang C (eds) An interdisciplinary response to mine water challenges. International Mine Water Association, Xuzhou, pp 645–649

Gees A (1990) Flow measurement under difficult measuring conditions: field experience with the salt dilution method, vol. 193. IAHS-AISH Publication, pp 255–262

Geisenheimer P (1913) Die Wasserversorgung des Oberschlesischen Industriebezirks. Selbstverl. des Oberschlesischen Berg- und Hüttenmännischen Vereins, Kattowitz, 97 p.

Geldenhuys AJ, Maree JP, de Beer M, Hlabela P (2003) An integrated limestone/lime process for partial sulfate removal. J S Afr Inst Min Metall 103:345–354

Geller W, Schultze M, Kleinmann R, Wolkersdorfer C (2013) Acidic pit lakes – the legacy of coal and metal surface mines. In: Environmental science and engineering. Springer, Heidelberg, 525 p. https://doi.org/10.1007/978-3-642-29384-9

Georgaki I, Dudeney AWL, Monhemius AJ (2004) Characterisation of iron-rich sludge: correlations between reactivity, density and structure. Miner Eng 17(2):305–316. https://doi.org/10.1016/j.mineng.2003.09.018

George DR, Ross JR (1970) Recovery of uranium from mine water by countercurrent ion exchange [abstract]. J Met 22(12):A25

Germer C (2001) Vergleichende Untersuchungen zur Sauerwasserprognose anhand verschiedener deutscher Erzproben. Clausthal-Zellerfeld, Papierflieger, p 131

Geroni JN (2010) The potential for semi-passive mine water treatment by $CO_2$ stripping at Ynysarwed, S.Wales. In: Wolkersdorfer C, Freund A (eds) Mine water and innovative thinking – International Mine Water Association symposium. Cape Breton University Press, Sydney, pp 123–127

Geroni JN (2011) Rates and mechanisms of chemical processes affecting the treatment of ferruginous mine water. Unpubl. PhD thesis. Cardiff University, Cardiff, 262 p.

Geroni JN, Sapsford D, Florence KM (2011) Degassing $CO_2$ from mine water: implications for treatment of circumneutral drainage. In: Mine water – managing the challenges – 11th International Mine Water Association Congress. RWTH Aachen University, Aachen, pp 319–324

Geroni JN, Sapsford DJ, Barnes A, Watson IA, Williams KP (2009) Current performance of passive treatment systems in South Wales, UK. In: Water Institute of Southern Africa, International

Mine Water Association (eds) Proceedings, International Mine Water Conference. Document Transformation Technologies, Pretoria, pp 486–496
Gerth J, Hirschmann G, Paul M, Jacobs P, Förstner U, Heise S, Barth J, Gocht T (2006) Ingenieurgeochemie im Boden- und Gewässerschutz – Praxisbeispiele und rechtlicher Rahmen. In: Förstner U, Grathwohl P (eds) Ingenieurgeochemie: Technische Geochemie – Konzepte und Praxis, 2nd edn. Springer, Berlin, pp 243–436. https://doi.org/10.1007/978-3-540-39512-6_3
Getliffe K (2002) European waste law: has recent case law impacted upon the mess? Environ Law Rev 4(3):171–178
Gezahegne WA, Planer-Friedrich B, Merkel BJ (2007) Obtaining stable redox potential readings in gneiss groundwater and mine water: difficulties, meaningfulness, and potential improvement. Hydrogeol J 15(6):1221–1229. https://doi.org/10.1007/s10040-007-0174-0
Gibert O, Rötting T, De Pablo J, Cortina JL, Bolzicco J, Carrera J, Ayora C (2004) Metal retention mechanisms for the Aznalcóllar permeable reactive Barrier (SW Spain). In: Jarvis AP, Dudgeon BA, Younger PL (eds) Mine water 2004 – proceedings International Mine Water Association symposium. 2. University of Newcastle, Newcastle upon Tyne, pp 61–68
Glasser O (1995) Wilhelm Conrad Röntgen and die Geschichte der Röntgenstrahlen, 2nd edn. Springer, Berlin, 381 p. https://doi.org/10.1007/978-3-642-79312-7
Glover HG (1975) Acidic and ferruginous mine drainages. In: Chadwick MJ, Goodman GT (eds) The ecology of resource degradation and renewal. Blackwell, Oxford, pp 173–195
Gnielinski V, Mersmann A, Thurner F (1993) Verdampfung, Kristallisation, Trocknung – mit 30 Übungsbeispielen. Braunschweig, Viehweg, 260 p. https://doi.org/10.1007/978-3-642-58073-4
Goette H (1934) Das Henry-Verfahren zum Klären von Ab- und Schlammwässern. In: Technische Blätter – Wochenschr z Dtsch Bergwerksztg, vol. 42, pp 683–684
Golden DM, Gray RE, Meiers RJ, Turka RJ (1996) Use of coal combustion by-products in abandoned mine land reclamation. In: Proceedings, West Virginia surface mine drainage task force symposium 17 (Electronic document)
Golder Associates Inc. (2009) Literature review of treatment technologies to remove selenium from mining influenced water, report no. 08-1421-0034 Rev. 2. Golder Associates, Lakewood, p 31
Gougar MLD, Scheetz BE, Roy DM (1996) Ettringite and C-S-H Portland cement phases for waste ion immobilization – a review. Waste Manage 16(4):295–303. https://doi.org/10.1016/S0956-053X(96)00072-4
Govind R, Kumar U, Puligadda R, Antia J, Tabak HH (1997) Biorecovery of metals from acid mine drainage. In: Emerging technologies in hazardous waste management, vol 7. Plenum, New York, pp 91–101. https://doi.org/10.1007/978-1-4615-5387-8_8
Gozzard E, Bowden LI, Younger PL (2005) Permeable reactive barriers for mine water treatment in the UK – lessons from laboratory-scale applications. Mine water 2005 – mine closure. University of Oviedo, Oviedo, pp 619–625
Grab T, Storch T, Groß U (2018) Energetische Nutzung von Grubenwasser aus gefluteten Bergwerken. In: Bauer M, Freeden W, Jacobi H, Neu T (eds) Handbuch Oberflächennahe Geothermie. Springer, Heidelberg, pp 523–586. https://doi.org/10.1007/978-3-662-50307-2_17
Gray JE (1997) Environmental geochemistry and mercury speciation of abandoned mercury mines in Southwestern Alaska. Open-File Rep US Geol Surv OF 97-0496:31
Gray JE (1998) Environmental geochemistry, mercury speciation, and effects to surrounding ecosystems of a belt of mercury deposits and abandoned mercury mines in Southwestern Alaska, USA. Open-File Rep US Geol Surv OF 98-0209:7–9
Gray JE, Kelley KD, Goldfarb RJ, Taylor CD (1997) Environmental studies of mineral deposits in Alaska. Open-File Rep US Geol Surv OF 97-0496:31. https://doi.org/10.3133/b2156
Grebenyuk VD, Pisaruk VI, Mukha SI, Penkalo II (1979) Electrodialysis of softened mine water. J Appl Chem USSR 52(6):1262–1266

Grehl S, Lösch R, Jung B (2018) Perfect match – combining robotics and IoT is revolutionising mining automation. World Mining Frontiers 1:21–24

Grischek T, Feistel U, Ebermann J, Musche F, Bruntsch S, Uhlmann W (2016) Field experiments on subsurface iron removal in the Lusatian mining region. In: Drebenstedt C, Paul M (eds) IMWA 2016 – mining meets water – conflicts and solutions. TU Bergakademie Freiberg, Freiberg, pp 292–297

Groß A, Knolle F (2011) Zur Hydrogeologie der Grube Einheit nach Flutung der Karst- und Bergbauhohlräume, Landkreis Harz, Sachsen-Anhalt. Mitt Verb dt Höhlen- u Karstforscher 57(3):78–86

Grothe H (1962) Lexikon des Bergbaues. Lueger Lexikon der Technik, Bd 4, 4th edn. Deutsche Verlags-Anstalt, Stuttgart, 727 p.

Grüschow A (1991) Bericht über Laboruntersuchungen zur Senkung der Schwermetallgehalte and der Acidität in Sickerwässern der Grube Straßberg. Kali Südharz AG, Sondershausen, 5 p.

Gunther P, Mey WS (2008) Selection of mine water treatment technologies for the Emalahleni (Witbank) water reclamation project. In: WISA biennal conference, Sun City, pp 1–14

Gusek J, Plocus V (2015) Case study – 19 years of acid rock drainage mitigation after a bactericide application. In: Brown A et al (eds) Agreeing on solutions for more sustainable mine water management. Gecamin, Santiago, pp. 1–11 (Electronic document)

Gusek JJ (2002) Sulfate-reducing bioreactor design and operating issues – is this the passive treatment technology for your mine drainage? In: Proceedings annual conference – National Association of abandoned mine land programs, Park City, pp 1–14 [CD-ROM]

Gusek JJ (2009) A periodic table of passive treatment for mining influenced water. In: Proceedings revitalizing the environment: proven solutions and innovative approaches, Billings, pp 550–562. 10.21000/JASMR09010550

Gusek JJ (2013) A periodic table of passive treatment for mining influenced water – revisited. In: Brown A, Figueroa L, Wolkersdorfer C (eds) Reliable mine water technology. International Mine Water Association, Golden, pp 575–580

Gusek JJ, Figueroa LA (2009) Mitigation of metal mining influenced water. Management technologies for metal mining influenced water, Vol 2. SME, Littleton, p 304

Gusek JJ, Waples JS (2009) A periodic table of passive treatment for mining-influenced water. In: Proceedings Securing the Future 2009 & 8th ICARD, Skellefteå, pp 1–10 (Electronic document)

Haber F (1927) Das Gold im Meerwasser. Angew Chem 40(11):303–314. https://doi.org/10.1002/ange.19270401103

Hahn F, Ignacy R, Bussmann G, Jagert F, Bracke R, Seidel T (2018) Reutilization of mine water as a heat storage medium in abandoned mines. In: Proceedings IMWA 2018 – risk to opportunity, Pretoria, vol. II, pp 1057–1062

Hall J, Glendinning S, Younger PL (2005) Is mine water a source of hazardous gas? Mine water 2005 – mine closure. University of Oviedo, Oviedo, pp 141–145

Halliburton Company (1970) New mine sealing techniques for water pollution abatement. Wat Poll Contr Res Ser 14010 DMO: 1–163

Hamai T, Okumura M (2010) A result of batch test to select effective co-precipitator of zinc containing mine drainage treatment. In: Wolkersdorfer C, Freund A (eds) Mine water and innovative thinking – International Mine Water Association symposium. Cape Breton University Press, Sydney, pp 127–131

Hamilton QUI, Lamb HM, Hallett C, Proctor JA (1999) Passive treatment systems for the remediation of acid mine drainage Wheal Jane, Cornwall, UK. Water Environ J 13(2):93–103. https://doi.org/10.1111/j.1747-6593.1999.tb01014.x

Hamm V, Collon-Drouaillet P, Fabriol R (2008) Two modelling approaches to water-quality simulation in a flooded iron-ore mine (Saizerais, Lorraine, France): a semi-distributed chemical reactor model and a physically based distributed reactive transport pipe network model. J Contam Hydrol 96(1):97–112. https://doi.org/10.1016/j.jconhyd.2007.10.004

Hammack RW, Dvorak DH, Edenborn HM (1994a) Bench-scale test to selectively recover metals from metal mine drainage using biogenic $H_2S$. In: Proceedings international land reclamation and mine drainage conference, Pittsburgh, vol. 1, pp 214–222

Hammack RW, Edenborn HM, Dvorak DH (1994b) Treatment of water from an open-pit copper mine using biogenic sulfide and limestone – a feasibility study. Water Res 28(11):2321–2329. https://doi.org/10.1016/0043-1354(94)90047-7

Hammack RW, Hedin RS (1989) Microbial sulfate reduction for the treatment of acid mine drainage: a laboratory study. In: Proceedings, West Virginia surface mine drainage task force symposium 10 (Electronic document). https://doi.org/10.21000/JASMR89010673

Hampson CJ, Bailey JE (1982) On the structure of some precipitated calcium alumino-sulfate hydrates. J Mater Sci 17(11):3341–3346. https://doi.org/10.1007/BF01203504

Hansen C (1998) Modellierung der hydrogeochemischen Entwicklung saurer Grubenwässer – Flutung der Schwefelkieslagerstätte Einheit bei Elbingerode (Harz). Unpubl. Dipl-A TU Clausthal, Clausthal-Zellerfeld, p 137

Harford AJ, Jones DR, Dam Rav (2012) Ecotoxicology of actively treated mine waters. International Mine Water Association symposium. Edith Cowan University, Bunbury, pp 615–622

Harries RC (1985) A field trial of seeded reverse-osmosis for the desalination of a scaling-type mine water. Desalination 56:227–236. https://doi.org/10.1016/0011-9164(85)85027-X

Harrington J (2016) Combining in situ treatment and active water treatment: case study at Schwartzwalder uranium mine. In: BC/MEND workshop, Vancouver, PowerPoint Presentation

Harrington J, Harrington J, Lancaster E, Gault A, Woloshyn K (2015) Bioreactor and in situ mine pool treatment options for cold climate mine closure at Keno Hill, YT. In: Brown A et al (eds) Agreeing on solutions for more sustainable mine water management. Gecamin, Santiago, pp 1–10 (Electronic document)

Härtel G, Haseneder R, Pukade B, Steinberger P, Rieger A, Riebensahm M (2007) Aufbereitung von Acid Mine Drainage (AMD) mittels Membranverfahren. Wiss Mitt Inst Geol 35:11–18

Hartinger L (2007) Handbuch der Abwasser- and Recyclingtechnik für die metallverarbeitende Industrie, 2nd ed (reprint). Hanser, München, 714 p.

Hartley FR (1972) Conventional processes to produce yellow cake. In: Proceedings AAEC symposium on uranium processing, Lucas Heights II-1–II-44. https://inis.iaea.org/search/search.aspx?orig_q=RN:4050498

Hasche A, Wolkersdorfer C (2004) Mine water treatment with a pilot scale RAPS-system. Wiss Mitt Inst Geol 25:93–99

Hasche-Berger A, Wolkersdorfer C (2005) Pilot scale RAPS-system in Gernrode/Harz Mountains. In: Merkel BJ, Hasche-Berger A (eds) Uranium in the environment. Springer, Heidelberg, pp 317–328. https://doi.org/10.1007/3-540-28367-6_32

Hasche-Berger A, Wolkersdorfer C, Simon J (2006) Laborexperimente als Grundlage für ein RAPS-System (Reducing and alkalinity producing system). Wiss Mitt Inst Geol 31:37–45

Hatch (2014) Study to identify BATEA for the management and control of effluent quality from mines. MEND report, MEND 3.50.1. The Mine Environment Neutral Drainage [MEND] Program, Ottawa, p 527

Hause DR, Willison LR (1986) Deep mine abandonment sealing and underground treatment to prelude acid mine drainage. In: Proceedings, West Virginia surface mine drainage task force symposium vol. 7, Paper 14, p 14

Häyrynen K, Pongrácz E, Väisänen V, Pap N, Mänttäri M, Langwaldt J, Keiski RL (2009) Concentration of ammonium and nitrate from mine water by reverse osmosis and nanofiltration. Desalination 240(1–3):280–289. https://doi.org/10.1016/j.desal.2008.02.027

Hebley HF (1953) Stream pollution by coal mine wastes. Min Eng 5:404–412
Hedin RS (1998) Recovery of a marketable iron product from coal mine drainage. In: Proceedings, West Virginia surface mine drainage task force symposium 19 (Electronic document)
Hedin RS (2002) Recovery of marketable iron oxide from mine drainage. In: Proceedings, West Virginia Surface Mine Drainage Task Force symposium 23, pp 1–6
Hedin RS (2003) Recovery of marketable iron oxide from mine drainage in the USA. Land Contam Reclam 11(2):93–97. https://doi.org/10.2462/09670513.802
Hedin RS (2016) Long-term minimization of mine water treatment costs through passive treatment and production of a saleable iron oxide sludge. In: Drebenstedt C, Paul M (eds) IMWA 2016 – mining meets water – conflicts and solutions. TU Bergakademie Freiberg, Freiberg, pp 267–1273
Hedin RS, Hyman DM, Hammack RW (1988) Implications of sulfate-reduction and pyrite formation processes for water quality in a constructed Wetland: preliminary observations. Bur Mines Inf Circ IC-9183: 382–388. 10.21000/JASMR88010382
Hedin RS, Nairn RW, Kleinmann RLP (1994a) Passive treatment of coal mine drainage. Bur Mines Inf Circ IC-9389: 1–35
Hedin RS, Watzlaf GR, Nairn RW (1994b) Passive treatment of acid-mine drainage with limestone. J Environ Qual 23(6):1338–1345. https://doi.org/10.2134/jeq1994.00472425002300060030x
Hedrich S, Lünsdorf H, Kleeberg R, Heide G, Seifert J, Schlömann M (2011) Schwertmannite formation adjacent to bacterial cells in a mine water treatment plant and in pure cultures of *Ferrovum myxofaciens*. Environ Sci Technol 45(18):7685–7692. https://doi.org/10.1021/es201564g
Heinrich B, Guderitz I, Neumann V, Pokrandt K-H, Benthaus F-C, Ulrich K-U (2011) In-lake neutralisation and post-rehabilitation treatment of a lignite mining pit lake – lessons learned. In: Mine water – managing the challenges – 11th International Mine Water Association Congress. RWTH Aachen University, Aachen, pp 343–347
Heinze G, Märten H, Schreyer J, Seeliger D, Sommer K, Vogel D (2002) Flood water treatment by improved HDS technology. In: Merkel BJ, Planer-Friedrich B, Wolkersdorfer C (eds) Uranium in the aquatic environment. Springer, Heidelberg, pp 793–801. https://doi.org/10.1007/978-3-642-55668-5_92
Heinzel E, Hedrich S, Janneck E, Glombitza F, Seifert J, Schlömann M (2009) Bacterial diversity in a mine water treatment plant. Appl Environ Microbiol 75(3):858–861. https://doi.org/10.1128/AEM.01045-08
Heitfeld M, Denneborg M, Rosner P, Müller F, Lieser U (2012) Signifikante Belastungsquellen des Erzbergbaus und mögliche Maßnahmen im Rahmen der Bewirtschaftungsplanung NRW. ARGE Erzbergbau, Aachen, 164 p.
Hellier WW (1998) Abatement of acid mine drainage by capping a reclamed surface mine with fluidized combustion ash. Mine Water Environ 17(1):28–40. https://doi.org/10.1007/BF02687242
Helms W (1995) Sauerwasser im Erzbergbau – Entstehung, Vermeidung und Behandlung. Bergbau 46(2):65–71
Hem JD (1985) Study and interpretation of the chemical characteristics of natural water. US Geol Surv Water Suppl Pap 2254:1–263. https://doi.org/10.3133/wsp2254
Hemm M, Schlundt A, Kapfer M, Nixdorf B (2002) Beispiele für Neutralisierungsversuche am Steinberger See (Bayern) und Zieselsmaar (Nordrhein-Westfalen) – aus der UBA-Studie "Tagebauseen in Deutschland". Gewässerreport 7:37–42
Henrikson L, Brodin Y-W (1995) Liming of acidified surface waters – a Swedish synthesis. Springer, Berlin, 458 p. https://doi.org/10.1007/978-3-642-79309-7
Herbert H-J, Sander W (1989) Verfahren zur verfälschungsfreien Messung und Probennahme von hochkonzentrierten Salzlösungen im Untertagebereich. Kali und Steinsalz 10(4/5):137–141
Herschy RW (1995) Streamflow measurement, 3rd edn. Taylor & Francis, Oxon, 507 p. https://doi.org/10.1007/1-4020-4497-6_214

Hessisches Ministerium für Wissenschaft und Kunst (2011) Water features and Hercules within the Bergpark Wilhelmshöhe – Nomination for Inscription on the UNESCO World Heritage List – Nomination dossier. Hessisches Ministerium für Wissenschaft und Kunst, Kassel, 291 p.

Hettersdorf F (1908) Über Selbstreinigung der Flüsse. Dtsch Vierteljahrsschr öffentl Gesundh-Pfl 40(4/1):615–636

Heviánková S, Bestová I (2007) Removal of manganese from acid mine drainage. J Min Metal 43A:43–52

Hicks WS, Bowmann GM, Fitzpatrick RW (1999) East trinity acid sulfate soils – part 1: environmental hazards. Technical report – CSIRO Land and Water 14/99, p 77

Hill RD (1968) Mine drainage treatment – state of the art and research needs, report no. BCR 68-150. Federal Water Pollution Control Administration, Cincinnati, p 99

Hilton T (1993) Technical information for fighting acid mine drainage. In: Proceedings, West Virginia surface mine drainage task force symposium 14 (Electronic document)

Hilton T, Adair J, Caruccio FT, Geidel G, Greskovich J, Faulkner B, Hajek J, Hall R, Keaveny G, Lilly R, O'Dell A, Tuckwiller E (1989) The magic of water treatment. In: Proceedings, West Virginia surface mine drainage task force symposium 10 (Electronic document)

Himsley A, Bennett JA (1985) Removal of toxic elements in mine effluents by CCIX. In: Proceedings, 2nd International Mine Water Association Congress 2, pp 661–671

Hobba WA Jr (1987) Underground coal mines as sources of water for public supply in Northern Upshur County, West Virginia. Water-Res Invest Rep 84–4115:1–38. https://doi.org/10.3133/wri844115

Hobiger G, Klein P, Denk J, Grösel K, Heger H, Klein P, Kohaut S, Kollmann WFH, Lampl H, Lipiarsky P, Pirkl H, Schedl A, Schubert G, Shadlau S, Winter P (2004) GEOHINT – Geogene Hintergrundgehalte oberflächennaher Grundwasserkörper, Report No. Zl. 70.215/08-VII 1/03. Geologische Bundesanstalt, Wien, 97 p.

Hofmann KA (2013) Anorganische Chemie, 21st edn. Vieweg+Teubner, Wiesbaden, 865 p.

Höglund LO, Herbert R, Lövgren L, Öhlander B, Neretnieks I, Moreno L, Malmström ME, Elander P, Lindvall M, Lindström B (2004) MiMi – performance assessment main report. MiMi Print, Luleå, p 345

Holgate R (1991) Prehistoric flint mines. Shire archaeology, Bd 67. Shire, Princes Risborough, p 56

Hölting B, Coldewey WG (2013) Hydrogeologie – Einführung in die Allgemeine und Angewandte Hydrogeologie, 8th edn. Springer, Berlin, 326 p. https://doi.org/10.1007/978-3-8274-2354-2

Hölting B, Coldewey WG (2019) Hydrogeology, Springer textbooks in earth sciences, geography and environment. Springer, Berlin

Holtzhausen L (2006a) From trapped to tap – mine water becomes a commodity. The Water Wheel 5(3):12–15

Holtzhausen L (2006b) World first – full scale biosure plant commissioned. The Water Wheel 5(3):19–21

Hoppe A (1996) Ein Boden ist ein Boden ist kein Boden – Plädoyer für eine einfache und einheitliche Begriffsbestimmung. Nachr Dt Geol Ges 1996:70–78

Horenburg D (2008) Fischsterben im Uhlenbach hat juristische Folgen. Mitteldtsch Ztg, 8 Aug 2008

Horova AI, Kolesnyk VY, Kulikova DV (2011) Udoskonalennya Sporud Mekhanichnoi Ochystky Shakhtnykh Vod [Mine water mechanical treatment facilities improvement]. Naukovyj visnyk Nacional'noho Hirnycoho Universytetu [Sci Bull Nat Min Univ] 5:107–113

Hubert E, Wolkersdorfer C (2015) Establishing a conversion factor between electrical conductivity and total dissolved solids in South African mine waters. Water SA 41(4):490–500. https://doi.org/10.4314/wsa.v41i4.08

Huisamen A (2017) Quantification methods and management of hydrogeochemistry in decommissioned collieries of the Mpumalanga Coalfields. Unpublished PhD thesis, University of Pretoria, p 191

Hung BQ (2017) VOxFlotation – future solution for water treatment. Helsinki Metropolia University of Applied Sciences, p 50

Hunger B, Weidemüller D, Westermann S, Stevens JM (2005) Transforming landscapes – recommendations based on three industrially disturbed landscapes in Europe. International Building Exhibition (IBA) Fürst Pückler Land, Großräschen, p 152

Hutton B, Kahan I, Naidu T, Gunther P (2009) Operating and maintenance experience at the Emalahleni Water Reclamation Plant. In: Water Institute of Southern Africa, International Mine Water Association (eds) Proceedings, International mine water conference. Document Transformation Technologies, Pretoria, pp 415–430

Hwang J-Y, Sun X (2012) Removal of ions from water with electrosorption technology. In: Water in mineral processing. SME, Littleton, pp 87–95

Hydrometrics Inc. (2001) A new process for sulfate removal from industrial waters. In: Water online. http://www.wateronline.com/doc/a-new-process-for-sulfate-removal-from-indust-0001. Accessed 24 Feb 2013

I. B. A. Internationale Bauausstellung Fürst-Pückler-Land (2010) New landscape Lusatia. jovis, Berlin, 304 p.

Iakovleva E, Sillanpää M (2013) The use of low-cost adsorbents for wastewater treatment in mining industries. Environ Sci Pollut Res 20(11):7878–7899. https://doi.org/10.1007/s11356-013-1546-8

IARC Working Group on the Evaluation of Carcinogenic Risks to Humans (2016) Polychlorinated biphenyls and polybrominated biphenyls. IARC Monogr 107:1–502

Ibanez JG, Hemandez-Esparza M, Doria-Serrano C, Fregoso-Infante A, Singh MM (2007) Environmental chemistry – fundamentals. Springer, Heidelberg, 334 p.

International Union of Pure and Applied Chemistry (2014) Compendium of chemical terminology – gold book, version 2.3.3 edn. International Union of Pure and Applied Chemistry, 1622 p. https://doi.org/10.1351/goldbook

Isaacson W (2011) Steve Jobs. Little Brown, London, 630 p.

Isgró MA, Basallote MD, Barbero L (2022) Unmanned aerial system-based multispectral water quality monitoring in the Iberian Pyrite Belt (SW Spain). Mine Water Environ 41(1):30–41. https://doi.org/10.1007/s10230-021-00837-4

Islam MS, Kashem MA, Osman KT (2016) Phytoextraction efficiency of lead by arum (*Colocasia esculenta* L.) grown in soil. Int. J Soil Sci 11(4):130–136. https://doi.org/10.3923/ijss.2016.130.136

Jackson CB, Hach C (2004) Report on the validation of proposed EPA method 360.3 (Luminescence) for the measurement of dissolved oxygen in water and wastewater. Hach, Loveland, p 42

Jacobs P, Pulles W (2007) Best practice guideline H4: water treatment. Department of Water Affairs and Forestry, Pretoria, p 77

Jäger B, Obermann P, Wilke FL (1991) Studie zur Eignung von Steinkohlebergwerken im rechtsrheinischen Ruhrkohlebezirk zur Untertageverbringung von Abfall- und Reststoffen – Kurzfassung. LWA Materialien 2(91):1–73

Jäger B, Obermann P, Wilke FL, Heidrich F, Rüterkamp P, Skrzyppek J (1990) Studie zur Eignung von Steinkohlebergwerken im rechtsrheinischen Ruhrkohlebezirk zur Untertageverbringung von Abfall- und Reststoffen. LWA Studie, Düsseldorf, 628 p.

Jambor JL, Blowes DW, Ritchie AIM (2003) Environmental aspects of mine wastes. Short Course Series, Bd 31. Mineralogical Association of Canada, Waterloo, 430 p. https://doi.org/10.1144/1467-7873/03-029

Jang A, Lee J-H, Bhadri PR, Kumar SA, Timmons W, Beyette FR Jr, Papautsky I, Bishop PL (2005) Miniaturized redox potential probe for in situ environmental monitoring. Environ Sci Technol 39(16):6191–6197. https://doi.org/10.1021/es050377a

Janneck E, Arnold I, Koch T, Meyer J, Burghardt D, Ehinger S (2010) Microbial synthesis of schwertmannite from lignite mine water and its utilization for removal of arsenic from mine waters and for production of iron pigments. In: Wolkersdorfer C, Freund A (eds) Mine water and innovative thinking – International Mine Water Association symposium. Cape Breton University Press, Sydney, pp 131–135. https://doi.org/10.13140/2.1.4353.9201

Janneck E, Burghardt D, Martin M, Damian C, Schöne G, Meyer J, Peiffer S (2011) From waste to valuable substance: utilization of schwertmannite and lignite filter ash for removal of arsenic and uranium from mine drainage. In: Mine water – managing the challenges – 11th International Mine Water Association Congress. RWTH Aachen University, Aachen, pp 359–364

Janneck E, Glombitza F, Terno D, Wolf M, Patzig A, Fischer H, Rätzel G, Herbach K-D (2007a) Umweltfreundliche biotechnologische Gewinnung von Eisenhydroxisulfaten aus der Bergbauwasserbehandlung und deren Verwertung als Roh- und Grundstoff in der keramischen, Baustoffe produzierenden sowie Farben und Pigmente herstellenden Industrie zur Kosten-, Rohstoff- und Ressourceneinsparung, Teilprojekt 1: Koordination sowie Anlagenbetrieb und Produktherstellung, report no. 01 RI05013. G.E.O.S. Freiberg Ingenieurgesellschaft mbH, Freiberg, 114 p.

Janneck E, Krüger HG (1999) Sanierung von Bergbauwässern aus Altablagerungen des Schieferbergbaus – Pilotanlage für die Behandlung saurer Bergbauwässer. Unpublished Ergebnisbericht G.E.O.S. Freiberg, Freiberg, p 75

Janneck E, Schröder A, Schlee K, Glombitza F, Rolland W (2007b) Senkung des Kalkverbrauches bei der Grubenwasserreinigung durch physikalische Entfernung der ungebundenen Kohlensäure. Wiss Mitt Inst Geol 35:27–34

Jarvis A, Fox A, Gozzard E, Hill S, Mayes W, Potter H (2007) Prospects for effective national management of abandoned metal mine water pollution in the UK. In: Water in Mining Environments. Mako Edizioni, Cagliari, pp 77–81

Jarvis AP (2006) The role of dissolved carbon dioxide in governing deep coal mine water quality and determining treatment process selection. In: ICARD 2006, 7. Proceedings 7th International Conference on Acid Rock Drainage (ICARD), St. Louis, pp 833–843 [CD-ROM]. https://doi.org/10.21000/JASMR06020833

Jarvis AP, Alakangas L, Azzie B, Lindahl L, Loredo J, Madai F, Walder IF, Wolkersdorfer C (2012) Developments and challenges in the management of mining wastes and waters in Europe. In: Proceedings 9th International Conference on Acid Rock Drainage (ICARD), Ottawa, pp 1–12 [USB flash drive]

Jarvis AP, Younger PL (1999) Design, construction and performance of a full-scale compost wetland for mine-spoil drainage treatment at quaking houses. Water Environ Manage 13(5):313–318. https://doi.org/10.1111/j.1747-6593.1999.tb01054.x

Jenk U, Frenzel M, Metschies T, Paul M (2014a) Flooding of the underground uranium leach operation at Königstein (Germany) – a multidisciplinary report. In: Sui W, Sun Y, Wang C (eds) An interdisciplinary response to mine water challenges. International Mine Water Association, Xuzhou, pp 715–719

Jenk U, Zimmermann U, Uhlig U, Schöpke R, Paul M (2014b) In Situ mine water treatment: field experiment at the flooded Königstein Uranium Mine (Germany). Mine Water Environ 33(1):39–47. https://doi.org/10.1007/s10230-013-0241-9

Jenk U, Zimmermann U, Ziegenbalg G (2005) The use of $BaSO_4$ supersaturated solutions for in-situ immobilization of heavy metals in the abandoned Wismut GmbH uranium mine at Königstein.

In: Merkel BJ, Hasche-Berger A (eds) Uranium in the environment. Springer, Heidelberg, pp 721–727. https://doi.org/10.1007/3-540-28367-6_73

Jenke DR, Diebold FE (1984) Electroprecipitation treatment of acid mine wastewater. Water Res 18(7):855–859. https://doi.org/10.1016/0043-1354(84)90269-0

Jenkins M, Skousen JG (1993) Acid mine drainage treatment with the aquafix system. In: Proceedings, West Virginia surface mine drainage task force symposium 14 (Electronic document)

Jensen WB (2004) The symbol for pH. J Chem Educ 81(1):21. https://doi.org/10.1021/ed081p21

Jeuken B, Märten H, Phillips R (2008) Uranium ISL operation and water management under the arid climate conditions at Beverley, Australia. In: Proceedings, 10th International Mine Water Association Congress, pp 487–490

Jewiss C, Craw D, Pope J, Christenson H, Trumm D (2020) Dilution Processes of Rainfall-Enhanced Acid Mine Drainage Discharges from Historic Underground Coal Mines, New Zealand. Mine Water Environ 39(1):27–41. https://doi.org/10.1007/s10230-019-00650-0

Jobst W, Rentzsch W, Schubert W, Trachbrod K (1994) Bergwerke im Freiberger Land, 2nd edn. Medienzentrum der TU Bergakademie Freiberg, Freiberg, 227 p.

Johnson DB, Hallberg KB (2002) Pitfalls of passive mine water treatment. Rev Environ Sci Biotechnol 1:335–343. https://doi.org/10.1023/A:1023219300286

Johnson DB, Hallberg KB (2005) Acid mine drainage remediation options – a review. Sci Total Environ 338:3–14. https://doi.org/10.1016/j.scitotenv.2004.09.002

Johnson RH, Tutu H (2016) Predictive reactive transport modeling at a proposed uranium in situ recovery site with a general data collection guide. Mine Water Environ 35(3):369–380. https://doi.org/10.1007/s10230-015-0376-y

Johnston D (2004) A metal mines strategy for Wales. In: Jarvis AP, Dudgeon BA, Younger PL (eds) Mine water 2004 – Proceedings International Mine Water Association symposium. 1. University of Newcastle, Newcastle upon Tyne, pp 17–23

Jones JB, Ruggeri S (1969) Abatement of pollution from abandoned coal mines by means of in-situ precipitation techniques. Am Chem Soc Div Fuel Chem Prepr 13(2):116–119

Jovanovic NZ, Annandale JG, Claassens AS, Lorentz SA, Tanner PD (2001) Modeling irrigation with gypsiferous mine water – a case study in Botswana. Mine Water Environ 20(2):65–72. https://doi.org/10.1007/s10230-001-8084-1

Joyce KE, Anderson K, Bartolo RE (2021) Of Course we fly unmanned – we're women! Drones 5(1). https://doi.org/10.3390/drones5010021

Juby GJG (1992) Membrane desalination of service water from gold mines. J S Afr Inst Min Metall 92(3):65–69

Juby GJG, Pulles W (1990) Evaluation of electrodialysis reversal for the desalination of brackish mine service water, Report no. 0 947447 94 6. Water Research Commission, Pretoria, p 54

Juby GJG, Schutte CF (2000) Membrane life in a seeded-slurry reverse osmosis system. Water SA 26(2):239–248

Juby GJG, Schutte CF, van Leeuwen J (1996) Desalination of calcium sulfate scaling mine water: design and operation of the SPARRO process. Water SA 22(2):161–172

Kabongo JD (2013) Waste valorization. In: Idowu SO, Capaldi N, Zu L, Gupta AD (eds) Encyclopedia of corporate social responsibility. Springer, Berlin, pp 2701–2706. https://doi.org/10.1007/978-3-642-28036-8_680

Kaksonen AH, Puhakka JA (2007) Sulfate reduction based bioprocesses for the treatment of acid mine drainage and the recovery of metals. Eng Life Sci 7(6):541–546. https://doi.org/10.1002/elsc.200720216

Kaksonen AH, Şahinkaya E (2012) Review of sulfate reduction based bioprocesses for acid mine drainage treatment and metals recovery. In: McCullough CD, Lund MA, Wyse L (eds) International Mine Water Association symposium. Edith Cowan University, Bunbury, pp 207–214

Kalayev VA, Kamentsev AV, Kozlov VM (2006) Способ очистки шахтных вод от вредных примесей [Sposob ochistki shakhtnykh vod ot vrednykh primesey – Methoden um Grubenwasser von Verunreinigungen zu befreien]. Ugol [Kohle](Dezember): 57–59

Kalin M (1993) The application of ecological engineering to Selminco Summit. Boojum Research, Sydney, p 68

Kalin M (2004a) Passive mine water treatment: the correct approach? Ecol Eng 22:299–304. https://doi.org/10.1016/j.ecoleng.2004.06.008

Kalin M (2004b) Slow progress in controlling acid mine drainage (AMD): a perspective and a new approach. Peckiana 3:101–112

Kalin M, Caetano Chaves WL (2003) Acid reduction using microbiology – treating AMD effluent emerging from an abandoned mine portal. Hydrometallurgy 71(1):217–225. https://doi.org/10.1016/S0304-386X(03)00159-2

Kalin M, Fyson A, Smith MP (1993) ARUM – acid reduction using microbiology. In: Proceedings Fossil Energy Materials Bioremediation, Microbial Physiology, Jackson Hole, II, pp 319–328

Kalin M, Smith MP (1991) Biological amelioration of acidic seepage streams, Kamloops, pp 351–362

Kalka H (2018) aquaC – aquatische Chemie. UIT report, Dresden, p 93

Karadeniz M (2005) Asit Maden (Kaya) Drenajinda Aktif ve Pasif Çözüm Yöntemler [Aktive und passive Lösungsmethoden für saueres Grubenwasser]. In: Proceedings, Madencilik ve Çevre Sempozyumu, pp 91–97

Karakatsanis K, Cogho V (2010) Drinking water from mine water using the Hipro® Process – optimum coal mine water reclamation plant. In: Wolkersdorfer C, Freund A (eds) Mine water and innovative thinking – International Mine Water Association symposium. Cape Breton University Press, Sydney, pp 135–139

Karickhoff S, Brown D, Scott T (1979) Sorption of hydrophobic pollutants on natural sediments. Water Res 13(3):241–248. https://doi.org/10.1016/0043-1354(79)90201-x

Kassahun A, Laubrich J, Paul M (2016) Feasibility study on seepage water treatment at a uranium TMF site by ion exchange and ferric hydroxide adsorption. In: Drebenstedt C, Paul M (eds) IMWA 2016 – mining meets water – conflicts and solutions. TU Bergakademie Freiberg, Freiberg, p 858

Kauppila P, Räisänen ML, Myllyoja S (2013) Best environmental practices in metal mining operations [Metallimalmikaivostoiminnan parhaat ympäristökäytännöt]. Suomen ympäristö, 29en/2011. Finlands miljöcentral [Finnish Environment Institute], Helsinki, p 219

Kegel K (1950) Bergmännische Wasserwirtschaft einschließlich Grundwasserkunde, Wasserversorgung und Abwasserbeseitigung. Berg- und Aufbereitungstechnik, Band III Geologische und technologische Grundlagen des Bergbaues, III, 3rd edn. Knapp, Halle, 374 p.

Kelly DP, Wood AP (2000) Reclassification of some species of *Thiobacillus* to the newly designated genera *Acidithiobacillus* gen. nov., *Halothiobacillus* gen. nov. and *Thermithiobacillus* gen. nov. Int J Syst Evol Microbiol 50:511–516. https://doi.org/10.1099/00207713-50-2-511

Kemmer H-G (1985) Knappen aus dem Labor – Mikroben gewinnen Metalle und entschwefeln Kohle. Die Zeit, 24 (7 Juni 1985), 20 p.

Kepler DA, McCleary EC (1994) Successive Alkalinity-Producing Systems (SAPS) for the treatment of acid mine drainage. In: Proceedings international land reclamation and mine drainage conference, Pittsburgh, 1, pp 195–204

Kepler DA, McCleary EC (1995) Successive Alkalinity-producing Systems (SAPS). In: Proceedings, West Virginia Surface Mine Drainage Task Force symposium 16 (Electronic document)

Kepler DA, McCleary EC (1997) Passive aluminum treatment successes. In: Proceedings, West Virginia surface mine drainage task force symposium 18 (Electronic document)

Kester DR, Byrne RH, Liang Y-J (1975) Redox reactions and solution complexes of iron in marine systems. ACS Symp Ser 18:56–79. https://doi.org/10.1021/bk-1975-0018.ch003

Kickuth R (1977) Degradation and incorporation of nutrients from rural wastewaters by plant hydrosphere under limnic conditions. In: Voorburg JH (eds) Utilization of manure land spreading. EUR 5672e. Europäische Gemeinschaften, Luxemburg, pp 335–343

Kimball BA, Runkel RL, Walton-Day K, Bencala KE (2002) Assessment of metal loads in watersheds affected by acid mine drainage by using tracer injection and synoptic sampling: Cement Creek, Colorado, USA. Appl Geochem 17(9):1183–1207. https://doi.org/10.1016/S0883-2927(02)00017-3

Kinčl J, Jiříček T, Fehér J, Amrich M, Neděla D, Toman F, Velen B, Cakl J, Kroupa J (2017) Electromembrane processes in mine water treatment. In: Wolkersdorfer C, Sartz L, Sillanpää M, Häkkinen A (eds) IMWA 2017 – mine water & circular economy, vol II. Lappeenranta University of Technology, Lappeenranta, pp 1154–1161

King DJ, Simmler JJ, Decker CS, Ogg CW (1974) Acid strip mine lake recovery. J Water Pollut Control Fed 42(10):2301–2316

Kirby CS, Cravotta CA III (2005a) Net alkalinity and net acidity 1: theoretical consideration. Appl Geochem 20(10):1920–1940. https://doi.org/10.1016/j.apgeochem.2005.07.002

Kirby CS, Cravotta CA III (2005b) Net alkalinity and net acidity 2: practical considerations. Appl Geochem 20(10):1941–1961. https://doi.org/10.1016/j.apgeochem.2005.07.003

Kirby CS, Decker SM, Macander NK (1999) Comparison of color, chemical and mineralogical compositions of mine drainage sediments to pigment. Environ Geol Water Sci 37(3):243–254. https://doi.org/10.1007/s002540050382

Kirby CS, Dennis A, Kahler A (2009) Aeration to degas $CO_2$, increase pH, and increase iron oxidation rates for efficient treatment of net alkaline mine drainage. Appl Geochem 24:1175–1184. https://doi.org/10.1016/j.apgeochem.2009.02.028

Kleinmann B, Skousen J, Wildeman T, Hedin B, Nairn B, Gusek J (2021) The early development of passive treatment systems for mining-influenced water: a North American perspective. Mine Water Environ 40(4):818–830. https://doi.org/10.1007/s10230-021-00817-8

Kleinmann RLP (1990a) Acid mine drainage treatment using engineered wetlands. Int J Mine Water 9(1–4):269–276. https://doi.org/10.1007/BF02503697

Kleinmann RLP (1990b) At-source control of acid mine drainage. Int J Mine Water 9(1–4):85–96. https://doi.org/10.1007/BF02503685

Kleinmann RLP, Erickson PM (1982) Control of acid mine drainage using anionic surfactants. In: Proceedings, 1st International Mine Water Congress, Budapest, Hungary C, pp 51–64

Klimant I, Meyer V, Kohls M (1995) Fibre-optic oxygen microsensors, a new tool in aquatic biology. Limnol Oceanogr 40(6):1159–1165. https://doi.org/10.4319/lo.1995.40.6.1159

Klímková Š, Cerník M (2008) Application of zero-valent nanoparticles for acid mine drainage remediation. In: Proceedings, 10th International Mine Water Association Congress, pp 281–284

Klímková Š, Cerník M, Lacinova L, Filip J, Jancik D, Zboril R (2011) Zero-valent iron nanoparticles in treatment of acid mine drainage from in situ uranium leaching. Chemosphere 82(8):1178–1184. https://doi.org/10.1016/j.chemosphere.2010.11.075

Klinger C, Hansen C, Rüterkamp P, Heinrich H (2000) In situ tests for interactions between acid mine drainage and ferrihydrite sludge in the pyrite mine “Elbingerode” (Harz Mts.; Germany). In: Proceedings, 7th International Mine Water Association Congress, pp 137–145

Klinger C, Jenk U, Schreyer J (2002) Processes in passive mine water remediation with zero-valent iron and lignite as reactive materials. In: Merkel BJ, Planer-Friedrich B, Wolkersdorfer C (eds) Uranium in the aquatic environment. Springer, Heidelberg, pp 569–576. https://doi.org/10.1007/978-3-642-55668-5_67

Knops F, de la Mata MG, Mendoza Fajardo C, Kahne E (2012) Seawater desalination of the Chilean coast for water supply to the mining industry. In: International Mine Water Association symposium. Edith Cowan University, Bunbury, pp 697–704

Koch C (2010) Einsatz von alkalischen Materialien und $CO_2$ zur Neutralisierung bergbaubedingt versauerter oberirdischer Gewässer und nachfolgender $CO_2$-Mineralisierung. In: Proceedings des Dresdner Grundwasserforschungszentrums eV 40, pp 1–186

Koch C, Graupner B, Werner F (2008) Pit Lake treatment using fly ash deposits and carbon dioxid. In: Proceedings, 10th International Mine Water Association Congress, pp 579–582

Koch C, Mazur K (2016) A new pit lake treatment technology using calcium oxide and carbon dioxide to increase alkalinity. In: Drebenstedt C, Paul M (eds) IMWA 2016 – mining meets water – conflicts and solutions. TU Bergakademie Freiberg, Freiberg, pp 284–291

König J (1899) Die Verunreinigung der Gewässer, deren schädliche Folgen sowie die Reinigung von Trink- und Schmutzwasser, 2nd edn. Springer, Berlin, 514 p. https://doi.org/10.1007/978-3-642-91824-7

Koren JPF, Syversen U (1995) State-of-the-art electroflocculation. Filtr Sep 32(2): 146, 153–156. https://doi.org/10.1016/S0015-1882(97)84039-6

Koros WJ, Ma YH, Shimidzu T (1996) Terminology for membranes and membrane processes (IUPAC Recommendations 1996). Pure Appl Chem 68(7):1479–1489. https://doi.org/10.1351/pac199668071479

Körtl K (2012) Neuer Ablauf für den Steinberger See. Mittelbayerische Zeitung, 19 Jan 2012, 34 p.

Koschorreck M, Wendt-Potthoff K, Bozau E, Herzsprung P, Geller W (2006) In situ Neutralisation von sauren Bergbaurestseen – Prozesse im Sediment und begrenzende Faktoren. Wiss Mitt Inst Geol 31:55–60

Kostenbader PD, Haines GF (1970) High-density sludge treats acid mine drainage. Coal Age 75(September):90–97

Kranz K, Dillenardt J (2010) Mine water utilization for geothermal purposes in Freiberg, Germany: determination of hydrogeological and thermophysical rock parameters. Mine Water Environ 29(1):68–76. https://doi.org/10.1007/s10230-009-0094-4

Kuipers J, Maest A (2006) Comparison of predicted and actual water quality at hardrock mines – the reliability of predictions in environmental impact statements. Earthworks, Washington, p 195

Kuit WJ (1980) Mine and tailings effluent treatment at Kimberley, B.C. operations of Cominco Ltd. CIM Bull 73:105–112

Kuyucak N, Lindvall M, Rufo Serrano JA, Oliva AF (1999) Implementation of a high density sludge "HDS" treatment process at the Boliden Apirsa Mine Site. In: Fernández Rubio R (ed) Mine, water & environment. II. International Mine Water Association, Sevilla, pp 473–479

Kuyucak N, Payant S, Sheremata T (1995) Improved lime neutralization process. In: Proceedings Sudbury '95 – mining and the environment. Sudbury, vol. 1, pp 129–137

Kwong YTJ, Nordstrom DK (1989) Copper-arsenic mobilization and attenuation in an acid mine drainage environment. In: Proceedings, International symposium on water-rock interaction 6, pp 397–399

Kwong YTJ, Van Stempvoort DR (1994) Attenuation of acid rock drainage in a natural wetland system. In: Environmental geochemistry of sulfide oxidation, Bd 550. American Chemical Society, New York, pp 382–392. https://doi.org/10.1021/bk-1994-0550.ch025

Ladwig KJ, Erickson PM, Kleinmann RLP, Posluszny ET (1984) Stratification in water quality in inundated anthracite mines, Eastern Pennsylvania. Bur Mines Rep Invest 8837:35

Landesamt für Natur Umwelt und Verbraucherschutz NRW, Rahm H, Obschernicat K, Dittmar M, Rosenbaum-Mertens J, Selent K (2018) Belastungen von Oberflächengewässern und von aktiven Grubenwassereinleitungen mit bergbaubürtigen PCB (und PCB-Ersatzstoffen) – Ergebnisse des LANUV-Sondermessprogramms. Landesamt für Natur, Umwelt und Verbraucherschutz NRW, Recklinghausen, 109 p.

Landesanstalt für Umweltschutz Baden-Württemberg (2002) Durchflussermittlung mit der Salzverdünnungsmethode. Arbeitsanleitung Pegel- und Datendienst Baden-Württemberg. Landesanstalt für Umweltschutz Baden-Württemberg, Karlsruhe, 66 p.

Lane A (2016) Development and validation of an acid mine drainage treatment process for source water. Battelle Memorial Institute, Columbus

Lang B (1999) Permanent sealing of tunnels to retain tailings or acid rock drainage. In: Fernández Rubio R (ed) Mine, water & environment. II. International Mine Water Association, Sevilla, pp 647–655

Langguth HR, Voigt R (2004) Hydrogeologische Methoden, 2nd edn. Springer, Berlin, 1005 p.

Langmuir D, Chrostowski P, Vigneault B, Chaney R (2005) Issue paper on the environmental chemistry of metals, 2nd corr edn. ERG – US EPA, Lexington, p 106

Lata S, Singh PK, Samadder SR (2015) Regeneration of adsorbents and recovery of heavy metals: a review. Int J Environ Sci Technol 12(4):1461–1478. https://doi.org/10.1007/s13762-014-0714-9

Lausitzer und Mitteldeutsche Bergbau-Verwaltungsgesellschaft mbH (2019) Merkblatt – Montanhydrologisches Monitoring in der LMBV mbH. Lausitzer und Mitteldeutsche Bergbau-Verwaltungsgesellschaft mbH, Senftenberg, 35 p.

Lawson DC, Sonderegger JL (1978) Geothermal data-base study – mine-water temperatures. Special Publication – State of Montana Bureau of Mines and Geology 79, RLO-2426-T2-3, pp 1–38. https://doi.org/10.2172/6031879

Leavitt BR (2011) Aeration of mine water using a TROMPE. In: Proceedings, West Virginia surface mine drainage task force symposium 32, pp 1–11 (Electronic document)

Leavitt BR, Danehy T, Mahony R, Page B, Neely C, Denholm C, Busler S, Dunn MH (2013) Trompe technology for mine drainage treatment. Reclam Matters 2013(Fall):31–33

Leavitt BR, Danehy T, Page B (2012) Passive mixing to improve calcium oxide dissolution. In: Proceedings, West Virginia surface mine drainage task force symposium 33, pp 1–15 (Electronic document)

Leblanc M, Morales JA, Borrego J, Elbaz-Poulichet F (2000) 4,500-year-old mining pollution in Southwest Spain: long-term implications from modern mining pollution. Econ Geol 95:655–662. https://doi.org/10.2113/gsecongeo.95.3.655

Leitch RD (1935) Sealing of coal mines – will reduce acidity of their effluent waters. Coal Age 40:323–326

Lengke M, Davis A, Bucknam C (2010) Improving management of potentially acid generating waste rock. Mine Water Environ 29(1):29–44. https://doi.org/10.1007/s10230-009-0097-1

Leupolt G, Hocker M (1999) Befahrerhandbuch – Streitschrift zu Arbeitsweisen der praktischen bergbauhistorischen Forschung. Eigenverlag, Dresden, 264 p.

Lewis GP, Coughlin LL, Jusko WJ, Hartz S (1972) Contribution of cigarette smoking to cadmium accumulation in man. Lancet 299(7745):291–292. https://doi.org/10.1016/S0140-6736(72)90294-2

Lewis NM, Wangerud KW, Park BT, Fundingsland SD, Jonas JP (2003) Status of in situ treatment of Anchor Hill pit lake, Gilt Edge Mine Superfund site, South Dakota, USA. Australian Inst of Mining and Metallurgy Publication Series, 3. In: Proceedings 6th International Conference on Acid Rock Drainage (ICARD), Carlton, pp 779–788

Li N, Wania F, Lei YD, Daly GL (2003a) A comprehensive and critical compilation, evaluation, and selection of physical-chemical property data for selected polychlorinated biphenyls. J Phys Chem Ref Data 32(4):1545–1590. https://doi.org/10.1063/1.1562632

Li T, Patel RU, Ramsden DK, Greene J (2003b) Ground water recovery and treatment. In: Moyer EE, Kostecki PT (eds) MTBE remediation handbook. Springer, New York, pp 289–327. https://doi.org/10.1007/978-1-4615-0021-6

Lieber D (2003) Säure- und Basekapazität – Vergleich unterschiedlicher vor-Ort-Bestimmungsmethoden bei unterschiedlichen Bergwerkswässern und Bearbeitern. Unpublished Studienarbeit, Freiberg, 33 S

Ließmann W (2010) Historischer Bergbau im Harz, 3rd edn. Springer, Heidelberg, 470 p. https://doi.org/10.1007/978-3-540-31328-1

Lilley T (2012) Membrane based water and wastewater treatment solutions. In: International Mine Water Association symposium. Edith Cowan University, Bunbury, pp 631–636

Linke S, Schiffer L (2002) Development prospects for the post-mining landscape in Central Germany. In: Remediation of abandoned surface coal mining sites. Springer, Berlin, pp 111–149. https://doi.org/10.1007/978-3-662-04734-7_5

Lipták BG (2003) Process measurement and analysis. In: Instrument engineers' handbook, vol I, 4th edn. CRC, Boca Raton, p 1828

Liu H, Zhao X, Qu J (2010) Electrocoagulation in water treatment. In: Comninellis C, Chen G (eds) Electrochemistry for the environment. Springer, Heidelberg, pp 245–262. https://doi.org/10.1007/978-0-387-68318-8_10

Lloyd JW, Heathcote JA (1985) Natural inorganic chemistry in relation to groundwater – an introduction. Clarendon, Oxford, p 296

Loop CM (2004) Lessons learned from full-scale, non-traditional placement of fly ash. In: Proceedings, state regulation of coal combustion by-product placement at mine sites: a technical interactive forum, pp 43–46

Loop CM, Scheetz BE, White WB (2003) Geochemical evolution of a surface mine lake with alkaline ash addition: field observations vs. laboratory predictions. Mine Water Environ 22(4):206–213. https://doi.org/10.1007/s10230-003-0023-x

Loop CM, Scheetz BE, White WB (2007) The big gorilla demonstration project. In: Coal ash beneficial use in mine reclamation and mine drainage remediation in Pennsylvania. The Pennsylvania Department of Environmental Protection, Harrisburg, pp 246–301

Lorax Environmental (2003) Treatment of sulfate in mine effluents. INAP – International Network of Acid Prevention, London, p 129

Losavio M, Lauf A, Elmaghraby A (2019) The internet of things and issues for mine water management. In: Proceedings Mine Water – Technological and Ecological Challenges (IMWA 2019), Perm, pp 678–683

Lu M (2004) Pit lakes from sulfide ore mining, geochaemical characterization before treatment, after liming and sewage sludge treatment – cases studies at Rävlidmyran and Udden, Sweden. Luleå University of Technology, p 31

Luckner L, Totsche O (2017) In-Lake-Neutralisation von Bergbaufolgeseen im Lausitzer und Mitteldeutschen Braunkohlerevier – Aktueller Stand und Bewertung der Technischen Entwicklung. LMBV, Senftenberg, 40 p.

Lundgren DG, Silver M (1980) Ore leaching by bacteria. Ann Rev Microbiol 34:263–283. https://doi.org/10.1146/annurev.mi.34.100180.001403

Luukkonen T, Runtti H, Niskanen M, Tolonen E-T, Sarkkinen M, Kemppainen K, Rämö J, Lassi U (2016) Simultaneous removal of Ni(II), As(III), and Sb(III) from spiked mine effluent with metakaolin and blast-furnace-slag geopolymers. J Environ Manage 166(Supplement C):579–588. https://doi.org/10.1016/j.jenvman.2015.11.007

Lydersen E, Löfgren S, Arnesen RT (2002) Metals in Scandinavian surface waters: effects of acidification, liming, and potential reacidification. Crit Rev Environ Sci Technol 32(2–3):73–295. https://doi.org/10.1080/10643380290813453

Macadam J, Shail R (1999) Abandoned pits and quarries: a resource for research, education, leisure and tourism. In: Spalding A (ed) The conservation value of abandoned pits and quarries in Cornwall. The Historic Environment Service, Truro, pp 71–80

Mair C (2002) Hydrogeologie, Hydrogeochemie und Isotopie der Grund- und Grubenwässer im Einzugsgebiet des Burgfeyer Stollens bei Mechernich, Eifel. Unpublished Dissertation. RWTH Aachen, Aachen, 270 p.

Makhathini TP, Mulopo J, Bakare BF (2020) Possibilities for acid mine drainage co-treatment with other waste streams: a review. Mine Water Environ 39(1):13–26. https://doi.org/10.1007/s10230-020-00659-w

Malisa R, Maree JP, Hardwick E, Oosthuizen F (2013) Resin freeze desalination process for acid recovery. In: Brown A, Figueroa L, Wolkersdorfer C (eds) Reliable mine water technology. International Mine Water Association, Golden, pp 655–659

Mallo JC, Marco SGD, Bazzini SM, Río JL (2010) Aquaculture: an alternative option for the rehabilitation of old mine pits in the pampasian region, Southeast of Buenos Aires, Argentina. Mine Water Environ 29(4):285–293. https://doi.org/10.1007/s10230-010-0120-6

Mamelkina M (2019) Electrochemical treatment of mining waters. In: Proceedings Mine Water – Technological and Ecological Challenges (IMWA 2019), Perm, Russia, pp 212–216

Man M, Sparrow B, Low M (2018) A cradle to grave treatment solution for mine waters. Paper presented at the IMWA 2018 – risk to opportunity, Pretoria, II, pp 1093–1098

Maneval DR (1968) The little scrubgrass creek AMD plant. Coal Min Process 5(9):28–32

Marchlewitz B (1959) Untersuchungen über die Mikrobenassoziation saurer Grubenwässer. Unpublished dissertation. Uni Greifswald, Greifswald, 57 p.

Marchlewitz B, Hasche D, Schwartz W (1961) Untersuchungen über das Verhalten von Thiobakterien gegenüber Schwermetallen. Z Allg Mikrobiol 1(3):179–191. https://doi.org/10.1002/jobm.19610010302

Marchlewitz B, Schwartz W (1961) Untersuchungen über die Mikroben-Assoziation saurer Grubenwässer. Z Allg Mikrobiol 1(2):100–114. https://doi.org/10.1002/jobm.19610010203

Maree JP, Hulse G, Dods D, Schutte CE (1991) Pilot-plant studies on biological sulfate removal from industrial effluent. Water Sci Technol 23(7–9):1293–1300. https://doi.org/10.2166/wst.1991.0581

Maree JP, van Tonder GJ, Millard P, Erasmus TC (1996) Pilot-scale neutralisation of underground mine water. Water Sci Technol 34(10):141–149. https://doi.org/10.2166/wst.1996.0250

Märten H (2006) Neueste Trends zur aktiven Wasserbehandlung und Anwendungsbeispiele. Wiss Mitt Inst Geol 31:13–22

Martikainen P, Korhonen K, Tarkiainen L (2021) Heavy metal toxicity and mortality – association between density of heavy metal bands and cause specific hospital admissions and mortality: population based cohort study. BMJ 375:e067633. https://doi.org/10.1136/bmj-2021-067633

Martinez-Olmos A, Capel-Cuevas S, López-Ruiz N, Palma AJ, de Orbe I, Capitán-Vallvey LF (2011) Sensor array-based optical portable instrument for determination of pH. Sens Actuator B Chem 156(2):840–848. https://doi.org/10.1016/j.snb.2011.02.052

Mast MA, Verplanck PL, Wright WG, Bove DJ (2007) Characterization of background water quality. US Geol Surv Prof Pap 1651(E7):347–386

Matschullat J, Ottenstein R, Reimann C (2000) Geochemischer Hintergrund – berechenbar? In: Bergbau und Umwelt – Langfristige geochemische Einflüsse. Springer, Berlin, pp 1–23. https://doi.org/10.1007/978-3-642-57228-9_1

Matthies R, Jarvis AP, Aplin AC (2009) Performance evaluation of two reducing and alkalinity producing systems for coal mine drainage remediation after 4 years of operation. In: Water Institute of Southern Africa, International Mine Water Association (eds) Proceedings, international mine water conference. Document Transformation Technologies, Pretoria, pp 531–538

Mayes WM, Gozzard E, Potter HAB, Jarvis AP (2008) Quantifying the importance of diffuse mine-water pollution in a historically heavily coal mined catchment. Environ Pollut 151(1):165–175. https://doi.org/10.1016/j.envpol.2007.02.008

Mayes WM, Jarvis AP, Younger PL (2005) Assessing the importance of diffuse mine water pollution: a case study from County Durham, UK. In: Mine water 2005 – mine closure. University of Oviedo, Oviedo, pp 497–505

McBain JW (1909) The mechanism of the adsorption ("sorption") of hydrogen by carbon. Philos Mag Ser 6 18(108):916–935. https://doi.org/10.1080/14786441208636769

McCloskey K, Almoric E, Bessbousse H, Mezailles N, Van Zutphen S (2010) Magpie polymers – selective metal capture. In: Wolkersdorfer C, Freund A (eds) Mine water and innovative thinking – International Mine Water Association symposium. Cape Breton University Press, Sydney, pp 143–147

McCullough CD, Lund MA (2011) Limiting factors for crayfish and finfish in acidic coal pit lakes. In: Mine water – managing the challenges – 11th International Mine Water Association Congress. RWTH Aachen University, Aachen, pp 35–40

McCullough CD, Schultze M, Vandenberg J (2020) Realizing beneficial end uses from abandoned pit lakes. Minerals 10(2):133. https://doi.org/10.3390/min10020133

McCullough CD, Steenbergen J, Beest CT, Lund MA (2009) More than water quality: environmental limitations to a fishery in acid pit lakes of Collie, South-West Australia. In: Water Institute of Southern Africa, International Mine Water Association (eds) Proceedings, international mine water conference. Document Transformation Technologies, Pretoria, pp 507–511

McDonald DM, Webb JA, Taylor J (2006) Chemical stability of acid rock drainage treatment sludge and implications for sludge management. Environ Sci Technol 40(6):1984–1990. https://doi.org/10.1021/es0515194

McKenzie R (2005) Software update to better predict costs of treating mine drainage. Mine Water Environ 24(4):213–215. https://doi.org/10.1007/s10230-005-0098-7

McLemore VT (2008) Basics of metal mining influenced water. Management Technologies for Metal Mining Influenced Water, Bd 1. SME, Littleton, p 103

McLemore VT, Smith KS, Russell CC (2014) Sampling and monitoring for the mine life cycle. In: Management technologies for metal mining influenced water, Bd 6. SME, Littleton, p 191

McPhilliamy SC, Green J (1973) A chemical and biological evaluation of three mine drainage treatment plants. Work document no. 47. U. S. Environmental Protection Agency, Wheeling, p 53

Melin T, Rautenbach R (2007) Membranverfahren – Grundlagen der Modul- und Anlagenauslegung, 3rd edn. Springer, Heidelberg, 584 p. https://doi.org/10.1007/978-3-540-34328-8

MEND (2000) MEND Manual – treatment – active and passive, 5.4.2e. The Mine Environment Neutral Drainage [MEND] Program, Ottawa, p 109

MEND (2001) MEND Manual – Sampling and Analysis, 5.4.2b. The Mine Environment Neutral Drainage [MEND] Program, Ottawa, p 111

Menendez R, Clayton JL, Zurbuch PE, Sherlock SM, Rauch HW, Renton JJ (2000) Sand-sized limestone treatment of streams impacted by acid mine drainage. Water Air Soil Pollut 124(3–4):411–428. https://doi.org/10.1023/A:1005264124166

Menezes JCSS, Silva RA, Arce IS, Schneider IAH (2009) Production of a poly-ferric sulfate chemical coagulant by selective precipitation of iron from acidic coal mine drainage. Mine Water Environ 28(4):311–314. https://doi.org/10.1007/s10230-009-0084-6

Mentz JW, Warg JB, Skelly Loy Inc. (1975) Up-dip versus down-dip mining – an evaluation. Environ Prot Technol Ser EPA-670/2-75-047:1–74

Merkel BJ (2002) Flooding of the Königstein uranium mine – aquifer reactivity versus dilution. In: Merkel BJ, Planer-Friedrich B, Wolkersdorfer C (eds) Uranium in the aquatic environment. Springer, Heidelberg, pp 267–275. https://doi.org/10.1007/978-3-642-55668-5_30

Merkel BJ (2005) Alkalinitätserhöhung in sauren Grubenwässern durch $CO_2$-Zugabe. Wiss Mitt Inst Geol 28:51–55

Merkel BJ, Planer-Friedrich B (2002) Grundwasserchemie – Praxisorientierter Leitfaden zur numerischen Modellierung von Beschaffenheit, Kontamination und Sanierung aquatischer Systeme. Springer, Heidelberg, 219 p.

Merkel BJ, Schipek M, Scholz G, Rabe W (2010) Optimierung der Kalkung von Tagebaufolgeseen. Wiss Mitt Inst Geol 42:51–59

Merkel W, Dördelmann O, Mauer C, Rieth U (2016) Gutachterlichen [*sic!*] Untersuchung/Recherche zu den technischen Möglichkeiten einer PCB-Elimination von Grubenwässern ("PCB-Gutachten"), Report No. 16/034.1 vom 21.04.2016. Ministerium für Klimaschutz, Umwelt, Landwirtschaft, Natur- und Verbraucherschutz des Landes Nordrhein-Westfalen, Mülheim an der Ruhr, 70 p.

Meschke K, Hansen N, Hofmann R, Haseneder R, Repke JU (2018) Characterization and performance evaluation of polymeric nanofiltration membranes for the separation of strategic elements from aqueous solutions. J Membr Sci 546:246–257. https://doi.org/10.1016/j.memsci.2017.09.067

Metschies T, van Berk W, Jenk U, Paul M (2018) Predicting natural attenuation for flooding of an Isl-Uranium Mine – potentials and limitations. In: Proceedings IMWA 2018 – risk to opportunity. Pretoria I, pp 411–416

Meyer AJ, Dick TP (2010) Fluorescent protein-based redox probes. Antioxid Redox Signal 13(5):621–650. https://doi.org/10.1089/ars.2009.2948

Mihok EA, Deul M, Chamberlain CE, Selmeczi JG (1968) Mine water research – the limestone neutralization process, 7191. Bureau of Mines Report of Investigations, Washington, p 20

Millen BMJ (2003) Aspects of the hydrogeology of a mining region with a focus on the antimony content of the spring-water, Eiblschrofen Massif, Schwaz, Tyrol, Austria. Mitt Österr Geol Ges 94:139–156

Miller A, Wildeman T, Figueroa L (2013) Zinc and nickel removal in limestone based treatment of acid mine drainage – the relative role of adsorption and co-precipitation. Appl Geochem 37:57–63. https://doi.org/10.1016/j.apgeochem.2013.07.001

Miller D, D'Souza G (2008) Using aquaculture as a post-mining land use in West Virginia. Mine Water Environ 27(2):122–126. https://doi.org/10.1007/s10230-008-0038-4

Miller SD, Jeffery JJ, Murray GSC (1990) Identification and management of acid generating mine wastes – procedures and practices in South-East Asia and the pacific regions. Int J Mine Water 9(1–4):57–67. https://doi.org/10.1007/BF02503683

Milošević Z, Suarez Fernandez RA, Dominguez S, Rossi C (2019) Guidance and navigation software for autonomous underwater explorer of flooded mines. In: Proceedings mine water – technological and ecological challenges (IMWA 2019), Perm, Russia, pp 690–695

Mischo H, Cramer B (2020) New mines in an old mining district – opportunities and challenges of the 4th mining boom in the Ore Mountains Region. Mining Rep Glückauf 156(1):40–45

Mishra AK (2014) Application of nanotechnology in water research. Scrivener, Salem, 522 p. https://doi.org/10.1002/9781118939314

Misra M (1990) Towards unobtainium [new composite materials for space applications]. Aerosp Compos Mater 2(6):29–32

Mitchell P, Anderson A, Potter C (2000) Protection of ecosystem and human health via silica micro encapsulation of heavy metals. In: Proceedings, 7th International Mine Water Association Congress, pp 146–156

Mitchell P, Wheaton A (1999) From environmental burden to natural resource; new reagents for cost-effective treatment of, and metal recovery from, acid rock drainage. In: Sudbury '99 – mining and the environment II, 2. Sudbury Environmental, Sudbury, pp 1231–1240

Mitko K, Noszczyk A, Dydo P, Turek M (2021) Electrodialysis of coal mine water. Water Resour Ind 25. https://doi.org/10.1016/j.wri.2021.100143

Moers W (2000) The 13½ lives of Captain Bluebear – being the demibiography of a seagoing bear, with numerous illustrations and excerpts from the "Encyclopedia of the marvels, life forms and other phenomena of Zamonia and its environs" by professor Abdullah Nightingale. Secker & Warburg, London, 702 p.

Mollah MYA, Morkovsky P, Gomes JAG, Kesmez M, Parga J, Cocke DL (2004) Fundamentals, present and future perspectives of electrocoagulation. J Hazard Mater 144(1–3):199–210. https://doi.org/10.1016/j.jhazmat.2004.08.009

Mollah MYA, Schennach R, Parga JR, Cocke DL (2001) Electrocoagulation (EC) – science and applications. J Hazard Mater B84:29–41. https://doi.org/10.1016/S0304-3894(01)00176-5

Monasmith R, Myhre G, Geer J, Moe R, Vasquez BC, Allred S, Cyr M, Wooldridge E, Swenson H, Capoccia S, Centurion DP, Young C, Zodrow KR (2020) Photothermal floats for evaporation enhancement and waterfowl deterrence. Mine Water Environ 39(4):716–723. https://doi.org/10.1007/s10230-020-00729-z

Monasmith R, Zodrow KR (2020) Proper adhesive choice increases photothermal float durability in mine water disposal applications. Mine Water Environ 39(4):724–734. https://doi.org/10.1007/s10230-020-00730-6

Monzyk BF, Wang M, Usinowicz PJ, Conkle HN, Fahnestock FMV, Beers TJ (2010) Prevention of acid mine drainage by avoiding acid formation through product recovery with F-LLX™. In: Proceedings hydroprocess conference, Santiago, pp 1–10

More KS, Wolkersdorfer C (2021) Application of artificial intelligence systems in mine water management – an introduction to two effective predictive models. Paper presented at the Mine Water Management for future generations, Wales, pp 365–367

More KS, Wolkersdorfer C, Kang N, Elmaghraby AS (2020) Automated measurement systems in mine water management and mine workings – a review of potential methods. Water Resour Ind:100136. https://doi.org/10.1016/j.wri.2020.100136

Morin KA, Hutt NM (2006) Case studies of costs and longevities of Alkali-based water-treatment plants for ARD. In: Proceedings 7th International Conference on Acid Rock Drainage (ICARD), St. Louis, pp 1333–1344

Mossad M, Zou L (2011) Effects of operational conditions on the electrosorption efficiencies of capacitive deionization. In: Proceedings Chemeca 2011 engineering a better world, Sydney, pp 1–11

Motyka I, Skibinski L (1982) Utilization of salt mine waters in the means of environmental protection. In: Proceedings, 1st International Mine Water Congress, Budapest, Hungary D, pp 37–47

Mtombeni T, Maree JP, Zvinowanda CM, Asante JKO, Oosthuizen FS, Louw WJ (2013) Evaluation of the performance of a new freeze desalination technology. Int J Environ Sci Technol 10(3):545–550. https://doi.org/10.1007/s13762-013-0182-7

Munro LD, Clark MW, McConchie D (2004) A Bauxsol™-based permeable reactive barrier for the treatment of acid rock drainage. Mine Water Environ 23(4):183–194. https://doi.org/10.1007/s10230-004-0061-z

Muraviev D, Gonzalo A, Valiente M (1995) Ion-exchange on resins with temperature-responsive selectivity. 1. Ion-exchange equilibrium of $Cu^{2+}$ and $Zn^{2+}$ on iminodiacetic and aminomethylphosphonic resins. Anal Chem 67(17):3028–3035. https://doi.org/10.1021/ac00113a043

Murdock DJ, Fox JRW, Bensley JG (1995) Treatment of acid mine drainage by the high density sludge process. In: Proceedings Sudbury '95 – mining and the environment, Sudbury, pp 431–439

Musche F, Paufler S, Grischek T, Uhlmann W (2016) Detection of iron-rich groundwater "hot spots" entering streams in Lusatia. In: Drebenstedt C, Paul M (eds) IMWA 2016 – mining meets water – conflicts and solutions. TU Bergakademie Freiberg, Freiberg, pp 616–623

Muste M, Vermeyen T, Hotchkiss R, Oberg K (2007) Acoustic velocimetry for riverine environments. J Hydraul Eng 133(12):1297–1298. https://doi.org/10.1061/(ASCE)0733-9429(2007)133:12(1297)

Musyoka NM, Petrik LF, Balfour G, Misheer N, Gitari W, Mabovu B (2009) Removal of toxic elements from brine using zeolite Na-p1 made from a South African coal fly ash. In: Water Institute of Southern Africa, International Mine Water Association (eds) Proceedings, International mine water conference. Document Transformation Technologies, Pretoria, pp 680–687

Naftz D, Morrison SJ, Davis JA, Fuller CC (2002) Handbook of groundwater remediation using permeable reactive barriers – applications to radionuclides, trace metals, and nutrients. Academic Press, Amsterdam, p 539

Naik TR, Ramme BW, Kolbeck HJ (1990) Filling abandoned underground facilities with CLSM Fly Ash Slurry. Concr Int Des Constr 12(7):19–25

Nancucheo I, Johnson DB (2012) Selective removal of transition metals from acidic mine waters by novel consortia of acidophilic sulfidogenic bacteria. Microb Biotechnol 5(1):34–44. https://doi.org/10.1111/j.1751-7915.2011.00285.x

Nariyan E, Sillanpää M, Wolkersdorfer C (2017) Electrocoagulation treatment of mine water from the deepest working European metal mine – performance, isotherm and kinetic studies. Sep Purif Technol 177:363–373. https://doi.org/10.1016/j.seppur.2016.12.042

Nathoo J, Jivanji R, Lewis AE (2009) Freezing your brines off: eutectic freeze crystallization for brine treatment. In: Water Institute of Southern Africa, International Mine Water Association (eds) Proceedings, International mine water conference. Document Transformation Technologies, Pretoria, pp 431–438

National Coal Board – Mining Department (1982) Technical management of water in the coal mining industry. National Coal Board, London, p 129

Naumann HE, Wiram VP (1995) Alkaline additions to the backfill – a key mining/reclamation component to acid mine drainage prevention. In: Proceedings, West Virginia surface mine drainage task force symposium 16 (Electronic document)

Neef T (2004) Hydrogeochemische Verhältnisse in einem natürlichen Feuchtgebiet zur passiven Grubenwasserreinigung. Unpubl. Dipl.-Arb. TU Bergakademie Freiberg, Freiberg, 141 p.

Neitzel PL, Schneider P, Hurst S (2000) Feldversuche zur in-situ Entfernung von $Uran_{(NAT.)}$ und Ra-226 aus Berge- und Flutungswässern. Freiberger Forsch-H C 482:196–206

Neumann C (1999) Zur Pedogenese pyrit- und kohlehaltiger Kippsubstrate im Lausitzer Braunkohlerevier. Cottbuser Schriften zu Bodenschutz und Rekultivierung 8:225

Neumann V, Nitsche C, Tienz B-S, Pokrandt K-H (2007) Erstmalige Neutralisation eines großen Tagebausees durch In-Lake-Verfahren – Erste Erfahrungen zu Beginn der Nachsorgephase. Wiss Mitt Inst Geol 35:117–124

Nielsen G, Janin A, Coudert L, Blais JF, Mercier G (2018) Performance of sulfate-reducing passive bioreactors for the removal of Cd and Zn from mine drainage in a cold climate. Mine Water Environ 37(1):42–55. https://doi.org/10.1007/s10230-017-0465-1

Nordstrom DK (2011) Mine waters: acidic to circumneutral. Elements 7(6):393–398. https://doi.org/10.2113/gselements.7.6.393

Nordstrom DK, Alpers CN (1995) Remedial investigation, decisions and geochemical consequences at Iron Mountain Mine, California. In: Proceedings Sudbury '95 – mining and the environment, II, pp 633–642

Nordstrom DK, Alpers CN, Ptacek CJ, Blowes DW (2000) Negative pH and extremely acidic mine waters from Iron Mountain, California. Environ Sci Technol 34:254–258. https://doi.org/10.1021/es990646v

Nordstrom DK, Bowell RJ, Campbell KM, Alpers CN (2017a) Challenges in recovering resources from acid mine drainage. In: Wolkersdorfer C, Sartz L, Sillanpää M, Häkkinen A (eds) IMWA 2017 – mine water & circular economy. II. Lappeenranta University of Technology, Lappeenranta, pp 1138–1146

Nordstrom DK, McCleskey RB, Ball JW (2009) Sulfur geochemistry of hydrothermal waters in Yellowstone National Park: IV Acid–sulfate waters. Appl Geochem 24(2):191–207. https://doi.org/10.1016/j.apgeochem.2008.11.019

Nordstrom DK, McCleskey RB, Ball JW (2010) Challenges in the analysis and interpretation of acidic waters. In: Wolkersdorfer C, Freund A (eds) Mine water and innovative thinking – International Mine Water Association symposium. Cape Breton University Press, Sydney, pp 379–383

Nordstrom DK, Munoz JL (1994) Geochemical thermodynamics, 2nd edn. Blackwell, Oxford

Nordstrom DK, Nicholson A, Weinig W, Mayer U, Maest A (2017b) Geochemical modeling for mine site characterization and remediation. In: Management technologies for metal mining influenced water, vol 4. SME, Englewood, p 159

Norton PJ (1992) The control of acid mine drainage with Wetlands. Mine Water Environ 11(3):27–34. https://doi.org/10.1007/BF02914814

Norton PJ (1995) Mine closure and its effects on groundwater and the environment related to Uranium Mining. In: Proceedings, Uranium-mining and hydrogeology, Freiberg, Germany, GeoCongress, vol 1, pp 415–421

Noubactep C (2010) The fundamental mechanism of aqueous contaminant removal by metallic iron. Water SA 36(5):663–670

Noubactep C, Meinrath G, Volke P, Peter HJ, Dietrich P, Merkel BJ (2002) Mechanism of uranium fixation by zero valent iron: the importance of co-precipitation. In: Merkel BJ, Planer-Friedrich B, Wolkersdorfer C (eds) Uranium in the aquatic environment. Springer, Heidelberg, pp 581–590. https://doi.org/10.1007/978-3-642-55668-5_68

Novak P (1994) Improvement of water quality in rivers by aeration at hydraulic structures. In: Hino M (ed) Water quality and its control. Balkema, Rotterdam, pp 147–168

Ødegaard H (2004) Sludge minimization technologies – an overview. Water Sci Technol 49(10):31–40. https://doi.org/10.2166/wst.2004.0602

Ofner C, Wieber G (2008) Geothermische Potenziale gefluteter Bergwerke. bbr Jahresmag 59(12):72–77

Oh C, Ji S, Cheong Y, Yim G, Hong JH (2016) Evaluation of design factors for a cascade aerator to enhance the efficiency of an oxidation pond for ferruginous mine drainage. Environ Technol 37(19):2483–2493. https://doi.org/10.1080/09593330.2016.1153154

Oleksiienko O, Wolkersdorfer C, Sillanpää M (2017) Titanosilicates in cation adsorption and cation exchange – a review. Chem Eng J 317:570–585. https://doi.org/10.1016/j.cej.2017.02.079

Olías M, Nieto JM, Sarmiento AM, Cerón JC, Cánovas CR (2004) Seasonal water quality variations in a river affected by acid mine drainage: the Odiel River (South West Spain). Sci Total Environ 333(1–3):267–281. https://doi.org/10.1016/j.scitotenv.2004.05.012

Olias M, Ruiz Cánovas C, Macias F, Nieto JM (2017) Water resources degradation by acid mine drainage: the Sancho Reservoir (Odiel River Basin, SW Spain). In: Proceedings IMWA 2017 – mine water & circular economy, Lappeenranta, III, pp 1389–1396

Olyphant GA, Harper D (1998) Hydrologic conditions in the coal mining district of Indiana and implications for reclamation of abandoned mine lands. In: Proceedings 15th annual national meeting – American Society for Surface Mining and Reclamation, Princeton, pp 283–288. https://doi.org/10.21000/JASMR98010283

Opara A, Peoples MJ, Adams DJ, Martin AJ (2014) Electro-biochemical reactor (EBR) technology for selenium removal from British Columbia's coal-mining wastewaters. Miner Metall Proc 31(4):209–214

Opitz J, Alte M, Bauer M, Peiffer S (2019) Testing iron removal in a trifurcated pilot plant for passive treatment of circum-neutral ferruginous mine water. In: Proceedings Mine Water – Technological and Ecological Challenges (IMWA 2019), Perm, Russia, pp 56–261

Opitz J, Alte M, Bauer M, Peiffer S (2021) The role of macrophytes in constructed surface-flow wetlands for mine water treatment: a review. Mine Water Environ 40(3):587–605. https://doi.org/10.1007/s10230-021-00779-x

Opitz J, Alte M, Bauer M, Schäfer W, Söll T (2017) Investigation of a pit lake acting as a large-scale natural treatment system for diffuse acid mine drainage. In: Wolkersdorfer C, Sartz L, Sillanpää M, Häkkinen A (eds) IMWA 2017 – mine water & circular economy. II. Lappeenranta University of Technology, Lappeenranta, pp 1095–1102

Opitz J, Alte M, Bauer M, Schäfer W, Söll T (2020) Estimation of self-neutralisation rates in a lignite pit lake. Mine Water Environ 39(3):556–571. https://doi.org/10.1007/s10230-020-00692-9

Orescanin V, Kollar R (2012) A combined CaO/electrochemical treatment of the acid mine drainage from the "Robule" Lake. J Environ Sci Health A Tox Hazard Subst Environ Eng 47(8):1186–1191. https://doi.org/10.1080/10934529.2012.668405

Oshinsky DM (2005) Polio – an American story. Oxford University Press, Oxford, p 342

Ott AN (1988) Dual-acidity titration curves – fingerprint, indicator of redox state, and estimator of iron and aluminum content of acid mine drainage and related waters. US Geological Survey Water-Supply Paper W 2330:19–33

Oyewo O, Onyango M, Wolkersdorfer C (2019) Synthesis and application of alginate immobilized banana peels nanocomposite in rare earth and radioactive minerals removal from mine water. IET Nanobiotechnol 13(7):756–765. https://doi.org/10.1049/iet-nbt.2018.5399

Özcan H, Ekinci H, Baba A, Kavdir Y, Yüksel O, Yigini Y (2007) Assessment of the water quality of Troia for the multipurpose usages. Environ Monit Assess 130:389–402. https://doi.org/10.1007/s10661-006-9406-3

Paikaray S (2021) Environmental stability of schwertmannite: a review. Mine Water Environ 40(3):570–586. https://doi.org/10.1007/s10230-020-00734-2

Panizza M (2010) Importance of electrode material in the electrochemical treatment ofwastewater containing organic pollutants. In: Comninellis C, Chen G (eds) Electrochemistry for the environment. Springer, Heidelberg, pp 25–54. https://doi.org/10.1007/978-0-387-68318-8_2

Paños NH (1999) Bio-deshydrometallurgy. In: Fernández Rubio R (ed) Mine, water & environment. I. International Mine Water Association, Sevilla, pp 409–415

Paños NH, Bellini MR (1999) Microbial degradation of cyanides. In: Fernández Rubio R (ed) Mine, water & environment. I. International Mine Water Association, Sevilla, pp 201–206

Park BT, Wangerud KW, Fundingsland SD, Adzic ME, Lewis NM (2006) In situ chemical and biological treatment leading to successful water discharge from anchor hill pit lake, gilt edge mine superfund site, South Dakota, U.S.A. In: Proceedings 7th International conference on acid rock drainage (ICARD), St. Louis, pp 1065–1069. https://doi.org/10.21000/JASMR06021065

Parkhurst DL, Appelo CAJ (2013) Description of input and examples for PHREEQC Version 3 – a computer program for speciation, batch-reaction, one-dimensional transport, and inverse geochemical calculations. US Geol Surv Tech Methods 6(A43):1–497

Parry WT, Forster CB, Solomon DK, James LP (2000) Ownership of mine-tunnel discharge. Ground Water 38(4):487–496. https://doi.org/10.1111/j.1745-6584.2000.tb00240.x

Pasche E (2013) Nutzwärme aus dem Bergbau hat Zukunft. VDI Nachrichten, 15 März 2013, 8 p.

Paul M, Metschies T, Frenzel M, Meyer J (2012) The mean hydraulic residence time and its use for assessing the longevity of mine water pollution from flooded underground mines. In: Merkel B, Schipek M (eds) The new uranium mining boom, Springer geology. Springer, Heidelberg, pp 689–699. https://doi.org/10.1007/978-3-642-22122-4_79

Paul M, Meyer J, Jenk U, Baacke D, Schramm A, Metschies T (2013) Mine flooding and water management at uranium underground mines two decades after decommissioning. In: Brown A, Figueroa L, Wolkersdorfer C (eds) Reliable mine water technology. International Mine Water Association, Golden, pp 1081–1087

Peters T (2010) Membrane technology for water treatment. Chem Eng Technol 33(8):1233–1240. https://doi.org/10.1002/ceat.201000139

Petritz KM, Gammons CH, Nordwick S (2009) Evaluation of the potential for beneficial use of contaminated water in a flooded mine shaft in Butte, Montana. Mine Water Environ 28(4):264–273. https://doi.org/10.1007/s10230-009-0083-7

Pflug W (1998) Braunkohlentagebau und Rekultivierung – Landschaftsökologie – Folgenutzung – Naturschutz. Springer, Berlin, 1068 p. https://doi.org/10.1007/978-3-642-58846-4

Phieler R, Merten D, Roth M, Büchel G, Kothe E (2015) Phytoremediation using microbially mediated metal accumulation in *Sorghum bicolor*. Environ Sci Pollut Res 22(24):19408–19416. https://doi.org/10.1007/s11356-015-4471-1

Pinto R, Oliveira Z, Diamantino C, Carvalho E (2016) Passive treatment of radioactive mine water in Urgeiriça uranium mine, Portugal. In: Drebenstedt C, Paul M (eds) IMWA 2016 – mining meets water – conflicts and solutions. TU Bergakademie Freiberg, Freiberg, pp 881–888

Piper AM (1944) A graphic procedure in the geochemical interpretation of water analyses. Trans Am Geophys Union 25:914–923. https://doi.org/10.1029/TR025i006p00914

Piper AM (1953) A graphic procedure in the geochemical interpretation of water analyses. US Geol Surv Ground Water Notes Geochem 12:1–14. https://doi.org/10.1029/TR025i006p00914

PIRAMID Consortium (2003) Engineering guidelines for the passive remediation of acidic and/or metalliferous mine drainage and similar wastewaters – “PIRAMID Guidelines”. University of Newcastle Upon Tyne, Newcastle Upon Tyne, p 166

Platner SB (1929) *Cloaca Maxima*. In: A topographical dictionary of Ancient Rome. Oxford University Press, Oxford, pp 126–127

Plaza-Cazón J, González E, Donati ER (2021) Parastrephia quadrangularis: a possible alternative to inhibit the microbial effect on the generation of acid mine drainage. Mine Water Environ 40(4):994–1002. https://doi.org/10.1007/s10230-021-00830-x

Plotnikov NI, Roginets II, Viswanathan S (1989) Hydrogeology of ore deposits. Russian translations series, Bd 72. Balkema, Rotterdam, p 290

Plumlee GS, Logsdon MJ (1999) The environmental geochemistry of mineral deposits, 6A-B. Society of Economic Geologists, Littleton, pp 1–583. https://doi.org/10.5382/Rev.06

Poinapen J (2012) Biological sulfate reduction of acid mine drainage using primary sewage sludge. In: International Mine Water Association symposium. Edith Cowan University, Bunbury, pp 237–244

Poirier PJ, Roy M (1997) Acid mine drainage characterization and treatment at La Mine Doyon. In: Proceedings of fourth international conference on acid rock drainage 4, pp 1485–1497

Pourbaix M (1966) Atlas of electrochemical equilibria in aqueous solutions. Pergamon, Oxford, p 644

Pourbaix M (1973) Lectures on electrochemical corrosion. Plenum, New York, p 336. https://doi.org/10.1007/978-1-4684-1806-4

Pouw K, Campbell K, Babel L (2015) Best Available Technologies Economically Achievable (BATEA) to manage effluent from mines in Canada. In: Brown A et al. (eds) Agreeing on solutions for more sustainable mine water management. Gecamin, Santiago, pp. 1–11 (Electronic document)

Prasad B, Sangita K, Tewary BK (2011) Reducing the hardness of mine water using transformed fly ash. Mine Water Environ 30(1):61–66. https://doi.org/10.1007/s10230-010-0130-4

Preuß V, Koch T, Schöpke R, Koch R, Rolland W (2007) Weitergehende Grubenwasserreinigung – Sulfatentfernung mittels Nanofiltration. Wiss Mitt Inst Geol 35:19–25

Preuß V, Riedel C, Koch T, Thürmer K, Domańska M (2012) Nanofiltration as an effective tool of reducing sulfate concentration in mine water. Archit Civil Eng Environ ACEE 3:127–132

Preuß V, Schöpke R, Koch T (2010) Reduction of sulfate load by nanofiltration – process development in bench scale. In: Wolkersdorfer C, Freund A (eds) Mine water and innovative think-

ing – International Mine Water Association symposium. Cape Breton University Press, Sydney, pp 171–175
Price KR (1997) Wetland design and construction beyond the concept – practicalities of implementation. In: Younger PL (ed) Minewater treatment using Wetlands. Chartered Institution of Water and Environmental Management, London, pp 133–138
Přikryl P, Černý R, Havlík V, Segeth K, Stupka P, Toman J (1999) Deposition of wastewater into deep mines. Environmetrics 10(4):457–466. https://shortdoi.org
Prinz H, Strauß R (2018) Ingenieurgeologie, 6th edn. Springer, Heidelberg, 898 p. https://doi.org/10.1007/978-3-662-54710-6
Pulles W, Heath R (2009) The evolution of passive mine water treatment technology for sulfate removal. In: Water Institute of Southern Africa, International Mine Water Association (eds) Proceedings, International mine water conference. Document Transformation Technologies, Pretoria, pp 2–14
Pulles W, Juby GJG, Busby RW (1992) Development of the Slurry Precipitation and Recycle Reverse-Osmosis (SPARRO) technology for desalinating scaling mine waters. Water Sci Technol 25(10):177–192. https://doi.org/10.2166/wst.1992.0246
Punkkinen H, Räsänen L, Mroueh U-M, Korkealaakso J, Luoma S, Kaipainen T, Backnäs S, Turunen K, Hentinen K, Pasanen A, Kauppi S, Vehviläinen B, Krogerus K (2016) Guidelines for mine water management. VTT Technology 266:1–157
Punz W (2014) "Erbe des Bergbaus" – zur Vegetation ostalpiner Schwermetallstandorte. GeoAlp 11:239–249
Punz W, Mucina L (1997) Vegetation on anthropogenic metalliferous soils in the Eastern Alps. Folia Geobot Phytotaxon 32(3):283–295. https://doi.org/10.1007/BF02804008
Pust C, Schüppel B, Kwasny J (2011) Advanced treatment of waters affected by AMD. In: Mine Water – managing the challenges – 11th International Mine Water Association Congress. RWTH Aachen University, Aachen, pp 437–440
Pust C, Schüppel B, Merkel BJ, Schipek M, Lilja G, Rabe W, Scholz G (2010) Advanced Mobile Inlake Technology (AMIT) – an efficient process for neutralisation of acid open pit lakes. In: Wolkersdorfer C, Freund A (eds) Mine water and innovative thinking – International Mine Water Association symposium. Cape Breton University Press, Sydney, pp 175–179
Puura E, D'Alessandro M (2005) A classification system for environmental pressures related to mine water discharges. Mine Water Environ 24(1):43–52. https://doi.org/10.1007/s10230-005-0070-6
Raff O, Wilken R-D (1999) Uranium removal from water by nanofiltration. In: Fernández Rubio R (ed) Mine, water & environment. I. International Mine Water Association, Sevilla, pp 321–324
Rajeshwar K, Ibanez J (1997) Environmental electrochemistry – fundamentals and applications in pollution sensors and abatement. Academic Press, Amsterdam, p 776
Randall D, Nathoo J, Lewis AE (2009) Seeding for selective salt recovery during eutectic freeze crystallization. In: Water Institute of Southern Africa, International Mine Water Association (eds) Proceedings, international mine water conference. Document Transformation Technologies, Pretoria, pp 639–648
Ranville JF, Schmiermund RL (1999) General aspects of aquatic colloids in environmental geochemistry. In: The environmental geochemistry of mineral deposits, 6A. Society of Economic Geologists, Littleton, pp 183–199. https://doi.org/10.5382/Rev.06.08
Rauner M (2011) Tief im Osten – Der Ex-Direktor eines DDR-Bergwerks sucht im Erzgebirge wieder nach Erz – und versetzt Sachsen in einen Rohstoffrausch. Die Zeit, 19 März 2011
Rawlings DE (2005) Characteristics and adaptability of iron- and sulfur-oxidizing microorganisms used for the recovery of metals from minerals and their concentrates. Microbial Cell Factories 4(13):1–15

Rawlings DE, Tributsch H, Hansford GS (1999) Reasons why '*Leptospirillum*'-like species rather than *Thiobacillus ferrooxidans* are the dominant iron-oxidizing bacteria in many commercial processes for the biooxidation of pyrite and related ores. Nat Rev Microbiol 145:5–13

Razowska-Jaworek L, Pluta I, Chmura A (2008) Mine waters and their usage in the Upper Silesia in Poland – examples from selected regions. In: Proceedings, 10th International Mine Water Association Congress, pp 537–539

Reddy ST, Kramer HJM, Lewis AE, Nathoo J (2009) Investigating factors that affect separation in a eutectic freeze crystallisation process. In: Water Institute of Southern Africa, International Mine Water Association (eds) Proceedings, International mine water conference. Document Transformation Technologies, Pretoria, pp 649–656

Reddy ST, Lewis AE (2010) Waste minimisation through recovery of salt and water from a hypersaline brine. In: Wolkersdorfer C, Freund A (eds) Mine water and innovative thinking – International Mine Water Association symposium. Cape Breton University Press, Sydney, pp 179–183

Reddy ST, Lewis AE, Witkamp GJ, Kramer HJM, van Spronsen J (2010) Recovery of $Na_2SO_4 \cdot 10H_2O$ from a reverse osmosis retentate by eutectic freeze crystallisation technology. Chem Eng Res Des 88(9):1153–1157. https://doi.org/10.1016/j.cherd.2010.01.010

Rees SB, Bowell RJ, Wiseman IM (2002) Influence of mine hydrogeology on minewater discharge chemistry. Spec Publ Geol Soc Lond 198(379–390). https://doi.org/10.1144/GSL.SP.2002.198.01.26

Rees SB, Bright P, Connelly R, Bowell RJ, Szabo E (2004) Application of the welsh mine water strategy: Cwmrheidol case study. In: Jarvis AP, Dudgeon BA, Younger PL (eds) Mine water 2004 – proceedings International Mine Water Association symposium. 1. University of Newcastle, Newcastle upon Tyne, pp 235–244

Rees SB, Connelly R (2003) Review of design and performance of the Pelenna Wetland systems. Land Contam Reclam 11(2):293–300. https://doi.org/10.2462/09670513.828

Regenspurg S, Meima JA, Kassahun A, Rammlmair D (2005) Krustenbildung in Bergbauhalden. Wiss Mitt Inst Geol 28:63–67

Rennie R, Law J (2016) Oxford dictionary of chemistry, 7th edn. Oxford University Press, Oxford, p 608. https://doi.org/10.1093/acref/9780198722823.001.0001

Renton JJ, Rymer T, Stiller AH (1988) A computer simulation probability model for geochemical parameters associated with coal mining operations. Bur Mines Inf Circ IC-9183:76–82:10.21000/JASMR88010076

Ridpath I (2012) A dictionary of astronomy, 2nd edn. Oxford University Press, Oxford, p 544. https://doi.org/10.1093/acref/9780199609055.001.0001

Rieger A, Steinberger P, Pelz W, Haseneder R, Härtel G (2010) Nanofiltration of acid mine drainage. Desalin Water Treat 21(1–3):148–161. https://doi.org/10.5004/dwt.2010.1316

Roadcap GS, Kelly WR, Bethke CM (2005) Geochemistry of extremely alkaline (pH > 12) ground water in slag-fill aquifers. Ground Water 43(6):806–816. https://doi.org/10.1111/j.1745-6584.2005.00060.x

Rodebaugh D (2012) Shock the mine waste – pilot project treats toxic metals flowing into Animas River. The Durango Herald, 24 Sep 2012

Rodriguez J, Stopić S, Friedrich B (2007) Continuous electrocoagulation treatment of wastewater from copper production. World Metall Erzmetall 60(2):81–87. https://doi.org/10.13140/RG.2.2.15988.86407

Roehl KE (2004) Passive in situ remediation of contaminated groundwater: Permeable Reactive Barriers – PRBS. In: Prokop G, Younger PL, Roehl KE (eds) Conference papers. 35. Umweltbundesamt, Wien, pp 57–70

Roenicke H, Schultze M, Neumann V, Nitsche C, Tittel J (2010) Changes of the plankton community composition during chemical neutralisation of the Bockwitz pit lake. Limnologica 40(2):191–198. https://doi.org/10.1016/j.limno.2009.11.005

Rohwerder T, Gehrke T, Kinzler K, Sand W (2003) Bioleaching review part A – progress in bioleaching – fundamentals and mechanisms of bacterial metal sulfide oxidation. Appl Microbiol Biotechnol 63:239–248. https://doi.org/10.1007/s00253-003-1448-7

Römpp online (2013) Römpp online – Der effizientere Zugriff auf das Wissen der Chemie. Thieme, Stuttgart

Roscher M (2021) Die Projekte der Saxore Bergbau GmbH – Smart and Green Mining – Bergbautradition und bekannte Erzlagerstätten neu gedacht. bergbau 9:412–414

Roschlau H, Heintze W (1975) Bergbautechnologie, 1st edn. VEB Deutscher Verlag für Grundstoffindustrie, Leipzig, 349 p.

Rose AW, Bisko D, Daniel A, Bower MA, Heckman S (2004a) An "autopsy" of the failed Tangaskootack #1 Vertical Flow Pond, Clinton CO., Pennsylvania. In: Proceedings 21st annual national meeting – American Society for Surface Mining and Reclamation, Lexington, pp 1580–1594. https://doi.org/10.21000/JASMR04011580

Rose AW, Dietz JM (2002) Case studies of passive treatment systems – vertical flow systems. In: Proceedings 19th annual national meeting – American Society for Surface Mining and Reclamation, Lexington, pp 776–797. https://doi.org/10.21000/JASMR02010776

Rose P, Corbett C, Neba A, Whittington-Jones K (2004b) Sewage sludge as an electron donor in biological mine wastewater treatment: development of the rhodes BioSURE Process®. In: Jarvis AP, Dudgeon BA, Younger PL (eds) Mine water 2004 – proceedings International Mine Water Association symposium. 2. University of Newcastle, Newcastle upon Tyne, pp 111–118

Roseneck R (1993) Der Harz als historische Kulturlandschaft. ICOMOS – Hefte des Deutschen Nationalkomitees 11:54–61. https://doi.org/10.11588/ih.1993.0.22592

Rosner P, Heitfeld M, Pabsch T, Denneborg M (2012) Abgrenzung signifikanter Belastungsquellen des Erzbergbaus in Nordrhein-Westfalen im Rahmen der EG-Wasserrahmenrichtlinie – Methodik und Ergebnisse 12. Altbergbau-Kolloquium. Glückauf, Goslar, pp 289–304

Rottmann D (1969) Untersuchungen zu Wasseraufkommen, Wasserverwendung und Wasserableitung der Betriebe des Steinkohlenbergbaus in der Bundesrepublik Deutschland, aufgeteilt nach Revieren, sowie zur Beschaffenheit der gehobenen Grubenwässer im Hinblick auf ihre mögliche Nutzung. Unpubl. Diss. RWTH Aachen, Aachen, 209 p.

Rowe OF, Johnson DB (2009) Enhanced rates of iron oxidation in mine waters by the novel acidophilic bacterium "*Ferrovum myxofaciens*" immobilized in packed-bed bioreactors. Skellefteå, pp 1–4 (Electronic document)

Ruhrkohle AG (2014) Konzept zur langfristigen Optimierung der Grubenwasserhaltung der RAG Aktiengesellschaft für das Saarland. Ruhrkohle AG, Herne, p 17

Rychkov VN, Kirillov EV, Kirillov SV, Bunkov GM, Mashkovtsev MA, Botalov MS, Semenishchev VS, Volkovich VA (2016) Selective ion exchange recovery of rare earth elements from uranium mining solutions. AIP Conf Proc 1767(1):020017. https://doi.org/10.1063/1.4962601

S. R. I. Consulting Business Intelligence (2008) Disruptive civil technologies – six technologies with potential impacts on US interests out to 2025, CR 2008-07. National Intelligence Council, Washington, p 34

Sächsisches Landesamt für Umwelt und Geologie (1997) Merkblätter zur Grundwasserbeobachtung, Grundwasserprobenahme. Mater Wasserwirtsch L II-1/6(4):1–20

Sächsisches Landesamt für Umwelt, Landwirtschaft und Geologie (2003) Handbuch Grundwasserbeobachtung – Grundwasserprobenahme. Mater Wasserwirtsch 5:68

Şahinkaya E, Yurtsever A, İşler E, Coban I, Aktaş Ö (2018) Sulfate reduction and filtration performances of an anaerobic membrane bioreactor (AnMBR). Chem Eng J 2:47–55. https://doi.org/10.1016/j.cej.2018.05.001

Sahoo PK, Tripathy S, Panigrahi MK, Equeenuddin SM (2012) Mineralogy of Fe-precipitates and their role in metal retention from an acid mine drainage site in India. Mine Water Environ 31(4):344–352. https://doi.org/10.1007/s10230-012-0203-7

Salkield LU (1987) A technical history of the Rio Tinto mines – some notes on exploitation from pre-Phoenician times to the 1950s. Institution of Mining and Metallurgy, London, 114 p. https://doi.org/10.1007/978-94-009-3377-4

Samborska K, Sitek S, Bottrell SH, Sracek O (2013) Modified multi-phase stability diagrams: an AMD case study at a site in Northumberland, UK. Mine Water Environ 32(3):185–194. https://doi.org/10.1007/s10230-013-0223-y

Sand W, Gehrke T (2006) Extracellular polymeric substances mediate bioleaching/biocorrosion via interfacial processes involving iron(III) ions and acidophilic bacteria. Res Microbiol 157:49–56. https://doi.org/10.1016/j.resmic.2005.07.012

Sapsford D, Brabham P, Crane R, Evans D, Stratford A, Wright A (2017) Mine waste resource assessment and recovery via mine water. In: Wolkersdorfer C, Sartz L, Sillanpää M, Häkkinen A (eds) IMWA 2017 – mine water & circular economy. II. Lappeenranta University of Technology, Lappeenranta, pp 956–963

Sapsford DJ (2013) New perspectives on the passive treatment of ferruginous circumneutral mine waters in the UK. Environ Sci Pollut Res 20(11):7827–7836. https://doi.org/10.1007/s11356-013-1737-3

Sapsford DJ, Barnes A, Dey M, Liang L, Williams KP (2005) A novel method for passive treatment of mine water using a vertical flow accretion system. In: Loredo J, Pendás F (eds) Mine water 2005 – mine closure. University of Oviedo, Oviedo, pp 389–394

Sapsford DJ, Barnes A, Dey M, Williams K, Jarvis A, Younger P (2007) Low footprint passive mine water treatment: field demonstration and application. Mine Water Environ 26(4):243–250. https://doi.org/10.1007/s10230-007-0012-6

Sapsford DJ, Barnes A, Dey M, Williams KP, Jarvis A, Younger P, Liang L (2006) Iron and manganese removal in a vertical flow reactor for passive treatment of mine water. In: Proceedings ICARD 2006, St. Louis, vol. 7, pp 1831–1843 [CD-ROM]. https://doi.org/10.21000/JASMR06021831

Sapsford DJ, Williams KP (2009) Sizing criteria for a low footprint passive mine water treatment system. Water Res 43(2):423–432. https://doi.org/10.1016/j.watres.2008.10.043

Sarmiento A, Nieto JM, Olías M, Cánovas C (2005) Environmental impact of mining activities in the Odiel river basin (SW Spain). In: Mine water 2005 – mine closure. University of Oviedo, Oviedo, pp 89–94

Sartz L (2010) Alkaline by-products as amendments for remediation of historic mine sites. Örebro Universitet, Örebro, p 350

Schabronath CA (2018) A novel sampling technique using a sediment collector (SECO) to collect particles from suspended matter in mine water with regard to long term monitoring. In: Proceedings IMWA 2018 – risk to opportunity. Pretoria I, pp 193–198

Schäfer AI, Fane AG, Waite TD (2006) Nanofiltration – principles and applications, Nachdruck. Elsevier Advanced Technology, Amsterdam, p 560

Schäfer T, Schwarz MA (2019) The meaningfulness of effect sizes in psychological research: differences between sub-disciplines and the impact of potential biases. Front Psychol 10:813. https://doi.org/10.3389/fpsyg.2019.00813

Schetelig K, Richter H (2013) Nutzung stillgelegter Bergwerke oder tiefliegender Grundwasservorkommen zur Wärme-/Kältegewinnung und -speicherung. Wasser als Energieträger. In: Proceedings 43. Internationales Wasserbau-symposium, Aachen, pp 36–50

Scheuchenstuel Kv (1856) Idioticon der österreichischen Berg- und Hüttensprache – Zum besseren Verständnisse des österr. Berg-Gesetzes und dessen Motive für Nicht-Montanisten. Braumüller, Wien, 270 p.
Schindler W, Gerlach F, Kaden H (1995) pH-Sensor für Präzisionsmessungen in Tiefenwässern. Sens Mag 3:6–9
Schipek M (2011) Treatment of acid mine lakes – lab and field studies. Freiberg Online Geol 29:1–381
Schipek M, Graupner B, Merkel BJ, Wolkersdorfer C, Werner F (2006) Neutralisationspotential von Flugaschen – Restseesanierung Burghammer. Wiss Mitt Inst Geol 31:125–132
Schipek M, Merkel BJ (2012) Liming of acid surface waters: effect of water constituents and material impurities on calcite dissolution kinetics. In: International Mine Water Association symposium. Edith Cowan University, Bunbury, pp 275–286
Schipek M, Merkel BJ, Scholz G, Rabe W, Clauß D, Lilja G (2011) Recent results of the research project OILL (Optimizing In-Lake Liming). In: Mine water – managing the challenges – 11th International Mine Water Association Congress. RWTH Aachen University, Aachen, pp 457–461
Schipek M, Unger Y, Merkel BJ (2007) Alkalitätsverbessernde Maßnahmen in Tagebaufolgeseen: Nutzung von $CO_2$ und anderen industriellen "Abfall" produkten. Wiss Mitt Inst Geol 35:125–131
Schlegel A, Dennis R, Simms J (2005) Altes Eisen Schluckt Arsen. Wiss Mitt Inst Geol 28:69–73
Schmidt K, Senger D, Schwark D, Dalaly H, Elgin International Inc (1969) Sul-BiSul ion exchange process: field evaluation on brackish waters. Res Dev Prog Rep 446:139
Schneider P (2016) 1000 years of mining: what means geogenic background of metals in the rivers of the Harz Mountains? In: Drebenstedt C, Paul M (eds) IMWA 2016 – mining meets water – conflicts and solutions. TU Bergakademie Freiberg, Freiberg, pp 45–53
Schneider P, Neitzel PL, Hurst S (2000) In-situ treatment of mining wastewaters using Fe(0) and Fe/Mn-Compounds (abstract). J Conf Abs 5(2):892
Schneider P, Neitzel PL, Osenbrück K, Noubacteb C, Merkel BJ, Hurst S (2001) In-situ treatment of radioactive mine water using reactive materials – results of laboratory and field experiments in uranium ore mines in Germany. Acta Hydrochim Hydrobiol 29(2–3):129–138. 10/ff64fb
Schoeman JJ, Steyn A (2001) Investigation into alternative water treatment technologies for the treatment of underground mine water discharged by Grootvlei Proprietary Mines Ltd into the Blesbokspruit in South Africa. Desalination 133(1):13–30. https://doi.org/10.1016/S0011-9164(01)00079-0
Schönaich-Carolath in Tarnowitz ABHPv (1860) Beschreibung des Verfahrens zur Entsäuerung der für die Speisung der Dampfkessel auf der Königshütte in Oberschlesien bestimmten Grubenwasser. Z Berg-, Hütten- und Salinen-Wesen 8:28–31
Schöpke R (1999) Erarbeitung einer Methodik zur Beschreibung hydrochemischer Prozesse in Kippengrundwasserleitern. Schriftenr Siedl-Wasserwirtsch Umw, Bd 2. Eigenverlag Lehrstuhl Wassertechnik und Siedlungswasserbau, Cottbus, 136 p.
Schöpke R, Koch R, Ouerfelli I, Striemann A, Preuß V, Regel R (2001) Anwendung des Neutralisationspotenzials bei der Bilanzierung von Säure-Base-Reaktionen im Umfeld des Braunkohlebergbaues. Grundwasser 6(1):23–29. https://doi.org/10.1007/PL00010387
Schöpke R, Preuß V (2012) Bewertung der Acidität von bergbauversauerten Wässern und Anwendung auf die Sanierung. Grundwasser 17(3):147–156. https://doi.org/10.1007/s00767-012-0189-x
Schöpke R, Preuß V, Koch R, Bahl T (2006) Einsatzmöglichkeiten passiver reaktiver Wände zur Entsäuerung potenziell saurer Kippengrundwasserströme und deren Nutzung in aktiven Untergrundbehandlungsverfahren. Wiss Mitt Inst Geol 31:165–172
Schubert JP, McDaniel MJ (1982) Using mine waters for heating and cooling. In: Proceedings, 1st International Mine Water Congress, Budapest, Hungary D, pp 63–82

Schüring J, Schulz HD, Fischer WR, Böttcher J, Duijnisveld WHM (2000) Redox: fundamentals, processes and applications. Springer, Berlin, p 251. https://doi.org/10.1007/978-3-662-04080-5
Schwedt G (1995) Analytische Chemie – Grundlagen, Methoden und Praxis. Thieme, Stuttgart, 442 p.
Schwertmann U (1985) The effect of pedogenic environments on iron oxide minerals. Advances in soil science, Bd 1. Springer, Heidelberg, pp 171–200. https://doi.org/10.1007/978-1-4612-5046-3_5
Schwertmann U (1999) Giftfänger im Bergbauabraum – Neues Mineral: Schwertmannit. TUM-Mitteilungen der Technischen Universität München für Studierende, Mitarbeiter, Freunde 4, p 29
Schwertmann U, Bigham JM, Murad E (1995) The first occurrence of schwertmannite in a natural stream environment. Eur J Mineral 7(3):547–552. https://doi.org/10.1127/ejm/7/3/0547
Scofield CS, Wilcox LV (1931) Boron in irrigation waters. USDA Tech Bull 264:1–65
Scott RL, Hays RM (1975) Inactive and abandoned underground mines – water pollution prevention and control, EPA-440/9-75-007. U.S. Environmental Protection Agency, Washington, p. 338
Sebastian U (2013) Die Geologie des Erzgebirges. Springer, Heidelberg, 268 p. https://doi.org/10.1007/978-3-8274-2977-3
Seed LP, Yetman DD, Pargaru Y, Shelp GS (2006) The DesEL system – capacitive deionization for the removal of ions from water. In: Proceedings Water Environment Federation WEFTEC 2006, pp 7172–7180. https://doi.org/10.2175/193864706783761671
Seewoo S, Van Hille R, Lewis A (2004) Aspects of gypsum precipitation in scaling waters. Hydrometallurgy 75(1–4):135–146. https://doi.org/10.1016/j.hydromet.2004.07.003
Seidel K (1952) Pflanzungen zwischen Gewässern und Land. Mitt Max-Planck-Ges Förd Wiss 8:17–21
Seidel K (1966) Reinigung von Gewässern durch höhere Pflanzen. Naturwissenschaften 53(12):289–297. https://doi.org/10.1007/BF00712211
Seidler C (2012) Deutschlands verborgene Rohstoffe: Kupfer, Gold und Seltene Erden. Hanser, München, 252 p.
Selent KD, Grupe A (2018) Die Probenahme von Wasser – Ein Handbuch für die Praxis. Deutscher Industrieverlag, München, 488 p.
Seltenerden Storkwitz AG (2013) Seltenerden Storkwitz AG – Gutachten bestätigt Schätzungen der einzigen bekannten Seltenerden-Lagerstätte Mitteleuropa. Pressemitteilung, 31 Jan 2013, pp 1–4
Senes Consultants Limited (1994) Acid mine drainage – status of chemical treatment and sludge management practices. MEND Report, 3.32.1. The Mine Environment Neutral Drainage [MEND] Program, Richmond Hill, 179 p.
Sersch W, Uhl O (2006) Vorbereitende Maßnahmen zur Flutung der Grubenbaue im Rahmen der Stilllegung des Standorts Warndt/Luisenthal. Glückauf 142(6):254–261
Sharma SK (2001) Adsorptive iron removal from groundwater. Wageningen University, Lisse, p 202
Shepherd NL, Denholm CF, Dunn MH, Neely CA, Danehy TP, Nairn RW (2020) Biogeochemical analysis of spent media from a 15-year old passive treatment system vertical flow bioreactor. Mine Water Environ 39(1):68–74. https://doi.org/10.1007/s10230-020-00652-3
Shepherd R (1993) Ancient mining. Elsevier, London, p 494
Shiller AM (2003) Syringe filtration methods for examining dissolved and colloidal trace element distributions in remote field locations. Environ Sci Technol 37(17):3953–3957. https://doi.org/10.1021/es0341182
Shone RDC (1987) The freeze desalination of mine waters. J S Afr Inst Min Metall 87(4):107–112
Sierra Florez DC, Gaona LS (2021) Prospective analysis of artisanal mining in the Santa Rita Village: Municipality of Andes, Antioquia. Mine Water Environ 40(3):783–792. https://doi.org/10.1007/s10230-021-00778-y
Sikora J, Szyndler K (2005) Debiensko, Poland desalination plant treats drainage for Zero Liquid Discharge (ZLD). Technical paper, TP1039EN 0601. General Electric, Trevose, pp 1–5

Silva RA, Castro CD, Petter CO, Schneider IAH (2011) Production of iron pigments (Goethite and Haematite) from acid mine drainage. In: Mine water – managing the challenges – 11th International Mine Water Association Congress. RWTH Aachen University, Aachen, pp 469–473

Sim PG, Lewin JF (1975) Potentially toxic metals in New-Zealand coals. NZ J Sci 18(4):635–641

Simate GS, Ndlovu S (2014) Acid mine drainage: challenges and opportunities. J Environ Chem Eng 2(3):1785–1803. https://doi.org/10.1016/j.jece.2014.07.021

Simon A, Rascher J, Hoth N, Drebenstedt C (2017) Facial and paleogeografic understanding of tertiary sediments as basis to predict their specific AMD release. In: Wolkersdorfer C, Sartz L, Sillanpää M, Häkkinen A (eds) IMWA 2017 – mine water & circular economy. I. Lappeenranta University of Technology, Lappeenranta, pp 244–250

Sincero AP, Sincero GA (2002) Physical-chemical treatment of water and wastewater. CRC Press, Boca Raton, p 856. https://doi.org/10.1201/9781420031904

Singer PC, Stumm W (1969) Oxygenation of ferrous iron in mine drainage waters. Am Chem Soc Div Fuel Chem Prepr 13(2):80–87

Singer PC, Stumm W (1970) Acidic mine drainage – rate-determining step. Science 167(3921):1121–1123. https://doi.org/10.1126/science.167.3921.1121

Singh G, Rawat NS (1985) Removal of trace elements from acid mine drainage. Int J Mine Water 4(1):17–23. https://doi.org/10.1007/BF02505377

Siringi DO, Home P, Chacha JS, Koehn E (2012) Is Electrocoagulation (EC) a solution to the treatment of wastewater and providing clean water for daily use. J Eng Appl Sci 7(2):197–204

Sisler FD, Senftle FE, Skinner J (1977) Electrobiochemical neutralization of acid mine drainage. J Water Pollut Control Fed 49(3):369–374

Skelly and Loy, Penn Environmental Consultants (1973) Process, procedures, and methods to control pollution from mining activities, EPA-430/9-73-011. U.S. Environmental Protection Agency, Washington, p 390

Skinner SJW (2009) Groundwater clean-up: assessing mining impacts and rehabilitation/management alternatives. In: Water Institute of Southern Africa, International Mine Water Association (ed) Proceedings. International Mine Water Conference. Document Transformation Technologies, Pretoria, pp 713–722

Skousen J, Zipper CE, Rose A, Ziemkiewicz PF, Nairn R, McDonald LM, Kleinmann RL (2017) Review of passive systems for acid mine drainage treatment. Mine Water Environ 36(1):133–153. https://doi.org/10.1007/s10230-016-0417-1

Skousen JG, Demchak J, McDonald L (2002) Longevity of mine discharges from above-drainage underground mines. Reclamation 2002 – partnerships & project implementation. In Proceedings National Association of Abandoned Mine Land Programs, Park City, pp 1–15

Skousen JG, Lilly R, Hilton T (1988) Special chemicals for treating acid mine drainage. Green Lands 18(3):36–40

Skousen JG, Rose A, Geidel G, Foreman JW, Evans R, Hellier W (1998) Handbook of technologies for avoidance and remediation of acid mine drainage. The National Mine Land Reclamation Center, Morgantown, p 130

Skousen JG, Sexstone A, Ziemkiewicz PF (2000) Acid mine drainage control and treatment. In: Barnhisel RI, Darmody RG, Daniels WL (eds) Reclamation of drastically disturbed lands. American Society of Agronomy, Madison, pp 131–168

Skousen JG, Ziemkiewicz P (2005) Performance of 116 passive treatment systems for acid mine drainage. In: National Meeting of the American Society of Mining and Reclamation, Breckenridge, pp 1100–1130. https://doi.org/10.21000/JASMR05011100

Smetana R, Muzák J, Novák J (2002) Environmental impact of uranium ISL in Northern Bohemia. In: Merkel BJ, Planer-Friedrich B, Wolkersdorfer C (eds) Uranium in the aquatic environment. Springer, Heidelberg, pp 703–712. https://doi.org/10.1007/978-3-642-55668-5_82

Smit JP (1999) The treatment of polluted mine water. In: Fernández Rubio R (ed) Mine, water & environment. II. International Mine Water Association, Sevilla, pp 467–471
Smith GC, Steinman HE, Young EF Jr (1970) Clean water from coal mines. Min Eng 22:118–119
Smith KS (1999) Metal sorption on mineral surfaces – an overview with examples relating to mineral deposits. In: The environmental geochemistry of mineral deposits, 6A. Society of Economic Geologists, Littleton, pp 161–182. https://doi.org/10.5382/Rev.06.07
Smith MW, Brady KBC (1998) Alkaline addition. In: Coal mine drainage prediction and pollution prevention in Pennsylvania. Pennsylvanian Bureau of Mining and Reclamation, Harrisburg, pp 13.1–13.13. https://doi.org/10.5382/Rev.06.07
Society for Mining Metallurgy and Exploration (1998) Remediation of historical mine sites – technical summaries and bibliography. Society for Mining, Metallurgy, and Exploration, Littleton, p 118
Soldenhoff K, McCulloch J, Manis A, Macintosch P (2006) Nanofiltration in metal and acid recovery. In: Nanofiltration – principles and applications, Nachdruck. Elsevier Advanced Technology, Amsterdam, pp 459–477
Sørensen SPL (1909) Über die Messung und die Bedeutung der Wasserstoffionenkonzentration bei enzymatischen Prozessen. Biochem Z 21:131–200
Sparrow B, Man M, Zoshi J (2012) Novel salt extractor system for salt removal from mine water and calcium sulfate removal on the International Space Station. In: International Mine Water Association symposium. Edith Cowan University, Bunbury, pp 651–660
Spence C, McPhie M (1997) Streamflow measurement using salt dilution in tundra streams, Northwest Territories, Canada. J Am Water Resour Assoc 33(2):285–292. https://doi.org/10.1111/j.1752-1688.1997.tb03509.x
Spieth WF (2015) Verantwortung des Bergbaus für Stilllegung und Nachsorge – Neue Anforderungen nach Berg- und Wasserrecht als Herausforderung. In: Paul M (ed) Sanierte Bergbaustandorte im Spannungsfeld zwischen Nachsorge und Nachnutzung – WISSYM 2015. Wismut GmbH, Chemnitz, pp 45–52
Starr RC, Cherry JA (1994) In situ remediation of contaminated ground water: the funnel-and-gate system. Ground Water 32(3):465–476. https://doi.org/10.1111/j.1745-6584.1994.tb00664.x
Stemke M, Gökpinar T, Wohnlich S (2017) Erkundung stillgelegter Erzbergwerkschächte mittels Unterwasserkamera. Bergbau 11:511–513
Stengel-Rutkowski W (1993) Trinkwasserversorgung aus Grubengebäuden des ehemaligen Roteisensteinbergbaus im Lahn-Dill-Kreis (Rheinisches Schiefergebirge). Geol Jb Hessen 121:125–140
Stengel-Rutkowski W (2002) Trinkwasserversorgung aus Grubengebäuden des ehemaligen Bergbaus im Rheingau-Taunus-Kreis (Rheinisches Schiefergebirge). Jb Nass Ver Naturk 123:125–138
Šterbenk E, Pavšek Z, Petkovšek SAS, Mazej Grudnik Z, Kugonič Vrbič N, Poličnik H, Pokorny B, Ramšak R, Rošer Drev A, Mljač L, Bole M, Glinšek A, Mavec M, Druks P, Flis J, Kotnik K, Zorko V, Goltnik V, Vrhovšek D, Zupančič M, Urana D (2011) Šaleška jezera – vodni vir in razvojni izziv – Konèno poroèilo [Šales lake – Water source and development challenge – final report]. ERICo Velenje, Velenje, 246 p.
Šterbenk E, Ramšak R (1999) Pokrajinski Vidiki Rabe Premogovniškega Ugrezninskega Velenjskega Jezera [Regional aspects on the subsidence of Lake Velenje due to the use of coal]. Sonaravni razvoj v slovenskih Alpah in sosedstvu, Dela. Oddelek za geografijo Filozofske fakultete, Laibach, pp 215–223
Sterner JJ, Conahan HA (1968) Ion exchange treatment of acid mine drainage. Purdue Univ Eng Bull Ext Ser 53(2):101–110
Stevanović Z, Obradovic L, Marković R, Jonović R, Avramović L, Bugarin M, Stevanović J (2013) Mine wastewater management in the bor municipality in order to protect the bor river water.

In: Wastewater – treatment technologies and recent analytical developments. InTech, Shanghai, pp 41–62. https://doi.org/10.5772/51902

Stoddard CK (1973) Abatement of mine drainage pollution by underground precipitation, EPA-670/2-73-092. U. S. Environmental Protection Agency, Washington, p 125

Stolp W, Kiefer R (2009) Sulfatabtrennung aus schwefelsaurem Grubenwasser durch Elektrolyse. ThyssenKrupp Innovationswettbewerb 2009. ThyssenKrupp AG, Essen, pp 46–49

Stookey LL (1970) Ferrozine – a new spectrophotometric reagent for iron. Anal Chem 42(7):779–781. https://doi.org/10.1021/ac60289a016

Strathmann H (2012) Membranes and membrane separation processes, 1. Principles. In: Ullmann's encyclopedia of industrial chemistry. Wiley, Weinheim, pp 413–456. https://doi.org/10.1002/14356007.a16_187.pub3

Stropnik B, Tamše M, Ramšak R, Drev Rošer A, Stegnar P (1991) The effects of coal mining and energy production in the Šalek valley, Slovenia, on surface water bodies. In: Proceedings, 4th International Mine Water Association Congress, vol. 1, pp 213–221

Strosnider WHJ, Hugo J, Shepherd NL, Holzbauer-Schweitzer BK, Hervé-Fernández P, Wolkersdorfer C, Nairn RW (2020) A snapshot of coal mine drainage discharge limits for conductivity, sulfate, and manganese across the developed world. Mine Water Environ 39(2):165–172. https://doi.org/10.1007/s10230-020-00669-8

Strzodka M, Claus R, Preuss V, Thürmer K, Viertel K (2016) Advanced treatment of pit lakes using limestone and carbon dioxide. In: Drebenstedt C, Paul M (eds) IMWA 2016 – mining meets water – conflicts and solutions. TU Bergakademie Freiberg, Freiberg, pp 209–215

Stumm W, Lee GF (1961) Oxygenation of ferrous iron. Ind Eng Chem 53(2):143–146. https://doi.org/10.1021/ie50614a030

Stumm W, Morgan JJ (1996) Aquatic chemistry – chemical equilibria and rates in natural waters, 3rd edn. Wiley, New York, p 1022

Sumi L, Gestring B (2013) Polluting the future – how mining companies are contaminating our nation's waters in perpetuity. Earthworks, Washington, p 52

Sun X, Hwang J-Y (2012) Desalination of coal mine water with electrosorption. In: Water in mineral processing. SME, Littleton, pp 237–245

Svanks K, Shumate KS (1973) Factors controlling sludge density during acid mine drainage neutralization, report no. 392X – contract number A-022-OHIO. State of Ohio Water Resources Center – Ohio State University, Columbus, p 156

Systat Software I (2002) TableCurve 2D 5.01 for windows user's manual – automated curve fitting and equation discovery. SYSTAT Software/Marketing Department, Richmond, pp 1.1–13.14

Szilagyl G (1985) Rebuilding of original water balance in karstic reservoirs. In: Proceedings, 2nd International Mine Water Association Congress, vol. 2, pp 841–847

Tabak HH, Scharp R, Burckle J, Kawahara FK, Govind R (2003) Advances in biotreatment of acid mine drainage and biorecovery of metals: 1. Metal precipitation for recovery and recycle. Biodegradation 14(6):423–436. https://doi.org/10.1023/A:1027332902740

Takeno N (2005) Atlas of Eh-pH diagrams – intercomparison of thermodynamic databases. Geol Surv Jpn Open File Rep 419:1–285

Tarutis WJ, Stark LR, Williams FM (1999) Sizing and performance estimation of coal mine drainage wetlands. Ecol Eng 12(3–4):353–372. https://doi.org/10.1016/S0925-8574(98)00114-1

Taylor J, Waring C (2001) The passive prevention of ARD in underground mines by displacement of air with a reducing gas mixture: GaRDS. Mine Water Environ 20(1):2–7. https://doi.org/10.1007/s10230-001-8074-3

Teaf C, Merkel BJ, Mulisch HM, Kuperberg M, Wcislo E (2006) Industry, mining and military sites. In: World Health Organisation (eds) Protecting groundwater for health: managing the quality of drinking-water sources. IWA Publishing, London, pp 11.1–11.27, 23.1–23.18

Teiwes K (1916) Die Wasserhaltungsmaschinen. Die Bergwerksmaschinen, Bd 5. Springer, Berlin, p 490. https://doi.org/10.1007/978-3-642-90836-1

Temporärer Arbeitskreis Geobiotechnologie in der Dechema e. V. (2013) Geobiotechnologie – Stand und Perspektiven. DECHEMA Gesellschaft für Chemische Technik und Biotechnologie e.V., Frankfurt, 49 p.

Terwelp T (2013) Grubenwasserhaltung – Änderung der Grubenwasserhaltung im Ruhrrevier im Zuge der Stilllegung des Steinkohlenbergbaues. bergbau 64(3):102–105

The Pennsylvanian Department of Environmental Protection (1998) Coal mine drainage prediction and pollution prevention in Pennsylvania. Pennsylvanian Bureau of Mining and Reclamation, Harrisburg, p 398

Thiruvenkatachari R, Francis M, Cunnington M, Su S (2016) Application of integrated forward and reverse osmosis for coal mine wastewater desalination. Sep Purif Technol 163:181–188. https://doi.org/10.1016/j.seppur.2016.02.034

Thomas CW (1959) Errors in measurement of irrigation water. Trans Am Soc Civ Eng 124:319–332

Tillmans J (1919) Über die quantitative Bestimmung der Reaktion in natürlichen Wässern. Z Unters Nahr Genußm Gebrauchsgegenstände 38(1/2):1–16

To TB, Nordstrom DK, Cunningham KM, Ball JW, McCleskey RB (1999) New method for the direct determination of dissolved Fe(III) concentration in acid mine drainages. Environ Sci Technol 33(5):807–813. https://doi.org/10.1021/es980684z

Tönjes B (2016) Nachbergbau – Aufgaben und Verantwortung. Veröff Dt Bergbau-Mus Bochum 217:133–136

Torre MLDL, Grande JA, Santisteban M, Valente TM, Borrego J, Salguero F (2014) Statistical contrast analysis of hydrochemical parameters upstream of the tidal influence in two AMD-Affected Rivers. Mine Water Environ 33(3):217–227. https://doi.org/10.1007/s10230-013-0258-0

Totsche O, Fyson A, Kalin M, Steinberg CEW (2006) Titration curves – a useful instrument for assessing the buffer systems of acidic mining waters. Environ Sci Pollut Res 13(4):215–224. https://doi.org/10.1065/espr2005.09.284

Tracy LD (1921) Mine-water neutralizing plant at calumet mine. Trans Am Inst Min Met Eng 66:609–623

Train R, Breidenbach AW, Schaffer RB, Jarrett BM (1976) Development document for interim final effluent limitations guidelines and new source performance standards for the coal mining point source category, EPA 440/1-76/057-a. US Environmental Protection Agency, Washington, p 288

Treskatis C, Hein C, Peiffer S, Herrmann F (2009) Brunnenalterung – Sind Glaskugeln eine Alternative zum Filterkies nach DIN 4924? BBR Fachmag Leit-Bau Brunnenbau Geotherm 60(4):36–44

Triplett J, Filippov A, Pisarenko A (2001) Coal mine methane in Ukraine – opportunities for production and investment in the Donetsk Coal Basin. U.S. Environmental Protection Agency, Partnership for Energy and Environmental Reform, Washington, p 127

Trivedi P, Axe L, Tyson TA (2001) An analysis of zinc sorption to amorphous versus crystalline iron oxides using XAS. J Colloid Interface Sci 244:230–238. https://doi.org/10.1006/jcis.2001.7971

Trudinger PA (1979) The biological sulfur cycle. In: Trudinger PA, Swaine DJ (eds) Biogeochemical cycling of mineral-forming elements. Studies in environmental science, Bd 3. Elsevier, Amsterdam, pp 293–313. https://doi.org/10.1016/S0166-1116(08)71062-8

Trumm D, Pope J, Newman N (2009) Passive treatment of neutral mine drainage at a metal mine in New Zealand using an oxidising system and slag leaching bed. In: Proceedings securing the future 2009 & 8th ICARD, Skellefteå, pp 1–10 (Electronic document)

Turek M (2003) Recovery of NaCl from saline mine water in an electrodialysis-evaporation system. Chem Pap 57(1):50–52

Turek M (2004) Electrodialytic desalination and concentration of coal-mine brine. Desalination 162(1–3):355–359. https://doi.org/10.1016/S0011-9164(04)00069-4

Turek M, Dydo P, Klimek R (2005) Salt production from coal-mine brine in ED-evaporation-crystallization system. Desalination 184(1–3):439–446. https://doi.org/10.1016/j.desal.2005.03.047

Turek M, Gonet M (1997) Nanofiltration in the utilization of coal-mine brines. Desalination 108:171–177. https://doi.org/10.1016/S0011-9164(97)00024-6

Turner AJM, Braungardt C, Potter H (2011) Risk-based prioritisation of closed mine waste facilities using GIS. In: Mine water – managing the challenges – 11th International Mine Water Association Congress. RWTH Aachen University, Aachen, pp 667–672

U. S. Department of the Interior – Bureau of Reclamation (2001) Water measurement manual, 3rd edn. U.S. Government Print. Office, Washington, p 272

U. S. Department of the Interior – Bureau of Reclamation (2015) Technical evaluation of the gold king mine incident San Juan County, Colorado, Denver, p 89

U. S. Environmental Protection Agency (1983) Design manual – neutralization of acid mine drainage, EPA-600/2-83-001. U.S. Environmental Protection Agency – Office of Research and Development, Cincinnati, p 231

U. S. Environmental Protection Agency (2000) Abandoned mine site characterization and cleanup handbook, EPA 910-B-00-001. U.S. Environmental Protection Agency, Denver, San Fancisco, Seattle, p 408

U. S. Environmental Protection Agency (2007a) Assessment for non-radionuclides including arsenic, cadmium, chromium, copper, lead, nickel, nitrate, perchlorate, and selenium. Monitored natural attenuation of inorganic contaminants in ground water 2, EPA/600/R-07/140. U.S. Environmental Protection Agency, Cincinnati, p 108

U. S. Environmental Protection Agency (2007b) Technical basis for assessment. Monitored natural attenuation of inorganic contaminants in ground water 1, EPA/600/R-07/139. U.S. Environmental Protection Agency, Cincinnati, p 78

U. S. Environmental Protection Agency (2010) Assessment for radionuclides including tritium, radon, strontium, technetium, uranium, iodine, radium, thorium, cesium, and plutonium-americium. Monitored natural attenuation of inorganic contaminants in ground water 3, EPA/600/R-10/093. U.S. Environmental Protection Agency, Cincinnati, p 127

U. S. Environmental Protection Agency (2012 [2015]) Guidelines establishing test procedures for the analysis of pollutants under the clean water act – analysis and sampling procedures – 40 CFR Part 136. Federal Register 77(97):29758–29834

U. S. Environmental Protection Agency (2014) Reference guide to treatment technologies for mining-influenced water, EPA 542-R-14-001. U. S. Environmental Protection Agency, Washington, p 94

U. S. Geological Survey (2015) National field manual for the collection of water-quality data techniques of water-resources investigations. TWRI 9. U.S. Geological Survey, Washington

Uhlig U, Radigk S, Uhlmann W, Preuß V, Koch T (2016) Iron removal from the Spree River in the Bühlow pre-impoundment basin of the Spremberg reservoir. In: Drebenstedt C, Paul M (eds) IMWA 2016 – mining meets water – conflicts and solutions. TU Bergakademie Freiberg, Freiberg, pp 182–190

Uhlmann W, Nitsche C, Neumann V, Guderitz I, Leßmann D, Nixdorf B, Hemm M (2001) Tagebauseen – Wasserbeschaffenheit und wassergütewirtschaftliche Sanierung – Konzeptionelle Vorstellungen und erste Erfahrungen. Stud Tag-Ber 35:1–77

Uhlmann W, Theiss S, Franke C (2010) Untersuchung der hydrochemischen und ökologischen Auswirkungen der Exfiltration von eisenhaltigem, saurem Grundwasser in die Kleine Spree (nördlich Speicher Burghammer) und in die Spree (Ruhlmühle) – Abschlussbericht Teil 1:

Erkundung (Mai 2010), report no. 45042952. Institut für Wasser und Boden Dr. Uhlmann, Dresden, 58 p.
Ulbricht S (2013) Hightech im alten Stollen – Quellwärme für das Krankenhaus – Europaweit einmaliges Projekt kann starten. Wochenspiegel – Zeitung für Freiberg und Umgebung, 13 März 2013, 11 p.
Ulrich K-U, Bethge C, Guderitz I, Heinrich B, Neumann V, Nitsche C, Benthaus F-C (2012) In-Lake neutralization: quantification and prognoses of the acid load into a conditioned pit lake (Lake Bockwitz, Central Germany). Mine Water Environ 31(4):320–338. https://doi.org/10.1007/s10230-012-0206-4
Unger Y, Wolkersdorfer C (2006) Sideritbildung in Tagebaurestseen – Mögliche Sanierungsstrategie des Restlochs Spreetal Nordost. Wiss Mitt Inst Geol 31:119–124
Valanko R, Shestakova M, Pekonen P, Hesampour M, Hansen B, Halttunen S, Hofmann R, Pretorius R, Penttinen M, Recktenwald M, Karpova T, Rossum R, Grönfors O, Mattsson E, Ahlgren J, Nilsson B, Leen P, Havansi H, Abinet R (2020) About water treatment. Kemira Oyj, Helsinki
Valente TM, Gomes P, Sequeira Braga MA, Pamplona J, Antunes IM, Ríos Reyes CA, Moreno F (2019) Mineralogical attenuation processes associated with the evolution of acid mine drainage in sulfide-rich mine wastes. In: Proceedings Mine Water – Technological and Ecological Challenges (IMWA 2019), Perm, Russia, pp 146–153
Valenzuela F, Basualto C, Tapia C, Sapag J (1999) Application of hollow-fiber supported liquid membranes technique to the selective recovery of a low content of copper from a Chilean mine water. J Membr Sci 155(1):163–168. https://doi.org/10.1016/S0376-7388(98)00321-4
van der Ent A, Baker AJ, Reeves RD, Chaney RL, Anderson CW, Meech JA, Erskine PD, Simonnot MO, Vaughan J, Morel JL, Echevarria G, Fogliani B, Rongliang Q, Mulligan DR (2015) Agromining: farming for metals in the future? Environ Sci Technol 49(8):4773–4780. https://doi.org/10.1021/es506031u
van der Ham F, Witkamp GJ, de Graauw J, van Rosmalen GM (1998) Eutectic freeze crystallization: application to process streams and wastewater treatment. Chem Eng Process Process Intensif 37(2):207–213. https://doi.org/10.1016/S0255-2701(97)00055-X
Van der Sloot HA, Van Zomeren A (2012) Characterisation leaching tests and associated geochemical speciation modelling to assess long term release behaviour from extractive wastes. Mine Water Environ 31(2):92–103. https://doi.org/10.1007/s10230-012-0182-8
van der Walt A, Wolkersdorfer C (2018) Indicator parameters for metals of potential concern in South African acid and circum-neutral mine waters. In: Meier G et al (eds) 18. Altbergbau-Kolloquium – Tagungsband. Wagner, Nossen, pp 120–124
van Emmerik T, Popp A, Solcerova A, Müller H, Hut R (2018) Reporting negative results to stimulate experimental hydrology: discussion of "The role of experimental work in hydrological sciences – insights from a community survey". Hydrol Sci J J Sci Hydrol 63(8):1269–1272. https://doi.org/10.1080/02626667.2018.1493203
Van Houten L (2018) Mushroom compost – what is it and how is it used? Reclam Matters 2018(Spring):28–30
van Iterson FKT (1938) The clarification of coal-washery effluent. Proc Kon Ned Akad Wet 41(2):81–94
van Zyl HC, Maree JP, van Niekerk AM, van Tonder GJ, Naidoo C (2001) Collection, treatment and re-use of mine water in the Olifants River Catchment. J S Afr Inst Min Metall 101(1):41–46
Vankelecom IFJ, de Smet K, Gevers LEM, Jacobs PA (2006) Nanofiltration membrane materials and preparation. In: Nanofiltration – principles and applications, Nachdruck. Elsevier Advanced Technology, Amsterdam, pp 33–65

Veil J (2013) Water associated with oil and gas production – produced water management and water needs. In: Brown A, Figueroa L, Wolkersdorfer C (eds) Reliable mine water technology. International Mine Water Association, Golden, pp 1229–1233

Veith H (1871) Deutsches Bergwörterbuch. Korn, Breslau, p 600

Vepsäläinen M (2012) Electrocoagulation in the treatment of industrial waters and wastewaters. VTT Science 12:1–96

Verburg R, Bezuidenhout N, Chatwin T, Ferguson K (2009) The Global Acid Rock Drainage Guide (GARD Guide). Mine Water Environ 28(4):305–310. https://doi.org/10.1007/s10230-009-0078-4

Vidal C, Broschek U, Zuñiga G, Bravo L (2009) Zeotreat: multifunctional adsorbents for metal and sulfate removal in mining wastewaters from copper industry, Chilean case. In: Water Institute of Southern Africa, International Mine Water Association (eds) Proceedings, international mine water conference. Document Transformation Technologies, Pretoria, pp 454–461

Vidal-Legaz B (2017) EU raw materials information system and raw materials scoreboard: addressing the data needs in support of the EU policies – an example for water use in mining. In: Wolkersdorfer C, Sartz L, Sillanpää M, Häkkinen A (eds) IMWA 2017 – mine water & circular economy. II. Lappeenranta University of Technology, Lappeenranta, pp 854–861

Vinci BJ, Schmidt TW (2001) Passive, periodic flushing technology for mine drainage treatment systems. In: Barnhisel RI, Buchanan BA, Peterson D, Pfeil JJ (eds) Land reclamation – a different approach. 2. Proceedings 18th annual national meeting – american society for surface mining and reclamation, Lexington, pp 611–625

Viollier E, Inglett PW, Hunter K, Roychoudhury AN, Van Cappellen P (2000) The ferrozine method revisited – Fe(II)/Fe(III) determination in natural waters. Appl Geochem 15:785–790. https://doi.org/10.1016/S0883-2927(99)00097-9

Visser TJK, Modise SJ, Krieg HM, Keizer K (2001) The removal of acid sulfate pollution by nanofiltration. Desalination 140(1):79–86. https://doi.org/10.1016/S0011-9164(01)00356-3

Vogel D, Paul M, Sänger H-J, Jahn S (1996) Probleme der Wasserbehandlung am Sanierungsstandort Ronneburg. Geowissenschaften 14:486–489

Volckman OB (1963) Operating experience on a large scale electrodialysis water-demineralization plant. In: Advances in chemistry (saline water conversion – II), 38. Proceedings American Chemical Society, Washington, pp 133–157. https://doi.org/10.1021/ba-1963-0038.ch010

Vories KC, Harrington A (2004) State regulation of coal combustion by-product placement at mine sites – a technical interactive forum (Proceedings). U.S. Department of the Interior, Alton, p 222

Vories KC, Harrington A (2006) Flue Gas Desulfurization (FGD) by-products at coal mines and responses to the national academy of sciences final report "Managing Coal Combustion Residues in Mines" – a technical interactive forum (Proceedings). U.S. Department of the Interior, Alton, p 237

Voss L, Hsiao IL, Ebisch M, Vidmar J, Dreiack N, Böhmert L, Stock V, Braeuning A, Loeschner K, Laux P, Thünemann AF, Lampen A, Sieg H (2020) The presence of iron oxide nanoparticles in the food pigment E172. Food Chem 327:127000. https://doi.org/10.1016/j.foodchem.2020.127000

Voznjuk GG, Gorshkov VA (1983) Mine water utilization in the USSR national economy. Int J Mine Water 2(1):23–30. https://doi.org/10.1007/BF02504618

Vranesh G (1979) Mine drainage – the common enemy. In: Mine Drainage – Proceedings of the first international mine drainage symposium. Freeman, San Francisco, pp 54–97

Vymazal J (2011) Constructed wetlands for wastewater treatment: five decades of experience. Environ Sci Technol 45(1):61–69. https://doi.org/10.1021/es101403q

Wacker-Theodorakopoulos C (2000) Zehn Jahre Duales System Deutschland. Wirtschaftsdienst 80(10):628–630

Wagner B, Töpfner C, Lischeid G, Scholz M, Klinger R, Klaas P (2003) Hydrogeochemische Hintergrundwerte der Grundwässer Bayerns. GLA Fachber 21:1–250

Walder IF, Nilssen S, Räisänen ML, Heikkinen PM, Pulkkinen K, Korkka-Niemi K, Salonen V-P, Destouni G, Hasche A, Wolkersdorfer C, Witkowski AJ, Blachére A, Morel S, Lefort D, Midžić S, Silajdžić I, Coulton RH, Williams KP, Rees B, Hallberg KB, Johnson DB (2005) Contemporary reviews of mine water studies in Europe, part 2. Mine Water Environ 24(1):2–37. https://doi.org/10.1007/s10230-005-0068-0

Walter T (2006) Identifizierung der Grundwasserkörper nach EU-WRRL im Saarland und Ermittlung regionaler hydrogeochemischer Hintergrundwerte. Wiss Mitt Inst Geol 31:339–346

Walton-Day K (2003) Passive and active treatment of mine drainage. In: Jambor JL, Blowes DW, Ritchie AIM (eds) Environmental aspects of mine wastes, Bd 31. Mineralogical Association of Canada, Waterloo, pp. 335–359

Wang F, Wang Y, Jing C (2021) Application overview of membrane separation technology in coal mine water resources treatment in Western China. Mine Water Environ 40(2):510–519. https://doi.org/10.1007/s10230-021-00781-3

Wang Z, Zheng J, Tang J, Wang X, Wu Z (2016) A pilot-scale forward osmosis membrane system for concentrating low-strength municipal wastewater: performance and implications. Sci Rep 6:21653. https://doi.org/10.1038/srep21653

Warkentin D, Chow N, Nacu A (2010) Expanding sulfide use for metal recovery from mine water. In: Wolkersdorfer C, Freund A (eds) Mine water and innovative thinking – International Mine Water Association symposium. Cape Breton University Press, Sydney, pp 195–199

Wasserchemische Gesellschaft – Fachgruppe in der Gesellschaft Deutscher Chemiker, in Gemeinschaft mit dem Normenausschuss Wasserwesen (NAW) im DIN Deutsches Institut für Normung e.V. (2021) Deutsche Einheitsverfahren zur Wasser-, Abwasser- und Schlammuntersuchung – Physikalische, chemische, biologische und bakteriologische Verfahren, Stand 2021-10. Verlag Chemie, Weinheim, 9508 p.

Watten BJ, Schwartz MF (1996) Carbon dioxide pretreatment of AMD for limestone diversion wells. In: Proceedings, West Virginia Surface Mine Drainage Task Force symposium 17, J-1–10

Watzlaf GR, Schroeder KT, Kairies CL (2000) Long-term perpormance of alkalinity-producing passive systems for the treatment of mine drainage. In: Proceedings National Meeting of the American Society for Surface Mining and Reclamation, Tampa, pp 262–274. https://doi.org/10.21000/JASMR00010262

Waybrant KR, Blowes DW, Ptacek CJ (1995) Selection of reactive mixtures for the prevention of acid mine drainage using porous reactive walls. In: Proceedings Sudbury '95 – mining and the environment, Sudbury, vol. 3, pp 945–953

Webster JG, Nordstrom DK, Smith KS (1994) Transport and natural attenuation of Cu, Zn, As, and Fe in the acid mine drainage of leviathan and bryant creeks. In: Environmental geochemistry of sulfide oxidation, Bd 550. American Chemical Society, New York, pp 244–260. https://doi.org/10.1021/bk-1994-0550.ch017

Wehrli B (1990) Redox reactions of metal ions at mineral surfaces. In: Stumm W (ed) Aquatic chemical kinetics – reaction rates of processes in natural waters. Wiley, New York, pp 311–336

Weichgrebe D, Kayser K, Zwafink R, Rosenwinkel K-H (2014) Umweltauswirkungen von Fracking bei der Aufsuchung und Gewinnung von Erdgas insbesondere aus Schiefergaslagerstätten – Arbeitspaket 3 – Flowback – Stand der Technik bei der Entsorgung, Stoffstrombilanzen. Texte 53, pp AP3-I-AP3-126

Weilner C (2013) Entstehung und hydrobiologische Entwicklung der Tagebauseen im Oberpfälzer Seenland. Naturw Z Niederbayern 34:125–144

Weiner ER (2010) Applications of environmental aquatic chemistry – a practical guide, 2nd edn. CRC Press, Boca Raton, p 465

Welgemoed TJ (2005) Capacitive deionization technology™ – development and evaluation of an industrial prototype system. Unpublished master thesis University of Pretoria, Pretoria, p 90

Welgemoed TJ, Schutte CF (2005) Capacitive deionization technology™: an alternative desalination solution. Desalination 183(1):327–340. https://doi.org/10.1016/j.desal.2005.02.054

Werner A, Rieger A, Mosch M, Haseneder R, Repke JU (2018) Nanofiltration of indium and germanium ions in aqueous solutions: influence of pH and charge on retention and membrane flux. Sep Purif Technol 194:319–328. https://doi.org/10.1016/j.seppur.2017.11.006

Werner F, Graupner B, Merkel BJ, Wolkersdorfer C (2006) Assessment of a treatment scheme for acidic mining lakes using $CO_2$ and calcium oxides to precipitate carbonates. In: ICARD 2006, 7. Proceedings 7th International Conference on Acid Rock Drainage (ICARD), St. Louis, pp 2344–2353 [CD-ROM]. https://doi.org/10.21000/JASMR06022344

West Virginia Department of Health and Human Resources (2003) State of West Virginia source water assessment and protection program – source water assessment report – City of Gary McDowell County report no. WV3302420. Source Water Protection Unit, West Virginia, p 12

Westermann S, Goerke-Mallet P, Reker B, Dogan T, Wolkersdorfer C, Melchers C (2017) Aus Erfahrungen lernen: Evaluierung von Grubenwasseranstiegsprozessen zur Verbesserung zukünftiger Prognosen 17. Altbergbau-Kolloquium – Tagungsband. Wagner, Nossen, pp 259–273

Weston Solutions (2004) Draft preliminary wastewater treatment technology evaluation Tar Creek Superfund Site Oklahoma Picher mining district. U.S. Army Corps of Engineers, Tulsa, p 30

Weyer J (2010) $CO_2$ in underground openings and mine rescue exercise. In: Wolkersdorfer C, Freund A (eds) Mine water and innovative thinking – International Mine Water Association symposium. Cape Breton University Press, Sydney, pp 93–97

Whillier A (1977) Recovery of energy from the water going down mine shafts. J S Afr Inst Min Metall 78:183–186

Whitehead PG, Hall G, Neal C, Prior H (2005) Chemical behaviour of the Wheal Jane bioremediation system. Sci Total Environ 338(1–2):41–51. https://doi.org/10.1016/j.scitotenv.2004.09.004

Wiberg N, Wiberg E, Holleman AF (2016) Anorganische Chemie, 1+2, 103rd edn. De Gruyter, Berlin, p 2694

Wieber G, Pohl S (2008) Mine water: a source of geothermal energy – examples from the Rhenish Massif. In: Proceedings, 10th International Mine Water Association Congress, pp 113–116

Wiedemeier TH, Rifai HS, Newell CJ, Wilson JT (1999) Natural attenuation of fuels and chlorinated solvents in the subsurface. Wiley, New York, p 632. https://doi.org/10.1002/9780470172964

Wildeman T, Brodie GA, Gusek J (1993a) Wetland design for mining operations. BiTech, Richmond, p 408

Wildeman T, Gusek J, Dietz J, Morea S (1993b) Handbook for constructed wetlands receiving acid mine drainage, report no. CR 815325. Colorado School of Mines, Golden, p 233

Wilkin RT (2007) Metal attenuation processes at mining sites. Ground Water Issue EPA/600/R-07/092, pp 1–12

Wilkin RT (2008) Contaminant attenuation processes at mine sites. Mine Water Environ 27(4):251–258. https://doi.org/10.1007/s10230-008-0049-1

Wilmoth RC (1973) Applications of reverse osmosis to acid mine drainage treatment. U.S. Government Print. Office, Washington, p 159

Wilmoth RC (1977) Limestone and lime neutralization of ferrous iron acid mine drainage. Environmental Protection Technology Series EPA-600/2-77-101, p 95

Wilmoth RC, Scott RB, Harris EF (1977) Application of ion exchange to acid mine drainage treatment. In: Proceedings 32nd industrial waste conference, Lafayette, pp 820–829

Winner Global Energy, Environmental Services LLC. (2009) AMD value extraction process – liquid to liquid extraction final report, Report no. GR# 410003675. Winner Global Energy & Environmental Services LLC, Sharon, p 169

Wisotzky F (2001) Prevention of acidic groundwater in lignite overburden dumps by the addition of alkaline substances: pilot-scale field experiments. Mine Water Environ 20(3):122–128. https://doi.org/10.1007/s10230-001-8093-0

Wisotzky F (2003) Saure Bergbauwässer (Acid Mine Drainage) und deren Qualitätsverbesserung durch Zugabe von alkalisch wirkenden Zuschlagstoffen zum Abraum – Untersuchungen im Rheinischen Braunkohlenrevier. Dtsch Gewässerkd Jb Bes Mitt 61:167

Wisotzky F, Cremer N, Lenk S (2018) Angewandte Grundwasserchemie, Hydrogeologie und hydrogeochemische Modellierung – Grundlagen, Anwendungen und Problemlösungen, 2nd edn. Springer, Heidelberg, p 678. https://doi.org/10.1007/978-3-662-55558-3

Wisotzky F, Lenk S (2006) Darstellung und Evaluierung der Minderungsmaßnahmen zur Kippenwasserversauerung im Tagebau Garzweiler der RWE Power AG (Rheinisches Braunkohlenrevier, Germany). Wiss Mitt Inst Geol 31:133–138

Wissenschaftlich-Technische Werkstätten GmbH (1997) Leitfähigkeits-Fibel – Einführung in die Konduktometrie für Praktiker. Wissenschaftlich-Technische Werkstätten GmbH, Weilheim (unpubl.), 58 p.

Wohnlich S (2001) Natural Attenuation – Wiederkehr der natürlichen Selbstreinigung des Grundwassers? Grundwasser 6(1):1. https://doi.org/10.1007/pl00010385

Wolfers B, Ademmer C (2010) Grenzen der bergrechtlichen Nachsorgehaftung – Verhältnismäßigkeitsprüfung im Bergrecht nach dem Rammelsberg-Urteil. Dtsch Verwalt-Bl Abh 1:22–27

Wolkersdorfer C (1995) Die Flutung des ehemaligen Uranbergwerks Niederschlema/Alberoda der SDAG Wismut. Z Geol Wiss 23(5/6):795–808

Wolkersdorfer C (1996) Hydrogeochemische Verhältnisse im Flutungswasser eines Uranbergwerks – Die Lagerstätte Niederschlema/Alberoda. Clausthaler Geowiss Diss 50:1–216

Wolkersdorfer C (2008) Water management at abandoned flooded underground mines – fundamentals, tracer tests, modelling, water treatment. Springer, Heidelberg, p 465. https://doi.org/10.1007/978-3-540-77331-3

Wolkersdorfer C (2011) Tracer test in a settling pond – the passive mine water treatment plant of the 1 B mine pool, Nova Scotia, Canada. Mine Water Environ 30(2):105–112. https://doi.org/10.1007/s10230-011-0147-3

Wolkersdorfer C (2013) Management von Grubenwasser 3.0 – Blick in die Zukunft. Wiss Mitt Inst Geol 44:105–113

Wolkersdorfer C, Baierer C (2013) Improving mine water quality by low density sludge storage in flooded underground workings. Mine Water Environ 32(1):3–15. https://doi.org/10.1007/s10230-012-0204-6

Wolkersdorfer C, Göbel J, Blume C, Weber C (2004) Hydrogeologische Probenahmestellen in der Troianischen Landschaft. Stud Troica 14:169–200

Wolkersdorfer C, Göbel J, Hasche-Berger A (2016) Assessing subsurface flow hydraulics of a coal mine water bioremediation system using a multi-tracer approach. Int J Coal Geol 164:58–68. https://doi.org/10.1016/j.coal.2016.03.010

Wolkersdorfer C, Kubiak C (2008) Low density sludge storage in a flooded underground mine. In: Rapantova N, Hrkal Z (eds) Mine water environ. VSB – Technical University of Ostrava, Ostrava, pp 51–54

Wolkersdorfer C, Lopes DV, Nariyan E (2015) Intelligent mine water treatment – recent international developments. In: Paul M (ed) Sanierte Bergbaustandorte im Spannungsfeld zwischen Nachsorge und Nachnutzung – WISSYM 2015. Wismut GmbH, Chemnitz, pp 63–68. https://doi.org/10.13140/RG.2.1.2441.5849

Wolkersdorfer C, Mugova E, Daga VS, Charvet P, Vitule JRS (2021) Effects of mining on surface water – case studies. In: Irvine K, Chapman D, Warner S (eds) The encyclopedia of inland waters, 2nd edn. Elsevier, Oxford. https://doi.org/10.1016/B978-0-12-819166-8.00085-2

Wolkersdorfer C, Nordstrom DK, Beckie R, Cicerone DS, Elliot T, Edraki M, Valente TM, França SCA, Kumar P, Oyarzún Lucero RA, Soler AIG (2020) Guidance for the integrated use of hydrological, geochemical, and isotopic tools in mining operations. Mine Water Environ 39(2):204–228. https://doi.org/10.1007/s10230-020-00666-x

Wolkersdorfer C, Qonya B (2017) Passive mine water treatment with a full scale, containerized vertical flow reactor at the abandoned Metsämonttu mine site, Finland. In: Wolkersdorfer C, Sartz L, Sillanpää M, Häkkinen A (eds) IMWA 2017 – mine water & circular economy, vol I. Lappeenranta University of Technology, Lappeenranta, pp 109–116

Wolkersdorfer C, Tamme S, Hasche A (2003) Natural attenuation of iron rich mine water by a surface brook. In: Nel PJL (eds) Mine water environ. Proceedings 8th International Mine Water Association Congress, Johannesburg, pp 433–439

Wolkersdorfer C, Wackwitz T (2004) Antimony anomalies around abandoned silver mines in Tyrol/Austria. In: Jarvis AP, Dudgeon BA, Younger PL (eds) Mine water 2004 – Proceedings International Mine Water Association symposium. 1. University of Newcastle, Newcastle upon Tyne, pp 161–167

Wolkersdorfer C, Younger PL (2002) Passive Grubenwassereinigung als Alternative zu aktiven Systemen. Grundwasser 7(2):67–77. https://doi.org/10.1007/s007670200011

World Water Assessment Programme (2012) Managing water under uncertainty and risk – The United Nations world water development report 4. United Nations Educational Scientific and Cultural Organization, Paris, p 866

Yang D, Fan R, Greet C, Priest C (2020) Microfluidic Screening to Study Acid Mine Drainage. Environ Sci Technol 54(21):14000–14006. https://doi.org/10.1021/acs.est.0c02901

Yang JE, Skousen JG, Ok Y-S, Yoo K-Y, Kim H-J (2006) Reclamation of abandoned coal mine waste in Korea using lime cake by-products. Mine Water Environ 25(4):227–232. https://doi.org/10.1007/s10230-006-0137-z

Yeh S-J, Jenkins CR (1971) Disposal of sludge from acid mine drainage neutralization. J Water Pollut Control Fed 43(4):679–688

Yeheyis MB, Shang JQ, Yanful EK (2009) Long-term evaluation of coal fly ash and mine tailings co-placement – a site-specific study. J Environ Manage 91(1):237–244. https://doi.org/10.1016/j.jenvman.2009.08.010

Young EF, Steinman HE (1967) Coal mine drainage treatment. Purdue Univ Eng Bull Ext Ser 52(3):477–491

Younger PL (1995) Hydrogeochemistry of minewaters flowing from abandoned coal workings in County Durham. Q J Eng Geol 28(4):101–113. https://doi.org/10.1144/GSL.QJEGH.1995.028.S2.02

Younger PL (1997) The longevity of minewater pollution – a basis for decision-making. Sci Total Environ 194–195:457–466. https://doi.org/10.1016/S0048-9697(96)05383-1

Younger PL (1998) Design, construction and initial operation of full-scale compost-based passive systems for treatment of coal mine drainage and spoil leachate in the UK. In: Proceedings mine water and environmental impacts, Johannesburg, vol. 2, pp 413–424

Younger PL (2000a) Holistic remedial strategies for short- and long-term water pollution from abandoned mines. Trans Inst Min Metall Sect A 109:210–218. https://doi.org/10.1179/mnt.2000.109.3.210

Younger PL (2000b) Predicting temporal changes in total iron concentrations in groundwaters flowing from abandoned deep mines: a first approximation. J Contam Hydrol 44:47–69. https://doi.org/10.1016/S0169-7722(00)00090-5

Younger PL (2002a) Deep mine hydrogeology after closure – insights from the UK. In: Merkel BJ, Planer-Friedrich B, Wolkersdorfer C (eds) Uranium in the aquatic environment. Springer, Heidelberg, pp 25–40. https://doi.org/10.1007/978-3-642-55668-5_3

Younger PL (2002b) Mine water pollution from Kernow to Kwazulu-Natal: geochemical remedial options and their selection in practice. In: Proceedings, Ussher Society, vol. 10, pp 255–266

Younger PL (2007) Groundwater in the environment – an introduction. Blackwell, Oxford, p 318

Younger PL (2010) Where there is no pH meter: estimating the acidity of mine waters by visual inspection. In: Wolkersdorfer C, Freund A (eds) Mine water and innovative thinking – International Mine Water Association symposium. Cape Breton University Press, Sydney, pp 407–411

Younger PL, Adams R (1999) Predicting mine water rebound. Environment Agency, Bristol, p 109. https://doi.org/10.13140/2.1.4805.5681

Younger PL, Banwart SA (2002) Time-scale issues in the remediation of pervasively contaminated groundwaters at abandoned mines sites. In: Groundwater quality: natural and enhanced restoration of groundwater pollution, IAHS Publication, 275. IAHS Press, Wallingford, pp 85–591

Younger PL, Banwart SA, Hedin RS (2002) Mine water – hydrology, pollution. In: remediation. Kluwer, Dordrecht, p 464. https://doi.org/10.1007/978-94-010-0610-1

Younger PL, Coulton RH, Froggatt EC (2005) The contribution of science to risk-based decision-making: lessons from the development of full-scale treatment measures for acidic mine waters at Wheal Jane, UK. Sci Total Environ 338(1–2):137–154. https://doi.org/10.1016/j.scitotenv.2004.09.014

Younger PL, Curtis TP, Jarvis A, Pennell R (1997) Effective passive treatment of aluminium-rich, acidic colliery spoil drainage using a compost wetland at quaking houses, County Durham. Water Environ Manag 11:200–208. https://doi.org/10.1111/j.1747-6593.1997.tb00116.x

Younger PL, Jayaweera A, Elliot A, Wood R, Amos P, Daugherty AJ, Martin A, Bowden L, Aplin AC, Johnson DB (2003) Passive treatment of acidic mine waters in subsurface-flow systems – exploring RAPS and permeable reactive barriers. Land Contam Reclam 11(2):127–135. https://doi.org/10.2462/09670513.806

Younger PL, Jenkins DA, Rees SB, Robinson J, Jarvis AP, Ralph J, Johnston DN, Coulton RH (2004) Mine waters in Wales – pollution, risk management and remediation. Geol Ser Natl Mus Art Galleries Wales 23:138–154

Zhang Y, Charlet L, Schindler PW (1992) Adsorption of protons, Fe(II) and Al(III) on lepidocrocite (γ-FeOOH). Colloids and Surfaces 63(3):259–268. https://doi.org/10.1016/0166-6622(92)80247-Y

Zhao S, Zou L, Tang CY, Mulcahy D (2012a) Recent developments in forward osmosis: opportunities and challenges. J Membr Sci 396:1–21. https://doi.org/10.1016/j.memsci.2011.12.023

Zhao X, Zhang B, Liu H, Chen F, Li A, Qu J (2012b) Transformation characteristics of refractory pollutants in plugboard wastewater by an optimal electrocoagulation and electro-Fenton process. Chemosphere 87(6):631–637. https://doi.org/10.1016/j.chemosphere.2012.01.054

Zhuang J-M (2009) Acidic rock drainage treatment – a review. Recent Pat Chem Eng 2(3):238–252. https://doi.org/10.2174/2211334710902030238

Žibret G, Žebre M (2018) Use of robotics and automation for mineral prospecting and extraction. In: Joint conference of UNEXMIN, ¡VAMOS! and RTM projects, Bled. Geological Survey of Slovenia, p 57. https://doi.org/10.5474/9789616498579

Ziegenbalg G (1999) In-situ remediation of heavy metal contaminated soil or rock formations and sealing of water inflows by directed and controlled crystallization of natural occurring mineral. In: Fernández Rubio R (ed) Mine, water & environment. II. International Mine Water Association, Sevilla, pp 667–672

Ziemkiewicz PF (2006) Groundwater effects of coal combustion by-product placement in coal mines. In: Flue Gas Desulfurization (FGD) by-products at coal mines and responses to the national academy of sciences final report "Managing Coal Combustion Residues in Mines" – a technical interactive forum (Proceedings). U.S. Department of the Interior, Alton, pp 59–67

Ziemkiewicz PF, Ashby JC (2007) Wet FGD placement at the Mettiki mine in Maryland. In: Proceedings, West Virginia Surface Mine Drainage Task Force symposium 28, electronic resource 7 S

Ziemkiewicz PF, Brant DL, Skousen JG (1996) Acid mine drainage treatment with open limestone channels. In: Proceedings, West Virginia surface mine drainage task force symposium 17 (Electronic document). https://doi.org/10.21000/JASMR96010367

Ziemkiewicz PF, Skousen JG, Brant DL, Sterner PL, Lovett RJ (1997) Acid mine drainage treatment with armored limestone in open limestone channels. J Environ Qual 26(4):560–569. https://doi.org/10.2134/jeq1997.00472425002600040013x

Zinchenko D, Schauer PJ, Migchelbrink J, Liang HC, Tamburini JR, Nketia M, Rolston N, Hatley J, Lau T, Willis J (2013) Evaluation of high-rate clarification for water treatment at a Uranium mine – a case study. In: Brown A, Figueroa L, Wolkersdorfer C (eds) Reliable mine water technology. International Mine Water Association, Golden, pp 755–760

Zinck JMM (2005) Review of disposal, reprocessing and reuse options for acidic drainage treatment sludge, MEND Report 3.42.3. Mine Environment Neutral Drainage (MEND) Program, Ottawa, p 61

Zinck JMM (2006) Disposal, reprocessing and reuse options for acidic drainage treatment sludge. In: ICARD 2006, 7. Proceedings 7th International Conference on Acid Rock Drainage (ICARD), St. Louis, pp 2602–2617. https://doi.org/10.21000/JASMR06022604

Zinck JMM, Aubé BC (2000) Optimization of lime treatment processes. CIM Bull 93(1043):98–105

Zinck JMM, Wilson JL, Chen TT, Griffith W, Mikhail S, Turcotte AM (1997) Characterization and stability of acid mine drainage sludge, MEND Report 3.42.2a. Mine Environment Neutral Drainage (MEND) Program, Ottawa, 73 p.

Zipper CE, Skousen JG (2010) Influent water quality affects performance of passive treatment systems for acid mine drainage. Mine Water Environ 29(2):135–143. https://doi.org/10.1007/s10230-010-0101-9

Zoumis T (2003) Entwicklung aktiver Barrieren für die Entfernung von Schwermetallen aus Grubenwässern am Beispiel der Freiberger Grube, Sachsen. VDI-Verl, Düsseldorf, p 159

Zoumis T, Calmano W, Förstner U (2000) Demobilization of heavy metals from mine waters. Acta Hydrochim Hydrobiol 28(4):212–218. 10/dstsff

何绪文 (He Xu Wen), 贾建丽著 (Jia Jianli) (2009) 矿井水处理及资源化的理论与实践 [Theorie und Praxis der Aufbereitung und des Recyclings von Grubenwasser]. 煤炭工业出版社出版, 北京，234 p.

梁天成 (Liang Tiancheng) (2004) 矿井水处理技术及标准规范实用手册 [Praktisches Handbuch der Technologie und Standards zur Grubenwasseraufbereitung]. 当代中国音像出版社，北京

# Index


**A**
Abandoned mine, xi, xvii, 88, 166, 223, 234
  after use, 227
Aboriginal people, 229
Absorption, 26–28, 141
Acid capacity, 12, 13, 22, 25, 26, 70, 158, 162, 164, 180, 182, 185, 214
  definition, 69
  titration, 22
Acid-generating salts, *see* Secondary mineral
*Acidithiobacillus*, 30, 137, 198
  *ferrivorans*, 200
  *ferrooxidans*, 213
  *thiooxidans*, 30, 137
Acidity, 12–13, 15, 21, 22, 38, 68, 69, 86, 160, 213
  activity, 69
  calculation from analysis, 13
  juvenile, 12, 17, 214
  net acidity, 21, 22, 38
  titration, 69
  vestigial, 12
Acid-loving iron-oxidising sulfur rod bacterium, *see Acidithiobacillus, thiooxidans*
Acid mine leachate (AMD/ML), *see* Mine water, acid mine water
Acid rain, 134, 207
Acid rock drainage, *see* Mine water, acid mine water
Acid soil, 31, 32
Adit closure, 203
Adit Troll, *see* Walter Moers
Adopt Your Watershed (association), 191
Adsorption, 26–28, 141–143
AEG (company), 3
Aerogel, 116
Agglomeration, 21
Alfalfa, 186
Algae, 50, 122, 240
Alkalinity, xviii, 14, 21, 22, 25, 69, 70, 86, 156, 158, 161, 162, 171, 186, 187, 209, 235
  net, 21, 22, 38
Alternative fact, 3
Aluminium, 8, 28, 38, 39, 46, 61, 82, 101, 102, 108, 114, 132, 133, 146, 156, 160, 161, 168, 169, 173, 194, 200, 234, 235, 238
*Alyssum*, 240
Ammonium, 35, 84, 118, 128, 137, 142
  sulfate fertiliser, 118
Analysis, *see* Water analysis
Anglo American (company), 149
Animas River Stakeholders Group (association), 191
Anthropogenic, 24, 32, 84, 193, 216
Anthroposphere, 215, 243
Antimony, 8, 113, 130, 197, 234
Apatite, 213, 224
Apple (company), 5
AQM PalPower, 142
Aquaculture, 229
Aquaminerals Finland (company), 142
Aqua-Simon (company), 147
Arsenic, 2, 8, 9, 17, 53, 92, 130, 136, 139–142, 218, 234
Authority, 190, 197, 234
Autocatalysis, 179, 180
Avocado, 127




C. Wolkersdorfer, *Mine Water Treatment – Active and Passive Methods*,
https://doi.org/10.1007/978-3-662-65770-6

**B**
Background value, 67, 74, 91
Bacteria, 14, 22, 50, 52, 124, 134, 136, 137, 171, 197–199, 204, 206, 223, 240
Bactericide, 212, 213
Baffle sheet, 176
Barcode, 59
Barium sulfate, 218
Base capacity, 12, 13, 22, 26, 70, 89, 101, 118, 161, 163, 197, 210, 212
  definition, 12, 68
  titration, 22
  *See also* Acidity
Base metal, 29, 37
Bassanite, 99
BAT/BATA, *see* Best available technology economically achievable (BATEA)
Batch experiment, 141, 143
Bat conservation, 192, 224
BATEA, *see* Best available technology economically achievable (BATEA)
Bayoxid E33, 237
Bergakademie Freiberg, 49, 125
Best available technology economically achievable (BATEA), 6, 42, 102
BHP Billiton (company), 127
Biocide, 119
Biofilm, 57, 68, 85, 119, 134, 143
Biofouling, 57, 119, 130, 137, 232
  *See also* Fouling
Biogeochemistry, 200
Biomining, 198
Bioreactor
  compost, 14
  definition, 14
Biosorbent, 141, 239–241
Biosynthesis, 14
Blast furnace slag, 179
Bluetooth, 65
Boojum Research (company), 187
Brine, 4, 122, 147, 149, 236, 238

**C**
Cadmium, 48, 60, 115, 139, 144, 200, 235
Calcite, 99, 133, 211
Calcium, 53, 86, 99, 100, 105, 125, 128, 236
  hydroxide, 86, 99, 100, 105, 205, 212
Cameco Corporation (company), 131
Carbolime, 205
Carbon source, 136, 137, 170, 171
Carbonate, 25, 28, 31, 34, 44, 130
Cattails, 166
Chain of custody, 59
Changes in mining methods, 197
Charge balance, 55, 56
Chatham House rule, x
Chemical analysis, *see* Water analysis
Chemisorption, 26
*Cherax tenuimanus*, 229, 230
Chromium, 144, 218, 234
*Circa Instans*, ix
Circular economy, xi, 10, 14, 225, 234, 235
  definition, 14
Citizen participation, 191
Citizen's Volunteer Monitoring Program (association), 191
Clarivate Analytics, 120, 121, 142
Clean Air Task Force (association), 221
*Cloaca Maxima*, 189
$CO_2$, 50, 79, 87, 101, 108, 118, 123, 133, 167, 172, 185, 207, 209–211, 223
Coagulation, 21, 99, 101, 102, 177
Cobalt, 139
Colloid, 21, 28, 62, 63, 82, 102, 117, 120, 239
Coltan, xii
Column experiment, 87, 88, 141
Complexation, 26, 28, 104
Compost, 14, 87, 136, 169, 170, 172, 173, 179
Conceptual model, 43, 196
Conductivity, *see* Electrical conductivity
Contaminant of Potential Ecological Concern (COPEC), 8
Contaminated watercourse, 216
Cookbook, ix, 45, 244
Cooling water, 116, 147, 229
Copper, xi, 13, 29, 60, 79, 108, 114, 115, 123, 127, 139, 142, 198, 200, 234, 238
Coprecipitation, 26, 28, 130, 132, 142, 167, 194, 217
  *sensu lato*, 28
  *sensu stricto*, 28
Corporación Nacional del Cobre de Chile (company), 199
Cow dung, 14
Crayfish, 229, 230

Critical metal, 234
Cyanide, 123, 200

**D**

Dam, 175, 219, 222, 224
  dry, 222
  wet, 222
  *See also* Treatment method, mine sealing
Data
  evaluation, 43
  typo, 2
Dead-end filtration, 121, 125
Decant, *see* Mine water discharge
DeepL Translator, x
*Desulfobacter*, 206
*Desulfovibrio*, 206
Diadochic substitution, 132
Diamond, 113
Differential equation, 2
Directive 2006/21/EC of the European Parliament, 9, 10, 20
Disulfide weathering, *see* Pyrite oxidation
Dolomite, 170, 208
Dow Chemical (company), 125
Down-dip operation, 197
Draw solution, 129, 130
Drinking water, 2, 9, 19, 24, 34, 40, 121, 125, 133, 137, 140–142, 147, 149, 228, 236
Drone, 48–50
Duplicate research, 2, 243
Durov diagram, 36

**E**

Ecosphere, xi, 243
Efflorescent salt, 15, 194
  *See also* Secondary mineral
Electrical conductivity, 40, 46, 48, 54, 55, 58, 59, 66–68, 74, 112, 116
Electrochemical reaction, 27, 118
Electrode
  calibration, 47
  corrosion, 47
Electron donor, 170, 206
Elektrizitätswerke Reutte (company), 233
End-of-the-pipe, 190
Energeticon Museum, 232
Ensore Solutions (company), 219
EnvironOxide, 237
Erythropoietin, 14
Eternity liability, ix
Ethanol, 137, 205
Ettringite, 131, 132, 134, 141
  sludge, 134
Explosive, xi, 128

**F**

Fenton process, 144
Fermenter
  beer, 14
  Bionade, 14
  *See also* Treatment method, fermenter
Ferric, *see* Water sampling, ferric iron
Ferrous, *see* Water sampling, ferrous iron
*Ferrovum myxofaciens*, 135, 137
Fibreglass float, 177
Ficklin diagram, 36, 37, 79
Field
  book, 63, 64
  experiment, 179, 218, 221
  scale, 118, 130, 137, 220
First flush, 12, 15, 17, 43, 194
  definition, 15
  duration, 15
Fish
  culture, 229
  pond, 101, 176
Flerovium, 30
Flocculant, 99, 101, 102, 106, 109
  optimisation, 109
  polymer, 21, 109
Flocculation, 21, 102, 112, 134, 217, 218
Flooded mine, 12, 214
  alkaline injection, 216
  barium sulfate immobilisation, 218
  in-situ liming, 219
  lake, 212
  pit lake, 233
  sampling, 59
  tourist attraction, 228
  tracer, 64
    test, 215
  underground dam, 224
  uranium mine, 217
  *See also* Mine flooding

Flow measurement, 43, 70, 74, 76, 77
acoustic digital flow meter, 71
Acoustic Doppler Current Profiler, 74
area velocity flow meter, 77
calibration, 77
errors, 72, 75
gallons, 77
impeller flow meter, 70, 72–75
methods, 70
salt dilution method, 71, 74–77
weir, 70–74, 181, 184, 185
health and safety, 77
Flow rate, 43, 70, 122, 145, 153, 166–168, 181
Fly ash, 140, 179, 210–212, 214–216, 220, 221, 236
classification, 221
Fouling, 116, 119, 122, 123, 129
*See also* Biofouling
Fracking, 14, 130, 146
Freeboard, 167, 173
Freezing-out, 147
Fundación Chile (company), 140
Fungi, 30, 50, 52, 166, 240

**G**

GARD Guide (Global Acid Rock Drainage), x, 2, 26, 34, 141
column test, 88
Garimpeiros, xii
Geodetic datum, 63
Geographical information system (GIS), 196
Geothermal, 14, 229–233
well, 14
Geotube, 237
Gibbsite, 102, 200
GIS, *see* Geographical information system (GIS)
Glückauf, xiii, 7, 244
Gold Book, 28, 29
Green fluorescent protein (GFP), 85
Groundwater, xvii, 15, 17, 20, 24, 31, 40, 43, 47, 51, 52, 83, 86, 91, 175, 177–179, 190, 195, 209, 211, 212, 216–219, 221, 229
contaminated, 32, 34
recharge, 228
Groundwater Ordinance, 19, 90
Guiding foil, *see* Baffle sheet
Gypsum, 99, 128, 133, 233, 236, 239

**H**

Hach (company), 57, 65, 83
Hardness, 39, 86, 140
HDS, *see* Treatment method, high density sludge (HDS)
Healing water, 9
Heat exchanger, 229
Heavy metal, 8, 9, 29
astronomy, 8
meaningless term, 8
music, 29
plants, 240
Horse manure, 87, 171, 174
Horseshoe, 170
$H_2S$, 111, 186, 223, 238
Hydrated lime, 106, 108
Hydrogencarbonate, 17, 53, 207, 209
Hydraulic conditions, 203
Hydrogen gas, 218
Hydrogen peroxide, 144, 182
Hydrolysis, 12, 51, 53, 81, 97, 101, 104, 128, 164, 166, 169, 173, 181, 182, 184

**I**

Illegal mine explorer, 222
Indicator plants, 240
Inorganic acid, 12, 68
Inotec (company), 207
In situ experiment, 153, 219
Insulin, 241
Integrated Pollution Prevention and Control (IPPC), 42
Intelligent Mine of the Future, 123
Intelligent Mine Water Treatment, 123
Interested public, 214
Intermediate bulk container (IBC), 180, 181
International Mine Water Association, xvi, 191
International Space Station (ISS), 117
Internet of Mine Water, xvii, 5, 6
Internet of Things, 5, 49
Ion balance, 54–56
Iron species, 38, 60, 83
Irrigation, 233

J

JECO Corporation (company), 127
Jones and Laughlin Steel Corporation (company), 101, 213
*Juncus* sp., 166

K

Kärchert, 15
Keyplan Engineering (company), 239
Kiln dust, 100, 179
Kreislaufwirtschaft, *see* Circular economy

L

Lab-on-a-chip, 78
Laboratory
  experiment, 81, 162, 209, 213, 215, 218, 229, 238
  investigation, 13, 203, 214–216
  scale, 40, 114, 125, 141, 199
  test, xvii, 81, 87, 105, 215, 220, 235
Landfill, 166, 235
Lanxess (company), 138, 237
Lausitzer und Mitteldeutsche Bergbauverwaltungsgesellschaft, *see* LMBV (company)
Lawrence Livermore National Laboratory, 116
Leachate, *see* Mine water
Lead, 23, 60, 102, 113, 115, 139, 164
LEAG (company), 135
*Leptospirillum*, 200
LfULG (institution), xv, xxvi, 118
Lime, 86, 87, 101, 105, 131, 137, 185, 197, 204, 205, 210, 212, 216, 220
  quick, 100, 212
  slaked, 134, 207, 209, 211
Limestone, 86, 87, 105, 143, 161–163, 171, 179, 205, 208, 209, 212, 216, 220
Linear economy, 14
Living in a box, 110
LMBV (company), 118, 175, 176, 212
Load, 70
Locality
  Aachen, Germany, 39, 231
  Abzucht, Germany, 34, 175
  Afon Goch, United Kingdom, 224, 225
  Alberoda, Germany, 131
  Alexisbad, Germany, 81
  Alsdorf, Germany, 229, 232
  Animas River, USA, 115, 191, 226
  Bad Ems, Germany, 50
  Bad Gastein, Austria, 9, 229
  Bad Grund, Germany, 233
  Bad Schlema, Germany, 17, 131, 157
  Batán, Argentina, 229
  Bear Creek valley, USA, 218
  Beatrix gold mine, South Africa, 118
  Beerwalde, Germany, 233
  Bell Island, Newfoundland, 229
  Berkeley, USA, 140, 208, 236
  Berlin, Germany, 229
  Bernsteinsee, Germany, 208, 210, 211
  Biberwier, Austria, 233
  Big Gorilla, USA, 212
  Binchang, China, 127
  Bochum, Germany, 49, 232
  Bockwitz, Germany, 118, 210
  Bolesław Śmiały, Poland, 118
  Bor, Serbia, 116
  Bowden Close, United Kingdom, 172–174
  Brachmannsberg, Germany, 224
  Brazil, xii, 186
    small-scale mining, xii
  Brixlegg, Austria, 139, 199
  Brombach, Bavaria, 229
  Buckhannon, USA, 224
  Budelco, Netherlands, 137
  Burgfey, Germany, 141, 197
  Buttonwood adit, USA, 16
  Calumet Colliery, USA, 100
  Cape Breton Island, Canada, xv, 48, 62, 71, 73, 77, 82, 110, 153, 163, 167, 187, 217
  Carolina, South Africa, 79, 181
  Cement Creek, USA, 32, 115
  China, xii
  Chorzów, Poland, 99
  Clausthal, Germany, xv, 233
  Clough Foot, United Kingdom, 176
  Collie, Australia, 229
  Congo, xii
  Copper Cliff, Canada, 186
  County Durham, United Kingdom, 104, 105, 109, 165, 171, 172, 174
  Cunha Baixa, Portugal, 187
  Curley, USA, 185
  Cwm Rheidol, United Kingdom, 180, 181
  Dębieńsko, Poland, 118, 236

Locality (*cont.*)
- Delitzsch, Germany, xi
- Denniston Plateau, New Zealand, 194
- Dillenburg, Germany, 34
- Donets Basin, Ukraine, 115
- Donetsk, Ukraine, 115
- Drehna lake, Germany, 210
- Driscoll, USA, 214
- Drosen, Germany, 233
- Dunlap, USA, 220
- Dyffryn Adda, United Kingdom, 225
- Ecton, United Kingdom, 49
- Eden Project, United Kingdom, 227
- Edsån, Sweden, 208
- Ehrenfriedersdorf, Germany, 232
- Eichow, Germany, 176
- Eifel, Germany, 132, 141, 197
- Einheit mine, Germany, 215
- Elbingerode, Germany, 215
- El Salvador, Chile, 199
- eMalahleni, South Africa, xv, 120, 121, 123, 124, 126, 127, 129, 238–240
- Enos Colliery, USA, 152
- Erzgebirge, Germany, 135, 218, 225, 227
- Essen, Germany, 231, 232
- Ettringer Bellerberg, Germany, 132
- Fisher coal mine, USA, 213
- Flensburg, Germany, 147
- Frances, United Kingdom, 104
- Freiberg, Germany, xv, xvi, 17, 140, 225, 232, 233
- Fry Canyon, USA, 178
- Fünfkirchen, Hungary, 218
- Garzweiler, Germany, 31
- Geierswalder See, Germany, 208, 211
- Gera, Germany, 23
- German Democratic Republic (GDR), 233
- Germany
  - end of coal mining, xi
- Gernrode, Germany, 17, 76, 164, 173, 190
- Gessenhalde, Germany, 23
- Gessental, Germany, 216, 217
- Ghana, 84, 199
- Glasebach, Germany, 31
- Glubokaya, Ukraine, 115
- Gold Acres mine, USA, 221
- Gold King mine, USA, 116, 226
- Goslar, Germany, 10, 199
- Gottessegen, Germany, 99
- Grootvlei, South Africa, 137
- Großthiemig, Germany, 9
- Habichtswald, Germany, 228
- Hagenbachtal, Germany, 76
- Halifax, Canada, 108
- Harz Mountains, Germany, 75, 156, 173, 215, 233
- Heerlen, Netherlands, 231, 232
- Heiligkreuz, Poland, 241
- Heinrich Colliery, Germany, 231
- Helmsdorf, Germany, 142
- Helmstedt, Germany, 220
- Herkules Colliery, Germany, 228
- Hohe Warte, Germany, 17, 173
- Hope mine, Germany, 58
- Horden, United Kingdom, 104, 105, 109
- Howe-Wilburton, USA, 215
- Iberian pyrite belt, 32
- Ida-Bismarck mine, Germany, 156
- Idrija, Slovenia, 49
- Ilse lake, Germany, 118
- InterContinental Shanghai Wonderland Hotel, China, 227, 228
- Iron Mountain, USA, 24, 79
- Jänschwalde, Germany, 229
- Jarny, France, 125
- Jinping I, China, 224
- Johanngeorgenstadt, Germany, 218
- Kaatiala, Finland, 49
- Kassel, Germany, 228
- Kennecott, USA, 137
- Keno Hill, USA, 153
- Kentucky-Utah Tunnel, USA, 33
- Kilian adit, Germany, 200
- Kingston, USA, 231
- Kittilä, Finland, 201
- Kleinkogel, Austria, 199
- Knappensee, Bavaria, 194, 229
- Königsgrube mine, Upper Silesia, 99
- Königshütte, Upper Silesia, 99
- Königstein, Germany, 140, 217, 218, 234
- Koschen, Germany, 211
- Kotalahti, Finland, 153, 184
- Kromdraai, South Africa, 13, 121, 123, 127, 238, 240
- Krušnohoří, Czechia, 225, 227
- Lake Calumet, USA, 24
- Landusky, USA, 206
- Langban, Sweden, 229

Lehesten, Germany, 46, 157, 161
Leitzach, Bavaria, 183, 190
Lennestadt, Germany, 99
Lichtenauer See, Germany, 210
Logatec, Slovenia, 50
Lohsa, Germany, 208, 229
Lorraine, France, 17, 125
Löttringhausen, Germany, 99
Lusatia, Germany, xi, 8, 32, 39, 87, 107, 117, 118, 125, 135, 175, 179, 204, 205, 207, 208, 228, 229, 233, 238
Mansfeld, Germany, 200
Marsberg, Germany, 200
McArthur River, Canada, 131
Meggen, Germany, ix, 99, 153
Metsämonttu, Finland, 13, 73, 74, 83, 153, 181, 186
Mettiki, USA, 214
Meurthe-et-Moselle, France, 125
Middleburg, South Africa, 72
Mina da Passagem de Mariana, Brazil, 229
Mina de Campanema, Brazil, 164, 165
Minas Gerais, Brazil, 165
Minera Escondida, Brazil, 127
Mogale, South Africa, 133
Montevecchio, Italy, 194
Monticello, USA, 178
Morrison Busty Colliery, United Kingdom, 171
Mponeng, South Africa, 148
Munich, Bavaria, xv, 189
Mynydd Parys, United Kingdom, 224
Neville Street, Canada, 153, 167, 176, 183
Newcastle-upon-Tyne, United Kingdom, 99
New Vaal, South Africa, 120, 148, 149
New Zealand, 154, 162, 194
Nickel Rim, USA, 178
Niederschlag, Germany, 218
Niederschlema/Alberoda, Germany, 9, 31
Nöbdenitz, Germany, 233
North Fork, USA, 185
Nova Scotia, Canada, 48, 71, 73, 108, 110, 163, 167, 176, 183, 217, 230
Ojamo, Finland, 229
Oker, Germany, 34, 175
Othfresen, Germany, 156, 164
Paitzdorf, Germany, 157, 233
Partwitz lake, Germany, 212
Parys Mountain, United Kingdom, 222, 224
Pécs, Hungary, 140, 218, 234
Pelenna, United Kingdom, 173
Pennsylvania, USA, xv, 7, 16, 100, 140, 146, 163, 166, 169, 170, 172, 185, 197, 212–214, 221, 228, 231
Phillipstollen, Bavaria, 190
Plessa, Germany, 118, 205
Poços de Caldas, Brazil, 140
Pöhla, Germany, 17, 157
Quaking Houses, United Kingdom, 164, 165, 171
Rainitza, Germany, 118
Rammelsberg, Germany, ix, 10, 34, 175, 199, 227
Rehbach, Germany, 46
Reiche Zeche, Germany, 17, 59, 140, 232
residual lake 107, Germany, 118
residual lake 111, Germany, 205
Rio de Huelva, Spain, 32
Rio Odiel, Spain, 32
Rio Tinto, Spain, 32, 33, 227
River Gaunless, United Kingdom, 75
Robule lake, Serbia, 116
Rocky Flats, USA, 142
Ronneburg, Germany, 23, 216, 217, 219, 221
Rothschönberg adit, Germany, 16
Sabie, South Africa, 9
Sala, Sweden, 229
Šalektal, Slovenia, 216
Salem No. 2, USA, 219
Šaleška dolina, Slovenia, 24
Sarver, USA, 146
Scandinavia, 9, 153, 204, 209
Scheibe lake, Germany, 210, 211
Schlabendorf Nord, Germany, 182
Schlema (*see* Locality, Bad Schlema, Germany)
Schmölln, Germany, 233
Schönstein, Slovenia, 216
Schwandorf, Germany, 209
Schwartzwalder mine, USA, 219, 222
Schwarze Pumpe, Germany, 107, 228, 229
Schwaz, Austria, 197
Schwefelstollen, Germany, 81, 189
Sedlitz, Germany, 118, 209
Selminco, Canada, 187
Senftenberg lake, Germany, 212
Sequatchie County, USA, 220
Shaanxi, China, 127

Locality (*cont.*)
Shanghai, China, 227, 228
Shilbottle, United Kingdom, 179, 217
Shimao Quarry Hotel, China, 227
Sierra de Cartagena-La Unión, Spain, 164
Silverton, USA, 32
Skado Dam, Germany, 8, 179, 208
Smith Township, USA, 140
Songjiang, China, 227, 228
Soshanguve, South Africa, 148
Šoštanj, Slovenia, 216
South Korea, 154
Spree, Germany, 216, 233
Spreewald, Germany, 229, 233
Springhill, Canada, 230
St Aidan's, United Kingdom, 229
Stanley Burn, United Kingdom, 164
Steinbach, Germany, 164
Steinberger See, Bavaria, 194, 209, 229
St. Helena (island), xiii, xvi, 154
Stilfontein, South Africa, 133
St. Michael, USA, 146
Straßberg/Harz, Germany, 13, 22, 31, 97, 175, 215, 224, 236
Straž, Hungary, 134
Sykesville, USA, 146
Taff Merthyr, United Kingdom, 180
Tauferberg, Austria, 50
Tuna Hästeberg, Sweden, 229
Tzschelln, Germany, 118, 135
Ukraine, xv, 115
Umhausen, Austria, 50
Unden lake, Sweden, 208
Urgeiriça, Portugal, 49, 157, 187
Velenje lake, Slovenia, 24, 79, 216
Velenje, Slovenia, 216
Velenjsko jezero, Slovenia, 216
Vesta mine, USA, xv, 100, 101, 213, 214
Vetschau, Germany, 175, 176
Wackersdorf, Bavaria, 157, 194, 209, 229
Wallsend Rising Sun, United Kingdom, 99
Walvis Bay, Namibia, 154
Warndt, Germany, 224
Wasatch Mountains, USA, 33
Westmoreland County, USA, 100, 169, 170
West Rand, South Africa, 184
Wheal Jane, United Kingdom, 88, 109, 153
Wilhelm-Erbstollen, Austria, 197
Wilhelmshöhe, Germany, 228
Wilkes-Barr, USA, 231
Will Scarlet, USA, 205
Wohlverwahrt-Nammen, Germany, 9, 13
Wolf mine, Germany, 232
Wöllan, Slovenia, 216
Wüstenhain, Germany, 175
Wyoming Basin, USA, 16
Ynysarwed, United Kingdom, 22, 144, 154
Yorkshire colliery, USA, 197
Ypsilanta, Germany, 34
Zollverein, Germany, 232
Zwickau, Germany, 232
Глубокая, Ukraine, 115

**M**

Macrophyte, 157, 173, 186, 187
Magnesium, 53, 200, 236
Manganese, 23, 34, 53, 81, 82, 101, 102, 115, 127, 134, 142, 144, 161, 167, 168, 182, 234, 239
Marcasite, 29, 30, 43
Massachusetts Water Watch Partnership (association), 191
Matrix sorption, *see* Absorption
MaxPett restaurant, 189
Medicinal water, 19, 34
Melanterite, 25
Membrane
cross-flow, 121
filtration, 121
microorganism, 119
permeate, 118
retentate, 118, 120
size exclusion, 118
Mercury, 60, 115, 139, 144, 193, 218
Metalloid, *see* Semimetal
Metallophyte, 240
Metal of Concern (MOC), 8
Methane, 49, 172
Methanol, 219
MIBRAG (company), 198
Microbial catalysis, 14, 30, 212
Microorganism, 23, 30, 52, 111, 112, 136, 144, 154, 166, 167, 170, 179, 200, 204, 206, 213
Mine, 13
Mine 2030, 198
Mine drainage, 19
*See also* Mine water

Mine flooding, 12, 15, 18
active, 18
controlled, 18
definition, 18
first flush, 15
German hard coal mines, xi
Hope mine, Germany, 58
hydraulic potential, 18
hydraulics, 18
iron concentration, 17, 32
Mine Water Filling Model (MIFIM), 19
monitoring, 18
passive, 18
perectly mixed flow reactor, 17
prediction, 18, 19
rebound time, 15
time scale, 18
uncontrolled, 18
water level, 18
*See also* Flooded mine
Mine leachate, *see* Mine water
Mine pool, 214, 219, 235
Mineral water, 19, 239
Mine sealing, *see* Treatment method, mine sealing
Mine water
acid mine water, 9, 10, 19, 26, 29, 31, 32, 34, 35, 46, 52, 55, 77, 79, 82, 87, 99, 100, 115, 116, 118, 125, 126, 143, 146, 160, 164, 166, 169, 172, 175, 180, 181, 191, 209, 212, 213, 220, 221, 223–225
natural, 32
ammonium, 84
brine, 26
buffer mechanism, 29, 31, 34, 43
circumneutral, 26, 31, 78, 79, 115, 125, 180, 181, 194, 211, 221
classification, 9, 36, 91
cocktail of elements, 30
contamination avoidance, 34
definition, 9, 19
drinking water, 39
ferruginous, 10, 35, 38, 179, 227
fishing industry, 233
formation, 29
formation water, 19
hydroelectric power, 233
large waste stream, 243
leachate, 20
mining influenced water, 7, 9, 19
coining the term, 19
definition, 19
mining water, 20
monitoring, 90
natural source, 20
net acidic, 21, 22, 26, 39, 79, 169
net alkaline, 21, 22, 39, 169
neutral, 26
organic contaminant, 52, 114, 144, 177–178
ownerless, 20
precipitate, 19
pressure factor, 39
quality index, 39
radioactive contamination, 233
real one, 141–143, 149
saline, 14, 26, 131
seepage water, 20
sexy, 227
synthetic, 81, 115, 118, 137, 141, 143, 147, 200
titration, 39
turbine, 34, 233
Mine Water and the Environment (journal), 9, 10, 29
Mine water discharge, 15, 173, 190, 195, 197
Mine water rebound, *see* Mine flooding
Mine water treatment
active, 11, 88
column test, 86
definition, 10
liming, 86
costs, 89
design criteria, 88
endless treatment, 90
low cost system, 88
passive, 88
climate, 153
definition, 10, 22, 155
design errors, 90
incorrectly designed, 21
linguistic quibbles, 10
misconception, 154
road trip, 155
pilot plant, 1, 2, 72, 88, 115, 129, 131, 133, 135, 142, 146, 147, 153, 157, 164, 173, 206, 234 (*see also* Pilot test)

Mine water treatment (*cont.*)
  preliminary measures, 34
  semi passive, 22
  *See also* Treatment method
Mining Environment Neutral Drainage (MEND), 42, 88, 100, 141
  sampling, 50
Mining influenced water (MIW), *see* Mine water
Mining method, 197, 226
Mining waste, xi
  reuse, 15
Mining Waste Directive, *see* Directive 2006/21/EC of the European Parliament
Mining water, *see* Mine water
MinRoG, 234
MINTEK (organisation), 98, 133, 201
Modelling
  analogue, 203
  chemical-thermodynamic, 43, 70, 84, 215, 236
  numerical, 33, 43, 143, 195, 203, 207, 209, 215, 218
Molasses, 158, 219
Molybdenum, 131, 140
Monitoring, x, 10, 11, 18, 85, 145, 168, 187, 195, 207, 209, 219–221
  model-based, 51
Moonlighter, 192, 222
Multivariate statistics, 91
Munsell scale, 40
Mushroom compost, 14, 170
m-value, 12
Mycorrhization, 166
Myron L (company), 65

**N**
Nanosorbent, 141
Natural attenuation, *see* Treatment method, natural attenuation
Natural lake, 204, 209
Nickel, 61, 139, 199–201, 235
Nitrate, 35, 62, 128, 137, 173, 206
Noble metal, 29
Non-ferrous metal, 29
Non-governmental organization (NGO), 32
NP–$SO_4$ diagram, 39

**O**
Occlusion, 28
Ochre
  avoidance, 179
  deposits, 34
  geothermal, 232
  Nochten-Ochre, 238
  Nochten-Red, 238
  paint, 10, 175
  pH value, 10, 34
  pigments, 237
  precipitate, 175, 216
  reactive wall, 179
  scaling, 35, 66
  vertical flow reactor, 181
  yellow, 34
Olivine, 208
Open pit, 9, 10, 18, 19, 31, 32, 39, 49, 84, 121, 140, 149, 196, 204, 205, 207, 209, 210, 212, 214, 220, 229, 235, 236
Organic acid, 12–13
Organic contaminant, *see* Mine water, organic contaminant
Organic material, 85, 136, 167, 171, 196
Organic substrate, 14, 170, 172, 173, 186, 204
ORP, *see* Redox potential
Osmotic pressure, 125, 129
Oulipotent, 2
Overburden, 219, 220
  BAT, 42
  column test, 87
  fly ash, 221
  limestone, 220
  management, 198
  pyrite oxidation, 213
  renaturated, 32
  sulfate, 31
Ozone, 144, 145, 182

**P**
Pantone (company), 237
Paques Bio-Systems (company), 137
*Parastrephia quadrangularis*, 213
Patent
  DE 60028806 T2, 147
  DE 692 31 983.2, 179
  EP 0250626, 133
  EP 2 796 188 B1, 111

EP 2118020 A2, 146
RU 2315007 C1, 115
US 0176061 A1, 146
US 1862265A, 99
US 3,738,932, 95
US 5547588, 133
US 5,672,280, 102
US 6355221 B1, 111
US 7504030, 98
US 7504030 B2, 110
PCB, *see* Polychlorinated biphenyls
Pedogenic processes, 31
*See also* Acid soil
Perfectly mixed flow reactor (PMFR), *see* Mine flooding
Periodic table of the elements, 30
Permeability, 125, 160–162, 178, 179, 218
Permeate, 122, 129
Persons
Agricola, Georgius, xii, 10, 240
Carroll, Lewis, vii
da Vinci, Leonardo, 182, 185
Eliot, Thomas Stearns, vii
Ernst, Wilfried, 23, 200
Ficklin, Walter, xi
Fischer, Artur, 5
Gutenberg, Johannes, 1, 4, 5
Haber, Fritz, xi, 238
Jobs, Steve, 5
Kachan, Dallas, 5
Kostenbader, Paul D., 2, 95
Levy, Max, 3
Matschoss, Conrad, xii
Reis, Philipp, 5
Röntgen, Paul, 2
Salk, Jonas, 2
Simon, William, 116, 191
Sørensen, Søren Peter Lauritz, 24
Strachan, Robin, 227
v. Cotta, Bernhard, 230
v. Goethe, Johann Wolfgang, vii
v. Pettenkofer, Max, 189, 190, 192, 193
Vranesh, George, xi
Watt, James, 1, 4, 5
Wittgenstein, Ludwig, vii
Younger, Paul, v, 65, 164
Petroleum extraction, 14
Photothermal material, 177
*Phragmites* sp., 166
pH value, 24, 92
activity, 24
anthropogenic influence, 24
average, 79
calibration, 78
concepts, 77
definition, 24
electrodes, 78
high, 79
highest, 24
hydrogen ion exponent, 24
low, 32
lowest, 24
monitoring, 87
natural water, 79
negative, 24, 31
non existing pH-scale, 24
optical determination, 78
temperature compensation, 78
test strips, 78
unit, 24
visual determination, 40
Physisorption, 26
Phytomining, 198
Pigments, xv, 168, 235, 237, 238
Pilot test, 129, 147, 211
*See also* Mine water, pilot plant
Piper diagram, 36
Pitfall, x, 8, 45, 47
Pit lake
chemical treatment, 207
monitoring, 207
tourist attraction, 228
Polychlorinated biphenyls (PCB), x, 52, 62, 63, 144, 223
analysis, 62
congeners, 62
filtration, 62
microorganism, 23
Polymerisation, 27, 28
Portlandite, 211
Post-mining usage, 227
Potable water, 133
Potential literature, 232
Potentially dissolved metal, 61
Potentially toxic element (PTE), 8, 214
Potentially toxic metal (PTM), 8, 221
Potential–pH diagram, *see* Pourbaix diagram
Pourbaix diagram, 92, 93, 132

Power plant, 50, 116, 209, 211, 216, 221, 231, 233
Precious metal, 29, 238
Precipitation, 26, 28, 34, 64, 97, 102, 119, 126, 131, 136, 143, 156, 160, 187, 194, 195, 204, 218, 236, 237
  contaminants, 34
  heterogeneous, 180
Predominance diagram, *see* Pourbaix diagram
Premier Coal (company), 229, 230
Probe
  biofouling, 58 (*see also* Biofouling, fouling)
  scaling, 58
Process water, 20, 118, 121, 142, 147, 149
Produced water, 14
Project
  ARIDUA, 49
  BioHeap, 201
  BioMinE, 201
  ERMITE, 50, 54, 89
  MiMi, 153
  Minewater Project Heerlen, 231
  PIRAMID, 43, 156, 158, 161
  Re-Mining, 234
  UNEXMIN, 48, 49
  Vodamin, xv, xvii, xviii, 118
Prolotroll, *see* Stollentroll
Proxa (company), 148, 149
Publishing
  avoiding field experiments, 3
  errors in data, 3
  failed experiments, 3
  negative results, 4
  patents, 1
  plagiarism, 3
  problematic results, 90
  publication bias, 3
  typos, 2
  unpublished case study, 224
Pump-and-treat, 190
p-value, 25
Pyrite, 12, 29–32, 35, 37, 38, 43, 44, 93, 169, 170, 209
Pyrite oxidation, 12, 30, 32, 35, 79, 198, 203, 212, 222, 224
Pyrrhotine, 29, 43
Pyruvate, 205
Pyruvic acid, 205

**R**
Radio frequency identification (RFID), 5, 59
Radionuclide, 128, 142
Radium, 17, 131, 139, 157, 187
Rare earth element, 234, 241
Raw material recovery, 4, 115, 120, 141, 233, 243
Receiving watercourse, 67, 120, 164, 190, 194, 195, 212, 224, 226, 239
Redox potential, 83–85, 92, 144, 161, 164, 169, 178, 186, 187
  concept, 83
  correction, 83, 84
  definition, 83
  error, 84
  measurement, 83
  missing, 84
  standard hydrogen electrode (SHE), 83
Reed grass, 166
Residence time, 17, 91, 162, 164, 175–177, 181, 182, 185
Residual lake, 107, 204, 205, 208, 209, 211–213, 229
  *See also* Pit lake
Residues, 234
Retentate, 123, 126
Rhizosphere, 166
Rio Tinto (company), xi, 127
River pollution, 192
Root zone process, 156
Rushes, 166

**S**
Sampling
  acid capacity, 65
  alligator, bear, mountain lion, 49
  back-up equipment, 47
  base capacity, 65, 68
  checklist, 47
  danger, 50
  electrical conductivity, 66
  essential parameter, 64
  field measurement, 64
  flow through cell, 64, 65, 85
  gas danger, 49
  gas detector, 50
  gloves, 50

health and safety, 49
methods, 50
mine water, 45
on-site parameter, xvii, 51, 58, 64, 65
oxygen measurement, 46
restricted areas, 50
specific conductance, 66
temperature, 66
titration, 65
underground danger, 50
*See also* Water sampling
SansOx Oy (company), 110, 111
Saxon mine adit database, 192
Saxore Bergbau GmbH (company), xi
Scaling, 35, 58, 119, 128, 232
*See also* Biofouling
Schwertmannite, xv, 134, 135, 141, 194, 238
Science Citation Index, 120, 135, 142, 146, 178
SDAG Wismut (company), *see* Wismut GmbH (company)
Secondary mineral, 12, 15, 17, 91, 224
*See also* Efflorescent salt
Selective precipitation, 136, 200
Selenium, 130
Semimetal, 8, 28, 35, 167, 214, 216, 234
Sensor, 5, 48, 66, 85, 111, 243
Septic drain fields, 189
Silver, xi, 29, 134, 218
bullet, 134
SI unit prefixes, 11
Sludge, 4, 234
dewatering, 107, 109, 111, 134, 180, 237
recycled, 108
sewage, 14, 205
stability, 105
underground disposal, 101
Small-scale mining, xii
Soda ash, 100, 209, 210
Sodium fluorescein, *see* Uranine
Sodium hydroxide, 105, 115, 206–209
Software
ABATES, 88
AMDTreat+, 80, 90, 104
aqion, 55, 56
aquaC, 100, 104
ASPEN, 133
chemical-thermodynamic modelling, 43–45
FIDAP, 215
Geochemist's Workbench, 55, 93
Hydra/Medusa, 93
Mine Water Filling Model (MIFIM), 19
PHREEQC, 55, 56, 66, 67, 90, 104, 182, 209, 218, 220
Solid solution, 28
Sorbent, 141–143, 239
Sorption, 26–28, 101, 104, 141–143, 178, 241
adsorbate, 27
adsorbent, 27
matrix sorption, 27, 28
poor selectivity, 143
the small difference, 27
surface sorption, 27
*See also* Absorption; Treatment method, sorption
Source water, 14
Species, 12, 26, 30, 54, 60, 64, 80, 83, 85, 92, 93, 101, 167, 182, 217
Species protection, 224
Specific conductance, *see* Electrical conductivity
Speleologist, 192
Spray sprinkler, 211
Stability diagram, *see* Pourbaix diagram
Stakeholder, 20, 155, 227
State Water Acts, 19
Stollentroll, *see* Adit troll
Stone Age, xii
Straw, 136, 170, 186, 205
Sugar cane, 186
Sulfate, 9, 14, 17, 26, 28, 31, 32, 37–39, 53, 99, 101, 117–119, 122, 124, 125, 128, 129, 132, 133, 137, 140, 145–147, 155, 158, 169, 178, 194, 204, 212, 220, 221, 236, 238
reduction, 137, 169, 170, 178, 187, 194, 204–206
removal, 134, 136, 137, 145, 155
Sulfate-reducing bacteria, 136
Sulfide weathering, *see* Pyrite oxidation
Superfund, 219
Surface complexation, 26, 27
Surface water, xvii, 20, 67, 86, 216
Surface Water Ordinance, 19, 60, 90
Suspended solids, 35, 58, 99, 101, 102, 173, 239
Synthetic resin, 138, 140
Syringe, 48, 170

**T**
Tailings, 5, 23, 31, 37, 38, 42, 87, 134, 178, 187, 206, 213, 217, 221, 235
Talvivaara (company), 199, 201
Tannin, 13, 68
TDS, *see* Total dissolved solids (TDS)
Technetium, 218
Terrafame (company), 199, 201
*Thiobacillus thiooxidans, see Acidithiobacillus, thiooxidans*
*Thlaspi*, 200, 240
  *caerulescens*, 200
  *calaminare*, 200
Thorium, 139
Titration, 12, 13, 70
  alkalinity, 25
Total dissolved solids, 40, 67, 68, 117, 122, 125, 140, 163, 236
  calculation from electrical conductivity, 68
Total inorganic carbon (TIC), 70
Tourism, 127, 229
  tourist attraction, 227, 228
Tracer test, 58, 64, 71, 183, 203, 215, 235
Transition element, 200
Treatment method
  abatement, 189
  Acid Mine Drainage Demineralizer, 110
  Acid Reduction Using Microbiology (ARUM), 173, 176, 186, 187
  active, 95
  advanced oxidation, 144
  aeration, 11, 22, 97, 102, 106, 108, 110, 111, 134, 182, 184, 185
  aerobic limestone drain, 172
  agro-metallurgy, 198
  alkaline material mixing, 219
  alternative treatment method, 189
  anoxic limestone drain, 46, 156, 158, 160, 161, 163, 172, 173
  anti-fouling additive, 112
  AquaFix, 217
  biohydrometallurgy, 198
  bioleaching, 198, 200
  biometallurgy, x, 198
  biomimetics, 198
  biooxidation, 198
  bioreactor, 14, 136, 137, 156, 158, 198, 200, 241
  bioremediation, 5, 23, 198
  BioSURE, 137
  BIOX, 201
  Birtley-Henry process, 99
  boat supported, 212
  capacitive deionisation, 116
  cascade, 106, 108, 182, 184, 185
    stair type, 184
    weir type, 184
  Colloid Polishing Filter Method, 142
  compost wetland (*see* Treatment method, constructed wetland, anaerobic)
  condensation deionisation (*see* Treatment method, electrosorption)
  constructed wetland, xviii, 14, 20, 21, 23, 30, 136, 153, 157, 158, 164, 166, 167, 169, 170, 244
    aerobic, 11, 20, 21, 152, 153, 160, 162, 166–168, 173
    aerobic calculation, 168
    anaerobic, 14, 20, 21, 169–172
    municipal wastewater, 20
    plants, 166
    substrate, 169, 171
  contaminated watercourse, 216
  conventional mine water treatment (*see* Treatment method, low density sludge (LDS))
  desalination, 116, 120–122, 126, 127, 236
  DesEL process, 116
  dissolved air flotation, 126
  diversion, 217
  doing nothing, 189
  electric aerator, 106
  electro-biochemistry, 206, 207
  electrochemistry, xviii, 5, 27, 65, 85, 112, 117, 118, 122, 129, 141, 204, 206, 243
  electrocoagulation, 112–116, 144, 217
    electrode, 113
  electrodialysis, 117–121, 236
  electroflocculation, 112
  electrosorption, 116, 117
  enclosure, 197, 204, 205
  enhanced natural attenuation, 152, 197
  ettringite precipitation process (Walhalla), 133
  ettringite process, 133, 134
  $Fe^0$ (*see* Treatment method, zero-valent iron)
  fermenter, 136
  fixed-bed reactor, 136

F-LLX, xv, 111, 145, 146, 234
floating fixed-bed reactor, 138
flocculant, 99
flotation liquid-liquid extraction, 145
flow diagram, 6, 96, 158, 159
forward osmosis, 126, 129, 130
freeze crystallisation, 120, 147–149, 238
GaRD, 223
geobiotechnology, x, 198–201
glass beads, 110, 112
globular catalytic reaction, 110
GYP-CIX, 140
HARDTAC process, 110, 111
Hazleton Iron Removal System, 110
high density sludge (HDS), 2, 95, 97–99, 104, 107–110, 126, 213, 235
HiPRo, 238
HybridICE, 122, 147, 148
HydroFlex, 145
hydrogen contact reactor (*see* Treatment method, reducing and alkalinity producing system (RAPS))
hydrogen peroxide, 106
improved ettringite process, 133
in situ experimental facilities (*see* Treatment method, microcosm)
in situ remediation, xi, xvii, xviii, 14, 156, 203
  uranium mine, 217
in-lake liming, 204
in-lake process, 204, 207, 210
in-lake technology, 208
integrated treatment, 126
ion exchange, 42, 117, 122, 138–141, 147, 219, 234
  membrane, 117
iron sulfate precipitation (*see* Treatment method, schwertmannite process)
journey, 7, 243, 244
low density sludge (LDS), 97, 105–107, 110, 213, 236, 238
  pH values, 106
macrocosm, 204, 205
MAXI-STRIP, 110, 111
membrane, 116–123, 125, 128–130, 137, 234, 238, 243
  anion selective, 117
  cation selective, 117
  ceramic, 125
  cleaning, 125
  Cleaning in Place, 120
  concentrate, 121
  fouling, 117
  hydrolysis, 128
  material, 125
  microfiltration, 123
  planning, 122
  pre-treatment, 117
  size, 121
  ultrafiltration, 124
membrane electrolysis, 117, 118, 120
microbiological iron separation process (*see* Treatment method, schwertmannite process)
microbiological treatment, 135
microcosm, 196, 197, 204, 205
microfiltration, 118, 119, 121, 123, 129
mine sealing, 203, 222, 224, 226
modern treatment processes, 95
monitored natural attenuation, 151, 189, 190, 194, 195, 197
nanofiltration, 95, 118, 121–126, 128, 239
natural attenuating pit lake area (NAPA), 194
natural attenuation, xvii, xviii, 22, 34, 141, 151, 152, 189, 193–196, 209
neutralisation, 97 (*see also* Treatment method, low density sludge (LDS))
  alkali calculation, 104, 105
  alkaline material, 100
  first use, 99
  process, 98, 101
  sludge, 101
  target pH, 102
on-site chemical treatment, 212
on-site rehabilitation, 203
open limestone channel, 158, 163–165
oxic limestone drain, 158, 162, 163
OxTube, 110, 111, 181, 186
passive, 151, 152, 190
  design errors, 152
  installed system number, 152
  polishing pond, 154, 162, 186
  self-confidence, 152
passive oxidation system, 182
permeable reactive wall, 8, 177–179, 217, 218
  funnel-and-gate, 178, 179

Treatment method (*cont.*)
photocatalysis, 144
photolysis, 144
phytomining, 200
phytoremediation, 23, 239, 240
long way too go, 23
plasma technology, 144
precipitation, 130
preventative measure, 221
radiolysis, 144
reducing and alkalinity producing system (RAPS), 11, 14, 85, 153–155, 158, 166, 171–174, 179
short circuit, 173
reed bed (*see* Treatment method, constructed wetland, aerobic)
reinjection of residues, 213–215, 219, 224, 235, 236
negative effect, 216
residue reuse, 234
reverse osmosis, 95, 116–119, 121, 122, 125–129, 147, 149, 219, 236, 238, 239
Rigby Process, 115
RODOSAN process, 117, 118
sacrificial electrode, 112, 114
SAPS (*see* Treatment method, reducing and alkalinity producing system (RAPS))
SAVMIN Process, 98, 131, 133, 134
schwertmannite process, 134, 135
SCOOFI, 154, 180
seeded reverse osmosis, 129
semi-passive, 151
settlement lagoon (*see* Treatment method, settling pond)
settling basin (*see* Treatment method, settling pond)
settling pond, 101, 104, 110, 173, 175, 177, 182
physics, 177
Slurry Precipitation and Recycle Reverse Osmosis (SPARRO), 123, 127–129
sorption, 26, 28, 42, 106, 141, 167, 195
iron hydroxide, 142
staged neutralisation, 111
stimulated iron and sulfate reduction, 204
successive alkalinity producing system (*see* Treatment Method, reducing and alkalinity producing system (RAPS))
Sul-biSul, 140
sulfate reducing bioreactor (*see* Treatment Method, reducing and alkalinity producing system (RAPS))
Taff Merthyr, United Kingdom, 180, 181
THIOPAQ, 137
Trompe, 182, 185
ultrafiltration, 95, 118, 121, 124, 126, 127, 129, 137, 239
uranium mine, in situ remediation, 217
UV treatment, 144
vertical flow reactor (VFR), xvi, 111, 153, 179–181, 186
vertical flow wetlands (*see* Treatment method, reducing and alkalinity producing system (RAPS))
ZEOTREAT, 140
zero-valent iron (ZVI), 179, 217, 218
Treatment plant
planning, 41
preliminary investigations, 41
Tshwane University of Technology, 148
Turbulence, 75, 185
Turing Galaxy, 2, 146
*Typha* sp., 166

## U

Uhde GmbH (company), 118
University of Cape Town, 147
Unobtainium, 2
Unoccupied Aerial Vehicle (UAV), *see* Drone
Up-dip operation, 197
Uranine, 183
Uranium, 9, 13, 17, 31, 91, 92, 123, 125, 131, 139, 140, 157, 187, 193, 198, 212, 217–219

## V

Valorisation, 244
Vattenfall (company), 118, 135
Venturi, 71, 111, 186
VEOLIA (company), 133
Vineyards, 127
Vision 2030, 123

W

Walter Moers, *see* Prolotroll
Wastewater, 9, 19, 20, 189
  advanced oxidation, 144
  electrocoagulation, 112
  ion exchange, 139
  municipal, 20, 23, 129, 137, 144, 156, 244
  ore processing, 9
  probe, 57
  sludge, 137
  sorption, 141
Water analysis, xvii, 36, 43, 51, 86
  acid capacity, 25
  Al determination, 82
  analytical error, 55
  base capacity calculation, 69
  electrical conductivity, 55
  Fe determination, 80, 81
  filtration, 82
  ion balance, 55
  Mn determination, 81
  parameter, 54
  PHREEQC, 66
  preliminary investigation, 42
  procedures, 86
  quality control, 55
  seasonal changes, 43
  sludge, 211
  TDS, 68
  timing, 51
  validation, 57
  weather, 64
  *See also* Water sampling
Waterfowl, 177
Water Framework Directive (WFD), 19, 20, 90
Water level, 19, 72, 73, 177, 222
  flooded mine, 12, 18, 224
  measurement, 47
Water quality, 9, 11, 18, 22, 39, 43, 44, 67, 152, 153, 161, 164, 175, 176, 194, 196–198, 204, 208, 213–215, 217, 221, 224, 228, 235, 236
Water Resources Act, 19
Water sampling, 43, 47, 49–51
  acid-extractable metal, 61
  additional parameter, 53
  air, 51
  aluminium, 53, 61, 82
  blind samples, 57
  bottles, 53
  brines, 58
  carelessnesses, 47
  colloids, 52
  contamination, 53
  coordinates, 63
  coprecipitation, 61
  data processing, 63
  dissolved concentration, 60
  dissolved metals, 61
  documentation, 59, 63
  drone, 49
  dublicate samples, 57
  electrodes, 57
  Fe concentration, 60, 80
  Fe hydrolysis, 81
  Fe terminology, 80
  ferric iron, 80, 182, 238
  ferrous iron, 21, 29, 32, 80, 101, 123, 180, 182
  filter, 52
    residue, 51
  filtration, 52, 60–62, 193
    in field, 53
  main ions, 51
  membrane filter, 62
  mine water, 43
  naming, 57, 59
    recommendation, 59
  nitric acid, 61
  organics, 52
  oxygen
    Clark electrode, 57, 64, 85
    luminescent dissolved oxygen, 57, 65, 85
    saturation, 85
  parameter, 53
  PCB, 62
  photometry, 81
  potentially dissolved metal, 61
  precision, 58
  pre-treatment, 51
  probe, 57
    biofouling, 57
    calibration, 57
    cleaning, 57
  quantities, 51
  regulations, 47
  rinsing, 53
  sample mix up, 59

Water sampling (*cont.*)
smoking, 48
storage, 51
suspended metals, 61
total concentration, 60
total metals, 61
trace elements, 52
units, 63
waste, 48
*See also* Sampling; Water analysis
Weathering
pyrite (*see* Pyrite oxidation)
secondary minerals, 12, 15, 43
Western Deep Levels (mine), 148
Wetland
municipal wastewater, 23
Wismut GmbH (company), 131, 140, 142, 217, 221, 234
World Heritage Site, 225, 227, 228
WTW (company), 65

**Y**
Younger-Rees diagram, 37
YSI (company), 65, 84

**Z**
Zeolite, 138, 140, 141
Zero-discharge plant, 238
Zinc, xi, 28, 108, 114, 115, 139, 140, 164, 167, 198, 200, 234, 235

Printed in the United States
by Baker & Taylor Publisher Services